```
QH        125611
87.3    Dennison, Mark S.
.D45      Wetlands : guide
1993    to science, law,
        and technology
```

GAYLORD

Wetlands

Guide to Science, Law, and Technology

WETLANDS
Guide to Science, Law, and Technology

by

Mark S. Dennison

and

James F. Berry

np NOYES PUBLICATIONS
Park Ridge, New Jersey, U.S.A.

Copyright © 1993 by Mark S. Dennison
 No part of this book may be reproduced or utilized in
 any form or by any means, electronic or mechanical,
 including photocopying, recording or by any information storage and retrieval system, without permission
 in writing from the Publisher.
Library of Congress Catalog Card Number 93-5635
ISBN: 0-8155-1333-X
Printed in the United States

Published in the United States of America by
Noyes Publications
Mill Road, Park Ridge, New Jersey 07656

10 9 8 7 6 5 4 3 2 1

Library of Congress Cataloging-in-Publication Data

Dennison, Mark S.
    Wetlands : guide to science, law, and technology / by Mark S.
Dennison, James F. Berry.
        p.   cm.
    Includes bibliographical references (p.   ) and index.
    ISBN 0-8155-1333-X
    1. Wetlands. 2. Wetlands--Classification. 3. Wetlands--Law and
legislation--United States. 4. Wetland ecology. I. Berry, James
F. II. Title.
QH87.3.D45   1993
333.91'8--dc 20   93-5635
                                                           CIP

To

Tracey, my love and inspiration

M.S.D.

and

To

My daughter Jennifer, whose talent, determination, love, and encouragement have been my motivation.

J.F.B.

Preface

The past three decades have seen a massive increase in interest in environmental issues within the United States, but few have received as much attention as wetlands. Largely as a result of the efforts of modern wetland scientists, swamps, marshes, bogs, and coastal wetlands are no longer viewed as disease-ridden, dangerous places, or as mere obstacles to progress. It is now known that wetlands have many important functions and values. They serve as essential habitats for a variety of fish, wildlife, and plants, and are known as biological "nurseries" due to the large number of species (including many of great economic importance) that depend on wetlands for breeding and early development. Wetlands benefit humans directly by dissipating the effects of floods and storms, and filtering pollutants and other materials from surface water. Humans also depend on wetlands for an assortment of recreational activities, such as hunting, fishing, and bird watching. Refinements in wetland science not only inspired the currently-accepted "no net loss of wetlands" policy of the federal government, but also led to increasingly sophisticated technological innovations for restoring and creating lost wetland functions and values.

Wetlands are currently protected in various ways by numerous laws and regulations, but the cornerstone is § 404 of the federal Clean Water Act, which prohibits the discharge of dredged or fill material into wetlands. The federal, state, and local wetland protection laws have, however, created a quagmire of interrelated and, in some cases, contradictory restrictions on land use practices in wetland areas. Whether a particular area is considered a wetland under the current regulatory scheme, and whether a proposed activity in a wetland area is prohibited, critically impacts virtually every land use decision.

It is within the current setting that this book was written. It has become increasingly evident to us in our law practices and scientific experience that no single book currently available combines the background in wetland science necessary to understand wetland functions and values, with the framework of laws and regulations that affect wetlands. It is our hope that this volume will serve this purpose. In addition, we present practical applications of wetland science, laws, and regulations, combined with the latest technological methodology, which will allow regulators, the regulated community, and wetland conservationists to understand the complex issues involved. To assure that our goals for the book are met, we have been fortunate indeed to have enlisted the foremost experts in the field as contributors.

The emphasis in the book is on wetlands within the continental (coterminous) United States. The book begins in Chapter 1 with an overview of the major issues in wetlands protection, both scientific and regulatory. Chapter 2 provides a background in both historical and modern ecological approaches to wetlands study, and their impact on modern wetland issues. In Chapter 3, Dr. James Kushlan examines the characteristics of freshwater wetland types, including the important plants and animals of each, and Dr. Robert Livingston does the same for estuarine and marine wetlands in Chapter 4. In Chapters 5 and 6, Ralph Tiner provides a detailed, practical guide to the complex process of wetland determination and delineation, and examines important "problem" wetlands related to the process. In Chapter 7, we review the regulatory framework of wetlands protection, including an up-to-date examination of the important laws and regulations that affect wetlands. We examine the concept of wetlands "mitigation," and the necessity for replacing lost wetland values in Chapter 8. In Chapter 9, Dr. Thomas Burns provides a unique exploration of the concept of environmental "risk assessment" as it applies to wetlands. Finally, Dr. Robert Brooks provides a practical guide to wetland restoration and creation in Chapter 10.

The book is intended to meet the requirements of a variety of professionals. Wetland scientists will find the regulatory and legal chapters to be particularly useful; conversely, environmental attorneys will find the scientific and technological chapters to be helpful. Regional, state, and municipal planners will benefit by the coverage, as will other governmental representatives. Wetland conservationists will also find the book to be a useful guide to wetland protection programs and regulations. Finally, members of the regulated community will find in this volume an extremely useful review of the morass of regulatory and technical requirements associated with wetlands. Nevertheless, we are not sufficiently naive to believe that this book will serve all of the needs of everyone. Each chapter provides references and other sources of information by which an interested reader can locate additional information. The Literature Cited section refers to many current and historically important sources of information.

Finally, we wish to express our sincere thanks to the many individuals whose contributions have assisted us in the preparation of this book. Mark Dennison wishes to give his special thanks to Tracey J. Huff for all her help and support; and extend his thanks to Bob Noyes (Noyes Publications); William L. Want (Charleston, South Carolina); and Ken Ochab (Hackensack Meadowlands Development Commission). Jim Berry expresses his gratitude to Professor Fred P. Bosselman (IIT Chicago-Kent College of Law); Nancy E. Stroud, Lisa N. Mulhall, and Barbara A. Adams (Burke, Bosselman & Weaver, Boca Raton, FL and Chicago); William S. Sipple and Joseph DaVia (U.S. Environmental Protection Agency, Washington, D.C.); John Rogner (U.S. Fish & Wildlife Service, Barrington, IL); Linda Winter (National Wildlife Federation); Lindel L. Marsh (Siemon, Larsen & Marsh, Irvine, CA); Drs. Earl H. Meseth, Liane Cochran-Stafira, Paul F. Ries, and Martan Walters (Elmhurst College); Dr. Peter G. Merritt (Treasure Coast Regional Planning Council, Palm City, FL); to Mary N. Brust for her patience; and to Elmhurst College for support in the preparation of this book. Rob Brooks wishes to acknowledge the assistance of R. Brooks, T. Rightnour, T. Seafass, R. Tiner, and D. Wardrop. Tom Burns thanks B.W. Carnaby and G. Mowris. Ralph Tiner acknowledges Peter Veneman (University of Massachusetts, Amherst). Finally, Mark Dennison and Jim Berry wish to express their special gratitude to Jim "Detritus" Schwab (American Planning Association, Chicago) for his interest, encouragement, and assistance in the field.

Mark S. Dennison   *Ridgewood, New Jersey*
James F. Berry   *Elmhurst, Illinois*

About the Authors

Mark S. Dennison is an attorney and author or co-author of numerous books and articles on environmental, land use and zoning law issues. He has written seven books, including *Wetlands and Coastal Zone Regulation and Compliance*, with Steven Silverberg (Wiley Law Publications, 1993) and *Understanding RCRA Hazardous Waste Identification and Classification: Practical Guide for the Waste Generator* (Executive Enterprises, 1993), and has published more than twenty-five articles in various newsletters and trade journals. He is editor of the quarterly journal, *Remediation*, is contributing editor of the monthly newsletter *Environmental Liability in Real Estate Transactions*, and serves on the editorial advisory board for *Environmental Protection* magazine. Mr. Dennison has previously held positions as managing editor of the Environmental Law Series and the Zoning & Land Use Law Series at Clark Boardman Callaghan, and as acquisitions editor for the Environmental Law Library at Wiley Law Publications. He is in private practice in Ridgewood, New Jersey, specializing in environmental, land use and zoning law. He is admitted to practice in New Jersey and New York. Mr. Dennison earned his undergraduate degree, magna cum laude, in foreign languages at the State University of New York and completed an M.A. degree in German at Syracuse University. He earned his law degree at New York Law School, and is currently completing a Masters of Law degree in environmental law at Pace University School of Law.

James F. Berry is a biologist, college professor, and environmental attorney. He is currently an Associate Professor of Biology at Elmhurst College in Elmhurst, IL, and a Research Associate at the Carnegie Museum of Natural History in Pittsburgh, PA. As a biologist, Dr. Berry specializes in research on the fauna and ecology of wetlands and other sensitive areas, and on conservation biology. He has served as a biological consultant to a variety of government agencies and private organizations, including the U.S. Fish and Wildlife Service, California Department of Fish and Game, Florida Committee on Rare and Endangered Biota, Illinois Attorney General's Office, and the City of Chicago. As an attorney, Dr. Berry specializes in environmental, land use, and natural resources law. His legal clients have included local and state governments and agencies, such as the cities of Longboat Key, and Orlando, FL; Northbrook, IL; the Treasure Coast Regional Planning Council, FL; and the California Resources Agency. He has also represented or consulted for many private individuals and environmental organizations, such as the Florida Audubon Society, Izaak Walton League, and The Conservancy, Inc. (Naples, FL). He is licensed to practice law in all state courts in Florida and Illinois, as well as a number of federal district courts and courts of appeals. He is currently in private practice in Elmhurst, IL, having previously practiced with the law firm of Burke, Bosselman & Weaver in Chicago, IL and Boca Raton, FL. Dr. Berry's biological education includes B.S. and M.S. degrees from Florida State University, and a Ph.D. from the University of Utah. He received his law degree from IIT Chicago-Kent College of Law, having completed the law school's Program in Environmental and Energy Law. He is a member of many scientific and legal professional organizations, and has published over 60 scientific and legal articles and book chapters. He is a popular speaker on environmental issues for a variety of scientific, legal, and planning organizations and workshops, which have recently included the American Planning Association, Chicago Bar Association, Izaak Walton League of America, and the Midsouth Planning and Zoning Institute.

About the Contributors

Robert P. Brooks is currently an Associate Professor of Wildlife Ecology at The Pennsylvania State University, where his research focuses on wetlands, streams, and riparian areas in the northeastern and mid-Atlantic states. Dr. Brooks received his B.S. degree from Muhlenburg College, and M.S. and Ph.D. Degrees in wildlife biology from the University of Massachusetts. He has published over 50 technical papers, books, and book chapters, and has presented over 50 technical presentations at conferences and meetings. Dr. Brooks has over 17 years experience working with wetlands, streams, and water-dependent species, and has consulted with agencies, corporations, utilities, citizen groups, and individuals concerning issues in natural resources and management.

Thomas P. Burns is an Ecological Risk Scientist for the Environmental Analysis Division of Science Applications International Corp., in Oak Ridge, TN, where his research centers on ecological risk assessment, and on ecosystem dynamics and evolutionary ecology. He previously served as a Postdoctoral Fellow at Oak Ridge National Laboratory, and as a Visiting Research Scientist at the Department of Zoology, Kyoto University, Japan. Dr. Brooks received his B.S. degree from the University of Notre Dame, his M.S. from the University of Miami, and his Ph.D. from the University of Georgia's Institute of Ecology. He has published over 20 scientific articles and books, including *Theoretical Studies of Ecosystems: The Network Perspective* with M. Higashi (Cambridge University Press, 1991).

James A. Kushlan is Professor and Chair of the Department of Biology at the University of Mississippi, where he conducts research on wetland ecology, population and community ecology, and conservation biology. Previously, he served as Professor of Biology and

Director of the Center for Water Resources Studies at East Texas State University, and as Supervisory Research Biologist for the U.S. National Park Service in Everglades National Park. Dr. Kushlan received his B.S. (cum laude), M.S., and Ph.D. degrees from the University of Miami. He has published over 150 scientific papers, book chapters, and books, including *Storks, Ibises and Spoonbills of the World* with J. Hancock and M.P. Kahl (Academic Press, 1992), and *Freshwater Fishes of Southern Florida* with W.F. Loftus (Florida State Museum). He is Associate Editor of the journal *Wetlands*, and was previously the editor of the journals *Colonial Waterbirds*, and *Florida Field Naturalist*. He is a member of the Executive Board of the International Waterfowl and Wetlands Research Bureau, and has spoken at numerous national and international symposia and conferences.

Robert J. Livingston is Professor and Director of the Center for Aquatic Research and Resource Management at Florida State University, Tallahassee, where his research focuses on continuous, long-term analyses of various river and coastal aquatic systems across the Gulf of Mexico. His education includes an A.B. from Princeton University, and M.S. and Ph.D. degrees from the Institute of Marine and Atmospheric Sciences at the University of Miami. Dr. Livingston has published over 120 scientific papers and books, including *The Rivers of Florida* (Springer-Verlag, 1991), and *Ecological Processes in Coastal and Marine Systems* (Plenum Press, 1979). He has presented over 115 scientific papers and academic presentations.

Ralph W. Tiner is a wetland ecologist with more than 20 years experience in wetland classification, inventory, and delineation. He has directed wetland inventories in 15 states, for state and federal agencies. He is Adjunct Assistant Professor in the Plant and Soil Sciences Department at the University of Massachusetts, where he teaches wetland identification, classification, and delineation. He is author of numerous wetland publications, including *A Field Guide to Coastal Wetland Plants of the Northeastern United States* (Univ. of Massachusetts Press, 1987), *Field Guide to Nontidal Wetland Identification* (USFWS, 1987), *Maine Wetlands and Their Boundaries* (Maine Dept. of Economic and Community Development, 1991), and *Field Guide to Coastal Wetland Plants of the Southeastern United States* (Univ. of Massachusetts Press, 1993).

Condensed Contents

1. Overview 1
2. Ecological Principles of Wetland Ecosystems 18
3. Freshwater Wetlands 74
4. Estuarine Wetlands 128
5. Field Recognition and Delineation of Wetlands 153
6. Problem Wetlands for Delineation 199
7. The Regulatory Framework 213
8. Wetland Mitigation 278
9. Assessing Risks to Ecological Resources in Wetlands 304
10. Restoration and Creation of Wetlands 319
Literature Cited 352
Appendix: Federal and State Wetland Agencies and Offices 384
List of Abbreviations and Acronyms 406
Index to Scientific and Common Names 409
Index 424

Notice

To the best of the Publisher's knowledge the information contained in this publication is accurate; however, the Publisher assumes no responsibility nor liability for errors or any consequences arising from the use of the information contained herein. Final determination of the suitability of any information, procedure, or product for use contemplated by any user, and the manner of that use, is the sole responsibility of the user. Expert advice should be obtained at all times when implementation is being considered. Mention of trade names or commercial products does not constitute endorsement or recommendation for use by the Publisher.

Contents

1. **OVERVIEW** 1
   Mark S. Dennison, James F. Berry
   1.1 **Purpose and Scope of the Book** 1
   1.2 **Introduction to Wetlands** 3
       1.2.1 Wetland Definitions and Classification 3
               Definitions 4
               Classification 6
       1.2.2 Values of Wetlands 8
       1.2.3 Federal "No Net Loss of Wetlands" Policy 8
   1.3 **Introduction to Regulation** 10
       1.3.1 Overview of Regulatory Framework 10
       1.3.2 Controversy Over Federal Wetland Delineation Manual 12
   1.4 **Interplay Between Science and Regulation** 13
   1.5 **Jurisdictional Determinations** 16

2. **ECOLOGICAL PRINCIPLES OF WETLAND ECOSYSTEMS** 18
   James F. Berry
   Introduction 18
   2.1 **An Introduction to Ecosystems** 19
       2.1.1 Wetlands, and the Concept of an "Ecosystem" 19
       2.1.2 Communities 21
       2.1.3 Populations, Species, and Organisms 21
   2.2 **Community and Population Analysis** 23
       2.2.1 Community Classification and Ordination 24
               Classification 24
               Ordination 27
       2.2.2 Population Structure 31
               Species Richness and Diversity 31
               Competition 32
               Predation 38
               Measuring Population Dynamics 40
       2.2.3 Ecological Succession 42
               The "Climax" 43

		Autogenic and Allogenic Succession	45
	2.2.4	Food Webs, Nutrient Cycling, and Energy Flow	48
		Productivity	50
		Biogeochemical Cycles	52
		The Hydrologic Cycle	54
2.3	**Wetland Functions and Values**		54
	2.3.1	Fish and Wildlife Habitat	55
		Animal Reproduction and Development	55
		Wetland Animals	56
		Wetland Plants	57
		Recreation	61
	2.3.2	Groundwater Recharge and Water Quality	61
	2.3.3	Flood and Erosion Control	63
	2.3.4	Wetland Valuation	66
2.4	**Wetland Conservation**		67
	2.4.1	Wetland Losses	67
	2.4.2	Wetland Protection	70

3. FRESHWATER WETLANDS 74
James A. Kushlan

Introduction			74
3.1	**Swamps**		76
	3.1.1	Cypress Swamps	77
		Geophysical Characteristics	77
		Wetland Criteria Characteristics	79
		Characteristic Fauna	83
		Wetland Values	84
		Unusual Characteristics	85
	3.1.2	Palustrine Hardwood Swamps	85
		Geophysical Characteristics	86
		Wetland Criteria Characteristics	86
		Characteristic Fauna	89
		Wetland Values	89
		Unusual Characteristics	90
	3.1.3	Shrub Swamps	90
		Geophysical Characteristics	90
		Wetland Criteria Characteristics	91
		Characteristic Fauna	94
		Wetland Values	94
		Unusual Characteristics	94
	3.1.4	Floodplain Forests	95
		Geophysical Characteristics	95
		Wetland Criteria Characteristics	96
		Characteristic Fauna	99
		Wetland Values	99

			Unusual Characteristics	100
3.2	Marshes			100
	3.2.1	Palustrine Marshes		100
			Geophysical Characteristics	102
			Wetland Criteria Characteristics	102
			Characteristic Fauna	107
			Wetland Values	108
			Unusual Characteristics	108
	3.2.2	Floodplain Marshes		108
			Geophysical Characteristics	109
			Wetland Criteria Characteristics	110
			Characteristic Fauna	110
			Wetland Values	111
			Unusual Characteristics	111
	3.2.3	The Everglades		111
			Geophysical Characteristics	111
			Wetland Criteria Characteristics	112
			Characteristic Fauna	114
			Wetland Values	114
			Unusual Characteristics	115
	3.2.4	Prairie Potholes		115
			Geophysical Characteristics	116
			Wetland Criteria Characteristics	116
			Characteristic Fauna	117
			Wetland Values	118
			Unusual Characteristics	119
3.3	Bogs			119
			Geophysical Characteristics	120
			Wetland Criteria Characteristics	120
			Characteristic Fauna	122
			Wetland Values	123
			Unusual Characteristics	123
3.4	Deep Water and Other Inland Aquatic Ecosystems			123
	3.4.1	Rivers and Streams		123
			Geophysical Characteristics	124
			Ecological Characteristics	124
			Characteristic Fauna	125
3.5	Ponds and Lakes			126
			Geophysical Characteristics	126
			Ecological Characteristics	126
			Characteristic Fauna	127

4. **ESTUARINE WETLANDS** 128
   Robert J. Livingston
   Introduction 128
      4.1 Estuarine Wetland Functions and Values 130
         4.1.1   Overview of Estuarine Wetland

			Functions and Values	130
		4.1.2	Productivity	131
	4.2	Salt Marshes		134
		4.2.1	Geophysical Characteristics	135
		4.2.2	Wetland Criteria Characteristics	136
			Vegetation	136
			Soils	139
			Hydrology	139
		4.2.3	Characteristic Fauna	140
		4.2.4	Wetland Values	141
		4.2.5	Unusual Characteristics	141
	4.3	Mangrove Swamps		142
		4.3.1	Geophysical Characteristics	143
		4.3.2	Wetland Criteria Characteristics	143
			Vegetation	143
			Soils	144
			Hydrology	144
		4.3.3	Characteristic Fauna	144
	4.4	Wetland Values and the Future of Estuarine Wetlands		147

5. **FIELD RECOGNITION AND DELINEATION OF WETLANDS** 153
   Ralph W. Tiner

   Introduction 153
   5.1 Purpose and Use of the Corps Manual 154
   5.2 Wetland Indicators 159
       5.2.1 Hydrophytic Vegetation 161
             Basic Rule for Determining
             Hydrophytic Vegetation 162
             FAC Neutral Option 167
             Other Hydrophytic Vegetation
             Indicators 167
             Vegetation Inconclusive 167
       5.2.2 Hydric Soil 168
             Technical Criteria for Hydric Soil 168
             Soil Wetland Indicators 168
       5.2.3 Wetland Hydrology 173
             Wetland Hydrology Indicators 175
   5.3 Field Procedures for Wetland Delineation 175
       5.3.1 General Guidance for Field Work 177
       5.3.2 Corps Wetland Delineation
             Procedures 178
             Routine Determination Methods 179
             Comprehensive Determination
             Method 185
             Atypical Situations 186
             Problem Areas 187

		5.4	Major Wetland Data Sources	187
			5.4.1 Topographic Maps	188
			5.4.2 National Wetlands Inventory Maps	188
			5.4.3 Soil Surveys and Hydric Soil Lists	193
			5.4.4 Aerial Photographs	194
			5.4.5 Recorded Hydrologic Data	197

6. PROBLEM WETLANDS FOR DELINEATION ... 199
Ralph W. Tiner

Introduction ... 199
6.1 Problematic Wetland Plant Communities ... 200
6.2 Problematic Hydric Soils ... 206
6.3 Hydrologically Difficult Wetlands ... 210
6.4 Problematic Field Conditions ... 211

7. THE REGULATORY FRAMEWORK ... 213
Mark S. Dennison, James F. Berry

Introduction ... 213
7.1 Clean Water Act § 404 Program ... 213
    7.1.1 Background ... 213
    7.1.2 Wetland Jurisdiction ... 216
    7.1.3 Jurisdictional Determinations ... 217
    7.1.4 Jurisdiction Over "Adjacent" Wetlands ... 219
    7.1.5 Jurisdiction Over "Isolated" Wetlands ... 220
    7.1.6 Migratory Bird Link to Interstate Commerce ... 221
    7.1.7 Discharges Into Waters of the United States ... 223
    7.1.8 Federal Manual for Delineating Wetlands ... 224
    7.1.9 State Assumption of the § 404 Program ... 226
7.2 Dredge and Fill Permits ... 227
    7.2.1 The Application Process ... 228
        Who Must Apply? ... 228
        The Application Form ... 228
        Other Application Requirements ... 230
    7.2.2 Notice, Comment, and Conflict Resolution ... 232
        Public Notice ... 232
        Comments ... 233
        Conflict Resolution ... 234
7.3 Corps Responses to § 404 Permit Applications ... 234
    7.3.1 The Public Hearing ... 235
        Notice of Hearing ... 236
        Form and Conduct of the Hearing ... 236

	7.3.2	Issuing the Permit	237
	7.3.3	General and Nationwide Permits	238
		Nationwide Permits	239
		Nationwide Permit 26	240
		Other Exemptions	240
	7.3.4	The EPA Veto	241
7.4	**Remedies for § 404 Permit Denials**		243
	7.4.1	Lack of Administrative Appeals Mechanism	243
	7.4.2	Regulatory Takings Challenges	244
	7.4.3	Harm–Prevention Analysis	246
	7.4.4	Regulatory Takings in the Wetlands Context	247
7.5	**Challenging Issuance of a § 404 Permit**		248
	7.5.1	Challenges in Federal District Court	249
		Ripeness	249
		Statute of Limitations	250
		Right to Jury Trial	250
		Attorney's Fees	250
		Standard of Review	251
	7.5.2	The Administrative Record and *De Novo* Review	251
7.6	**Impact of Other Federal Environmental Laws**		253
	7.6.1	Overview of Other Environmental Laws That May Affect Wetland Activities	253
	7.6.2	National Environmental Policy Act (NEPA)	254
	7.6.3	River and Harbors Act of 1899 (RHA)	255
	7.6.4	Clean Water Act (CWA)	256
	7.6.5	Coastal Zone Management Act	257
	7.6.6	Endangered Species Act of 1973	258
	7.6.7	Marine Mammal Protection Act of 1972	261
	7.6.8	Marine Sanctuaries Act	261
7.7	**Advance Identification of Wetlands (ADID)**		262
	7.7.1	The ADID Process	262
	7.7.2	The Rookery Bay ADID	263
7.8	**Special Area Management Plans (SAMPS)**		264
	7.8.1	Purpose and Development of SAMPs	264
	7.8.2	SAMP for New Jersey's Hackensack Meadowlands District	266
7.9	**State and Local Wetland Regulations**		267
	7.9.1	State and Local Wetland Permit Requirements	267
	7.9.2	New York State Tidal Wetland Regulation	269
	7.9.3	New Jersey State Coastal Wetlands	

		Regulation	270
	7.9.4	New York State Wetlands Case	271
	7.9.5	United States Claims Court Wetlands Case	274
	7.9.6	Importance of Wetland Maps in the Case	275

8. WETLAND MITIGATION ... 278
James F. Berry, Mark S. Dennison

Introduction ... 278

8.1 The Regulatory Framework ... 280
- 8.1.1 The Origin of Wetland Mitigation Policy ... 281
- 8.1.2 Kinds of Mitigation Measures ... 283
- 8.1.3 The Role of Other Federal Agencies ... 284

8.2 Corps/EPA 404(b)(1) Mitigation Guidelines and Joint MOA ... 286
- 8.2.1 Overview ... 286
- 8.2.2 EPA/Corps Mitigation MOA ... 287
- 8.2.3 "Practicable Alternatives" ... 288
- 8.2.4 The "Sweedens Swamp" Case ... 290

8.3 Forms of Mitigation ... 291
- 8.3.1 Mitigation Alternatives ... 291
- 8.3.2 Sequencing ... 293
  - Avoidance ... 294
  - Minimization ... 294
- 8.3.3 Compensatory Mitigation ... 295
  - Exceptions to the Sequencing Requirement ... 296
  - Other Considerations ... 297

8.4 Wetland Banking ... 298
- 8.4.1 The Concept of Mitigation Banking ... 299
- 8.4.2 The Tenneco LaTerre Wetland Mitigation Bank ... 299
- 8.4.3 Advantages and Disadvantages of Mitigation Banks ... 301
  - Advantages ... 301
  - Disadvantages ... 301
  - Recommendations ... 302

9. ASSESSING RISKS TO ECOLOGICAL RESOURCES IN WETLANDS ... 304
Thomas P. Burns

Introduction ... 304

9.1 Endpoints and Understanding ... 306

9.2 The Ecological Risk Assessment ... 308
- 9.2.1 Problem Formulation ... 308
- 9.2.2 Exposure Assessment ... 311

xxii   Contents

    9.2.3 Effects Assessment 313
    9.2.4 Risk Characterization 314
  9.3 ERAs and Wetlands 316
    9.3.1 Examples of Wetland ERAs 316
    9.3.2 Special Problems with Wetland ERAs 317
  9.4 Summary 318

10. RESTORATION AND CREATION OF WETLANDS 319
  Robert P. Brooks
  Introduction 319
  10.1 **Definitions** 320
  10.2 **The State-of-the-Technology** 320
  10.3 **The Process** 322
    10.3.1 Functional Assessments 323
    10.3.2 Site Objectives 324
    10.3.3 Site Selection and Development of
       Conceptual Plans 325
    10.3.4 Prepare Construction Plans,
       Specifications, and Budget 331
    10.3.5 Implementation and Maintenance 334
    10.3.6 Monitoring and Evaluation 336
  10.4 **Case Study—Tipton Demonstration
     Wetlands** 341
    10.4.1 Background 341
    10.4.2 The Process 341
    10.4.3 Was the Tipton Demonstration Project
       a Success? 345
  10.5 **Cost Considerations and Conclusions** 347
  10.6 **Guide to Wetland Permit Sequencing and
     Project Management for Wetland
     Restoration or Creation** 348

LITERATURE CITED 352

**APPENDIX: FEDERAL AND STATE WETLAND
AGENCIES AND OFFICES** 384
  U.S. Army Corps of Engineers Division Offices 384
  U.S. Environmental Protection Agency
    National and Regional Offices 388
  Other Federal Offices 391
  State Wetland Agencies and Offices 392

LIST OF ABBREVIATIONS AND ACRONYMS 406

INDEX TO SCIENTIFIC AND COMMON NAMES 409

INDEX 424

Plates xxiii

I. Fringing cattail marsh and water lily bed in a pond.
II. Northern sedge meadow.
III. Northern bog in autumn.

IV. Louisiana cypress-gum swamp.
V. Hemlock swamp.
VI. Irregularly flooded New England salt marsh.
VII. Organic soil (muck).
VIII. Hydric mineral soil (histic epipedon).
IX. Hydric mineral soil.
X. Hydric sandy soil

1
Overview

Mark S. Dennison
and
James F. Berry

1.1 PURPOSE AND SCOPE OF THE BOOK

This book is intended as a practical guide to scientific, legal, and technical issues concerning wetlands. As such, it is written in the most practical terms, with numerous helpful examples and case studies of how specific issues should best be addressed. The book is organized in a way that exposes the reader in logical succession to the full gamut of complex scientific, legal, and technical aspects of wetlands. Because a broad spectrum of issues are covered in one comprehensive volume, space constraints necessarily limit full discussion of certain topics. Whenever additional information is available on a particular subject, the book points the reader toward additional sources. Wetlands have been the subject of countless articles and books - scientific, legal, and technical. The comprehensive listing of "literature cited" found at the end of the book contains references to many other helpful publications.

This book recognizes that wetland science, law, and technology are interdependent disciplines. Most other works focus on one of these disciplines while perhaps providing some cursory treatment of related disciplines. This book attempts to meld several different perspectives on the subject of wetlands and to show the interrelationships between the various professions that deal with wetland issues. Paramount is an understanding that no single discipline operates in a vacuum. For instance, the regulatory framework is driven by the state of the science of wetlands. Further, technological advances in the creation and restoration of wetlands are dependent on an understanding of the ecological principles that drive wetland functions and values.

The book is organized as a guide through the various scientific, legal, and technical components of wetlands. Within each individual chapter, extensive cross-referencing is provided to help the reader link related aspects of the issue being discussed. Further, within the presentation of each separate chapter is a discussion of how the various scientific, legal, and technical aspects of the subject interrelate. Each chapter has been written by an known authority with specialized experience in the topic being presented. Although a specific chapter may be consulted for a distinct explanation of a particular issue, the book is not a compilation of papers dealing with separate topics. Rather, the book has been developed as an integrated whole, as though one author had written the entire book.

The book begins with an overview of the issues, and explanation of the interdependence of wetland science, law, and technology. Chapter 2 provides a thoughtful discussion of the ecological principles that form the basis for wetland functions and values, which influence the state of the science. Chapter 3 offers a practical presentation of the various types of freshwater wetlands - the characteristics of each type, the values and functions of freshwater wetlands, and the hydrologic, vegetative and soil indicators used by scientists to classify each type of freshwater wetland. Chapter 4 continues with a discussion of marine and estuarine wetlands, presented in much the same fashion as the freshwater wetland types are examined in the preceding chapter. Chapter 5 is a practical guide to identifying and delineating wetlands, using various field procedures. The chapter explains the three wetland indicators - hydrology, hydrophytic vegetation, and hydric soils - and discusses how the three indicators relate to scientific delineation and legal classification of wetlands. Chapter 6 follows with specific discussion of problem wetland types, which are difficult to delineate because of problematic environmental conditions that influence the presence of the three wetland indicators. Chapter 7 outlines the legal framework for wetlands, explaining the regulatory requirements for activities undertaken in wetland areas, the authority of the U.S. Army Corps of Engineers, EPA, and other federal, state, and local agencies to protect wetlands, the impact of related environmental laws, and innovative wetland protection measures, such as ADIDs and SAMPs. Chapter 8 provides a practical discussion of the controversial issue of wetland mitigation, explaining Corps/EPA guidance concerning if, when, and how mitigation measures may be used. The innovative use of "wetland banking" is also examined. Chapter 9 takes a look at the recent use of ecological risk assessments (ERAs) to determine the possible risks to ecological resources in wetland areas. Finally, Chapter 10 offers a valuable discussion of wetland restoration and creation, describing the process in practical terms, with step-by-step procedures, cost estimates, and a helpful case study.

1.2 INTRODUCTION TO WETLANDS

The term "wetland" has been used to mean a variety of different habitats, which often have little in common. Certainly most people would recognize terms like "swamp," "marsh," or "bog" as representing some sort of area containing water during at least part of the year. Unfortunately, most people place these terms in a negative context. A "wetland" too often induces images of mosquito-ridden, snake-infested, smelly areas where disease is rampant and danger lurks (see Mitsch and Gosselink, 1993). Misperceptions that wetlands serve only as dangers to health and welfare, or as obstacles to progress, have resulted in the destruction or degradation of nearly half of the nation's wetlands resources as wetland areas were converted to agricultural uses, housing, industry, or for a variety of "reclamation" projects (see Section 2.4).

Nevertheless, at least a few positive values of wetlands were recognized quite early in our nation's history. For example, a few wetlands were preserved as habitats for migratory waterfowl, for fishing, or for human recreation as early as the nineteenth century. However, it has only been within the past half-century that wetland scientists and other interested people began studying the many positive attributes of wetlands in such a way that a full appreciation of the many values of wetlands were recognized. As a consequence, a new emphasis has been placed on the preservation and study of wetlands of all kinds.

1.2.1 Wetland Definitions and Classification

To those people who are unacquainted with the issues, it might seem that wetlands should be easy to define and classify. After all, how hard can it be to define a "swamp," a "marsh," or a "bog?" However, wetlands are neither easily defined nor classified.

Definitions. There is no single, formal definition of wetlands among wetland ecologists and managers, or even among government regulators. Wetland definitions often reflect the purposes for which they were created (e.g., regulation, scientific investigation, or conservation). Federal laws and regulations often contain different and contradictory definitions, while state and local laws and regulations often differ from each other and from federal definitions (see 56 Fed. Reg. 40,446 (Aug. 14, 1991)). There are over fifty federal and state wetland definitions (Willard et al., 1990). These definitions were developed to protect specific resources associated with water. Despite differences in actual wording, there is much agreement among scientists with respect to the types of areas or habitats that are wetlands (Environmental Defense Fund/World Wildlife Fund, 1992). Nevertheless, wetland definitions and technical criteria for their identification and delineation may be difficult for most of the general public to comprehend.

The problem of formal definitions is of particular interest to the regulated community (i.e., those persons who are subject to limitations and restrictions placed on them by federal, state, and/or local laws), and to those persons, organizations and agencies involved in wetland protection. Inconsistent wetland definitions place a severe burden on any private property owner who wishes to develop a parcel of land that may (or may not) contain wetlands which are potentially subject to regulation by a governmental agency. This burden is particularly harsh when the governmental agencies with jurisdiction have conflicting definitions.

Problems in defining wetlands stem from the nature of wetlands themselves. Wetlands vary enormously in their characteristics and functions (see Chapter 2). Many wetlands are transient, and may seem to disappear for long periods of time during droughts or when water levels are lowered. In addition, many wetlands are variously degraded by dredging and filling such that many of their natural functions are depleted. Willard et al. (1990) regarded wetland definitions as "concepts" while the wetlands themselves exist as a "physical reality." The correspondence between the concept and physical reality is imperfect.

Despite some disagreement regarding wetland definitions, a useful starting point is the regulatory definition of wetlands used by the U.S. Army Corps of Engineers and the U.S. Environmental Protection Agency in administering dredge and fill permitting under § 404 of the

Clean Water Act (33 CFR § 323.2, and 40 CFR § 230.3):

> "The term 'wetlands' means those areas that are inundated or saturated by surface or ground water at a frequency and duration sufficient to support, and that under normal circumstances do support, a prevalence of vegetation typically adapted for life in saturated soil conditions. Wetlands generally include swamps, marshes, bogs and similar areas."

Under this definition, an area is a wetland if it contains (1) wetland hydrology (the presence of water at or near the surface for a period of time), (2) hydrophytic vegetation (wetland plants), and (3) hydric soils (periodically anaerobic soils resulting from prolonged saturation or inundation). Implementation of this definition depends on the wetlands "delineation manual" in place at the time (see Chapters 5-7). Under most circumstances, an area must have *all three* attributes to qualify as a jurisdictional wetland (i.e., one subject to the permitting requirements of § 404 of the Clean Water Act). Many states have adopted versions of this definition (see review in Willard et al., 1990).

By contrast, the U.S. Fish and Wildlife Service (FWS; an agency of the Department of the Interior) uses a more expansive definition which defines a wetland as an area with *any one* of the attributes of wetland hydrology, hydrophytic vegetation, or hydric soils (Cowardin et al., 1979). The result of application of the FWS definition is that more areas would be considered wetlands than under the Corps/EPA definition. The more inclusive FWS definition reflects that agency's mandate to serve as a resource conservation agency, and also to facilitate its responsibility for a National Wetlands Inventory pursuant to § 208(i) of the Clean Water Act (33 USC § 1288(i)).

Both agency definitions reflect the scientific sophistication required to recognize and classify the various types of wetlands. For example, recognizing the various kinds of wetland plants requires an advanced knowledge of field botany.[1] Similarly, identification of the various kinds of hydric soils requires advanced knowledge of soil

---

[1] See Tiner, 1991; and Chapter 5. The USFWS has prepared plant lists for each of the regions of the country to facilitate wetland identification (see Reed, 1988a, and Chapters 5 and 6).

science, and wetland hydrology often requires training in geology and hydrology.

Classification. Most of the problems associated with inconsistent definitions of wetlands apply to classifications of wetlands as well, although there have been relatively few attempts at comprehensive classification of wetlands.

Perhaps the most serious impediment to classifying wetlands in a systematic fashion is that some wetlands do not form discrete tracts of easily identified habitat. Wetland types may overlap or grade into each other, creating a substantial transitional zone. For example, an estuarine embayment may gradually become a tidal marsh, then a brackish marsh, and finally freshwater marsh as one moves inland where freshwater inputs are higher. Moreover, one wetland type may develop from another through time in a process called "succession" (see Chapter 2).

The most significant wetland classifications have been those of the federal regulatory agencies, primarily the U.S. Fish and Wildlife Service. The earliest agency classification was the often-cited "Circular 39" classification, which organized wetlands into 20 categories of particular value for waterfowl management purposes (Shaw and Fredine, 1956). The Circular 39 classification was subsequently replaced by the comprehensive classification of Cowardin et al. (1979), shown in Figure 1.1. The Cowardin classification created the FWS definition of wetlands, and includes wetland types that may not exhibit all three wetland attributes as required by the Corps definition (for example, areas lacking wetland vegetation like mudflats). One particular value of the Cowardin classification has been its use for the National Wetlands Inventory by the FWS (see Chapter 5). It has also provided a standard classification which can be used to compare wetlands from different parts of the country.

Many wetland scientists and others prefer to use more conventional (but somewhat less objective) terms based on familiar wetland types such as "swamps," "marshes," "bogs," "fens," etc. These classifications are generally based on characteristic vegetation, soils, atmospheric and geological conditions, and/or water quality (see Mitsch and Gosselink, 1986). Most of these classifications recognize at least five categories of wetlands, based generally on water flow and salinity: marine, estuarine, lacustrine (lakes and ponds), riverine (rivers

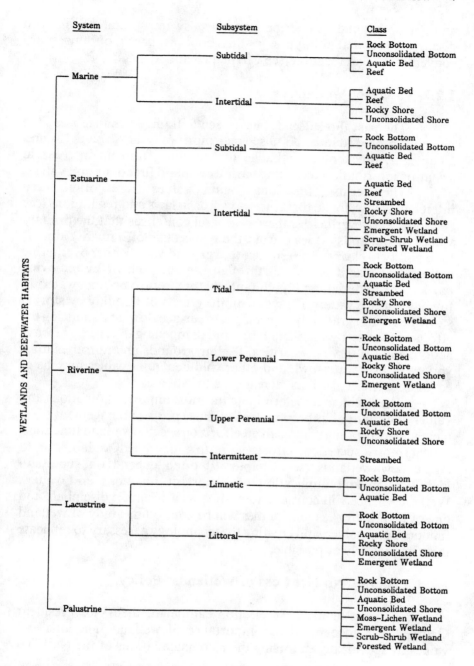

Figure 1.1: The classification hierarchy used by the U.S. Fish and Wildlife Service for wetlands and deepwater habitats (from Cowardin et al., 1979).

and streams), and palustrine (marshes, swamps, and bogs). The discussion of wetland types in this book is based on this more conventional classification (see Chapters 3 and 4).

1.2.2 Values of Wetlands

The past three decades have seen a dramatic change in public perception of wetlands. As discussed more fully in Section 1.4 and Chapter 2, much of this change has resulted from an increase in scientific information regarding various wetland functions and values.

The terms "functions" and "values" are often used interchangeably, in the literature as well as in laws and regulations (see Section 2.3). Nevertheless, there is general consensus that the primary values of wetlands (at least from the perspective of humans) are: (1) their ability to cleanse both surface and groundwater, either by filtering surface water as it percolates through wetland soils or by removing particulate material and pollutants before returning the water to flowing surface waters; (2) reducing the effects of flooding by storing stormwater and gradually returning it to surface flow, and reducing the effects of erosion by stabilizing soils and dampening the effects of wave action; and (3) serving as critical feeding grounds and nurseries for a variety of fish, waterfowl, and other wildlife. Other values may be of equal importance, such as recreation and esthetics.

The problem of determining the most important functions and values of wetlands has become critical, since proper management often means concentrating management efforts on specific wetland functions. Furthermore, successful efforts to restore degraded wetlands or to create new wetlands (see Chapter 9) often depend on successful identification and duplication of important functions and values. Wetland science will continue to mature as a scientific discipline, and one of its primary responsibilities will be continuing study of wetland functions and values, and creating the technology necessary to replicate them as precisely as possible.

1.2.3 Federal "No Net Loss of Wetlands" Policy

As part of his 1988 election campaign, then-Vice President George Bush recognized the importance of wetland functions and values by specifically endorsing the recommendations of the National

Wetlands Policy Forum, an intergovernmental, public and private coalition convened at the recommendation of EPA by The Conservation Foundation (see Conservation Foundation, 1988). This endorsement translated into the Vice President's campaign promise that there would be "no net loss" of the nation's wetlands (see Berry, 1992a).

After the 1988 election, the Bush administration moved quickly to implement aspects of a "no net loss of wetlands" policy. For example, the FWS released the National Wetlands Priority Conservation Plan in April, 1989, which implemented the federal Emergency Wetlands Resources Act of 1986 (see U.S. Fish and Wildlife Service, 1989), and specifically recognized wetland losses across the nation. It also established mechanisms for public-private cooperation in wetland protection, as well as intensified efforts for acquisition of wetlands by federal, state, and local governments (16 USC § 3901).[2]

Unfortunately, competing interests in the Bush administration soon led to departures from the Bush campaign pledge, and the charge by many environmentalists that the administration had abandoned its "no net loss" campaign pledge (for example, see Goldman-Carter, 1992). Examples of these alleged departures involved two controversial Memoranda of Agreement (MOA) in 1989 and 1990 between the Corps and EPA regarding wetlands mitigation requirements (see discussion in Chapter 8). Many environmentalists believed that conservative White House staff members and others outside Congress had been allowed to exert influence to weaken wetland protection by the agencies (for example, see Winter, 1990; but see Wilcher and Page, 1990, for the agency perspective; and see Albrecht, 1992, for a perspective from the regulated community).

However, many observers perceived the clearest signal of Bush administration abandonment of the "no net loss" policy to be proposed changes to the federal Wetlands Delineation Manual (see Sections 1.3.2 and 1.4). The proposed changes were the product of the White House's Council on Competitiveness (chaired by then-Vice President Dan Quayle), and would have reduced significantly the amount of wetlands that would receive federal protection under § 404 of the Clean Water Act (Environmental Defense Fund/World Wildlife Fund, 1992).

"No net loss of wetlands" remains federal policy under the

---

[2] For a review of Bush administration policies and contributions to "no net loss," see Deland (1992).

Clinton administration, despite heavy economic pressures to weaken it. It remains to be seen what specific programs will be put in place to protect wetlands under the Clinton administration, and what vitality the "no net loss" policy will retain under subsequent administrations.

1.3 INTRODUCTION TO REGULATION

1.3.1 Overview of Regulatory Framework

An environmental regulatory scheme for wetlands has developed slowly. In fact, unlike other areas of environmental regulation, including hazardous and solid waste regulation, endangered species protection, air pollution control, drinking water quality, and oil pollution control, no targeted and comprehensive legislation has been enacted to regulate wetland areas. Instead, wetlands have gradually been subject to regulatory oversight under some general provisions contained in § 404 of the Clean Water Act (33 USC Section 1344). Activities in wetlands were virtually uncontrolled until 1972 when the Section 404 program was adopted as part of the Clean Water Act. Until that time, dredging, filling, and human destruction of wetlands basically went unchecked. In fact, the U.S. Fish and Wildlife Service estimated that some 482,000 acres of saltwater wetlands were lost during the period from the mid-1950s to the mid-1970s (see Frayer et al., 1983; Dahl, 1990; Dahl et al., 1991).

Section 404 of the Clean Water Act provides the legal framework for regulation of wetlands. This section of the Clean Water Act gives authority to the U.S. Army Corps of Engineers to oversee dredge, fill and other activities in "navigable waters." The term "navigable waters" has been interpreted by the courts to mean "waters of the United States," regulation of which is not necessarily constrained by navigability.[3] In fact, under this interpretation of the term, wetlands are considered "waters of the United States," and need not be navigable for the Corps to exercise jurisdiction under the Section 404 Program. The Corps initially considered its jurisdiction to be limited by navigability in U.S. waters, however, following the expanded judicial

---

[3] See Natural Resources Defense Council, Inc. v. Callaway, 392 F. Supp. 685 (D.D.C. 1975). For a more detailed discussion of the meaning of "waters of the United States" see Section 7.1.3.

interpretation of the term navigable waters, the Corps prepared new regulations pursuant to the Clean Water Act to encompass regulation of various activities in wetland areas (40 Fed. Reg. 31,319 (1975)). Gradually, this jurisdiction has broadened to include wetlands that are "adjacent" to waters associated with interstate commerce, as well as certain intrastate "isolated" wetland areas located great distances from streams and navigable bodies of water.

The Corps and EPA have concurrent jurisdictional authority over wetlands under the Section 404 program. Although the Clean Water Act is essentially silent concerning which agency has authority to make jurisdictional determinations under the Section 404 Program, the EPA and Corps have formulated agreements detailing their respective jurisdictional responsibilities.[4] The Secretary of the Army, acting through the Chief of Engineers, is authorized to issue individual permits for the discharge of dredged or fill material into wetlands (33 USC § 1344(a); regulations at 33 CFR Part 323; see Section 7.2). In some circumstances, the Corps may issue "nationwide permits" for certain activities in jurisdictional wetlands that are deemed to have minimal environmental impacts.[5] EPA is responsible for formulating the guidelines used by the Corps to make the permit decisions (see Clean Water Act § 404(b)(1) Guidelines; Correction at 55 Fed. Reg., 9210, 9211 (Feb. 7, 1990)).

The Corps regulations set forth extensive procedures for the permit process (see Section 7.2; 33 CFR Parts 320, 323, and 325). Following submission of a permit application for activity in a wetland area, the Corps must decide whether to grant the permit and, if granted, whether any conditions should be placed on the permit (33 CFR § 320.4 lists the criteria for evaluating a permit application). In evaluating a permit application, the Corps is required to consider the recommendations of the FWS and the National Marine Fishery Service

---

[4] EPA/Department of Defense, Memorandum of Understanding on "Geographical Jurisdiction of the Section 404 Program (MOU)," 45 Fed. Reg. 45,018 (July 2, 1980); Department of the Army/EPA Memorandum of Agreement "Concerning the Geographic Jurisdiction of the Section 404 Program and the Application of the Exemptions under Section 404(f) of the Clean Water Act (MOA)" _ Fed. Reg. _ (Jan. 19, 1989).

[5] 33 USC § 1344(e). Corps regulations governing the Nationwide permit program are found at 33 CFR Part 330. See discussion of nationwide permits in Section 7.3.3.

(33 CFR § 320.4(c)), as well as comments and objections from other federal and state agencies.[6] In addition to Corps permits, an applicant may need to secure additional state and local wetland permit approvals (see Section 7.9), as well as various necessary approvals under other federal and state environmental laws (see Section 7.6).

1.3.2 Controversy Over Federal Wetland Delineation Manual

Today, regulation of wetlands is one of the most controversial areas of environmental protection law and policy. Private property owners, government regulators and wetland scientists are engaged in ongoing and vigorous debate concerning which wetlands should be regulated, how they should be regulated, and to what degree they should be regulated. Much of the controversy is driven by the conflicting economic interests of private property rights proponents and land developers who want to use wetlands, and the environmental protectionist goals of various public interest groups who want to preserve them.[7]

The conflict has been further compounded by congressional failure to adopt a comprehensive legislative scheme for regulating wetlands. Since amendment of the Clean Water Act in 1972, which instituted Corps and EPA oversight of wetland protection through the Section 404 program, Congress has opted to let the Corps, EPA, FWS and U.S. Department of Agriculture Soil Conservation Service (SCS) deal with the problem of how to define and regulate wetland areas. The result has been chaos and uncertainty for both the regulatory agencies and the regulated community. During the first seventeen years of the §

---

[6] For instance, the Coastal Zone Management Act permits a state to object to a proposed permit if the state has an approved Coastal Zone Management Program (CMP) and the state determines that issuance of the permit will be inconsistent with the goals of the state CMP. 16 USC § 1456(c)(3)(A).

[7] See Dennison (1992), describing the private property rights movement; and Environmental Defense Fund/World Wildlife Fund. 1992. How Wet Is a Wetland? The Impacts of the Proposed Revisions to the Federal Delineation Manual, Report by Environmental Defense Fund and World Wildlife Fund (Jan. 16, 1992), reproduced in Daily Env't Rep. (BNA) at E-1 (Jan. 17, 1992) (January 1992 report by the Environmental Defense Fund and the World Wildlife Fund indicating that the proposed changes to the manual would strip federal protection from approximately one-half the remaining wetlands in the United States).

404 program, the agencies couldn't even agree on how a wetland should be defined. Then they adopted the 1989 "Federal Manual for Identifying and Delineating Jurisdictional Wetlands" (1989 Manual; Federal Interagency Committee, 1989) without public input. Now the 1989 Manual has been shelved, and somewhat ironically in August 1992, its use even forbidden by congressional mandate (Pub. L. No. 102-377, 106 Stat. 1315, 1324 (1992)) while public comments on EPA's August 1991 proposed revisions were being considered (56 Fed. Reg. 40,446 (Aug. 14, 1991)).

The legal framework for regulating wetlands has primarily suffered its inadequacies because of a lack of consensus over what the legal definition should be in light of the scientific criteria. This is why revision of the 1989 Manual has been such a painstaking process. It's hardly surprising that EPA (after receiving over 80,000 public comments on proposed revisions to the manual) is putting the whole issue in the hands of the National Academy of Sciences. While a new manual is under consideration, it is important to note that the 1987 Corps of Engineers Wetlands Delineation Manual must be used for making wetland determinations (U.S. Army Corps of Engineers, 1987). In fact, in January 1993, EPA and the Corps signed a Memorandum of Agreement amending their January 19, 1989 MOA to provide that EPA and the Corps will use the 1987 Manual in making wetland determinations (58 Fed. Reg. 4995 (Jan. 19, 1993)).

1.4 INTERPLAY BETWEEN SCIENCE AND REGULATION

The science of wetland ecology has undergone an important transformation over the past thirty years (see Chapter 2). What was once a largely descriptive science whose primary function was to determine ways to "improve" wetlands for human purposes has grown into an imposing discipline that relies on the most technologically advanced biological, chemical, and mathematical procedures. A science that was once dominated by traditional natural historians is now replete with theoretical and applied ecologists, engineers, mathematicians and statisticians, and chemists.

This change has been driven at least in part by changes in public attitudes about environmental protection. Since the early 1960s, public awareness of the costs of environmental degradation and the destruction of natural resources, as well as explicit public recognition of the various values of wetlands (see Chapter 2), has resulted in public support for a greater federal and state role in protecting wetlands of all kinds. The result has been a series of changes in land use policies, and the environmental laws and regulations that implement these policies. Wetlands have received a share of the scrutiny, and have been affected in a variety of ways (see Chapter 7).

During the same period, the stakes in governmental decisions that affect wetlands have increased dramatically. The post-war baby boom of the 1950 (and escalating immigration from other countries) meant a rapidly increasing population in the U.S. which, coupled with a national increase in affluence, meant huge increases in demand for manufactured goods, agricultural goods, and housing. Added to a century of largely unchecked pollution, a variety of reclamation projects that altered natural ecosystems in many parts of the country, and federal, state, and local governments which had generally failed to become involved in environmental protection, an enormous burden was placed on land resources in many parts of the country. Within this climate of rapidly changing attitudes on environmental policy, coupled with dramatically improved technology, scientists began to identify the effects of wetland losses on a wide variety of human values with much greater precision (see Chapter 2). During the 1960s and 1970s it became increasingly evident that unchecked (and unregulated) destruction and degradation of wetlands could not continue without serious consequences. It was in this climate that the major wetlands protection statutes and regulations were enacted (see Chapter 7). Unfortunately, these regulations were controversial from the beginning, having been challenged from the outset by a variety of development and agricultural interests which argued that they were an unfair abridgement of fundamental property rights.

The result of the conflicting pressures of environmental awareness and conversion of land to human use can be seen virtually any time a developer attempts to develop for human use a parcel of property which even arguably contains wetlands. There are many of examples from around the country of multi-million dollar development projects that have been delayed or stopped altogether because a wetland may be present on the property. Although they do not receive the same degree of public interest, the same conflicts often arise when a farmer wishes to convert a wetland area to agricultural use, or when a private homeowner attempts to fill a small portion of a wetland to expand a home.

However, at the heart of these confrontations lies a fundamental contradiction in American values. Since its inception as a nation, people in the United States have cherished their rights to ownership in property. In fact, property rights of American citizens are carefully protected by many laws. The U.S. Constitution makes it very clear that property rights may not be abridged without due process of law (Amendment XIV, § 1), and private property may not be "taken" for a public purpose without payment of "just compensation" (Amendment V). Early common law in this country recognized that a person who owned real property in "fee simple absolute" (i.e., unencumbered by liens, easements, or other limitations) could do pretty much anything he pleased with the land short of creating abnormally dangerous conditions or nuisances that might affect neighbors (Cunningham et al., 1984).

Many Americans still feel that they may do as they please with their land without interference from the government. Unfortunately, this attitude frequently collides with the right of the government to restrict certain activities on privately owned land to protect societal values (e.g., water quality), or to reduce public harm (e.g., prevent flood damage). Many laws regulate activities on land, including those laws which specifically protect wetlands (see Chapters 7-8).

A recent example of this conflict arose in 1991 in the context of the proposed changes to the federal wetlands "delineation manual" in 1991 (discussed more fully in Chapters 5-7). Development and agricultural interests had successfully lobbied the Bush administration to withdraw the 1989 manual in use at the time, and propose a modified manual. The proposed manual would have reduced dramatically the number of wetlands which received protection under federal law by making the criteria for identification much more stringent (see Environmental Defense Fund/World Wildlife Fund, 1992). However, upon closer scrutiny it became evident that the proposed manual had not been conceived by persons with an adequate knowledge of wetland science, and it was criticized by scientists from all sides of the issue (including many from federal agencies) as being scientifically unworkable and ill conceived (see Huffman, 1991; Pierce, 1991; Shelley, 1991). The proposed 1991 manual apparently stands little chance of approval. Agency personnel are currently developing a new manual based on sound scientific principles.

The proposed 1991 delineation manual controversy serves as a useful warning to regulators, scientists, the regulated community, and the general population of the dangers of ignoring scientific information in formulating policies, laws, and regulations. At the same time, the controversy serves to remind the scientific community that it must strive to assure that the best scientific information is available to the public in general, and regulators and decision makers in particular.

1.5 JURISDICTIONAL DETERMINATIONS

One of the most costly and time consuming aspects of wetlands regulation is the determination of those situations under which the Corps might have "jurisdiction" over a potential wetland for purposes of § 404 and other federal laws and regulations.[8] Much of the controversy associated with wetlands protection reflects the high stakes in determination decisions. This controversy often places the Corps in the awkward position of trying to balance the protection of valuable natural resources against the interests of private citizens (often with substantial political clout) who wish to develop their land.

The Corps has developed a procedure for determining whether or not a "wetland" falls within the Corps' jurisdiction such that a § 404 permit is required (see Chapters 5 and 6). The *Corps of Engineers Wetlands Delineation Manual* (U.S. Army Corps of Engineers, 1987) contains the technical guidelines used to determine if an area contains jurisdictional wetlands. These guidelines are based on three wetland characteristics contained in the regulatory definition of wetlands, namely (1) hydrophytic (water-adapted) vegetation, (2) hydric soils, and (3) wetland hydrology.

If an area is determined to contain jurisdictional wetlands, and if the Corps determines that the proposed project will significantly degrade wetlands, then the Corps may deny the permit, require modifications of the project design, and/or require "mitigation" for the damage the project will cause (see Chapters 7-9). Moreover, the issuance of a § 404 permit has been determined to constitute the kinds of "federal action" that may trigger a variety of other federal statutes,

---

[8] The processes of wetland determination and delineation are discussed more fully in Chapters 5 and 6. See also Section 7.1 discussing Corps/EPA jurisdictional determinations.

such as the National Environmental Policy Act ("NEPA," 42 U.S.C. § 4321 *et seq.*), the Endangered Species Act (16 U.S.C. § 1531 *et seq.*), and the National Historic Preservation Act (16 U.S.C. § 470) among many others (see Want, 1993; Silverberg and Dennison, 1993; and Chapter 7). In addition, many states and local governments have additional regulations, all of which require compliance before construction on a wetland can begin.

Compliance with the many federal, state, and local wetlands regulations often places a tremendous burden on the potential developer. Some developers simply avoid any attempt at developing wetlands. A few developers attempt to circumvent the process by proceeding with their project, and later attempting to obtain an "after the fact" permit, or avoid detection altogether. The latter strategies are extremely dangerous and ill-advised since the criminal and civil penalties for failing to follow the procedures required under federal law may be quite severe (see Chapter 7).

With stakes so high, it is no wonder that a heavy emphasis is placed on the first step of the process, namely determining if an area contains jurisdictional wetlands. This part of the process has become so important to the regulated community and federal agencies that it often dominates the entire process. Indeed, much of the remainder of this book is devoted to various aspects of the determination process.

2

Ecological Principles of Wetland Ecosystems

James F. Berry

INTRODUCTION

The purpose of this chapter is to introduce the major concepts of ecology as they apply to the study of wetlands. The coverage is not intended to be comprehensive, and the reader who requires a more detailed discussion of these and related concepts is advised to consult the references contained in the following discussion.

In order to understand modern techniques for ecosystems analysis, it is important for the reader to be familiar with basic ecosystems ecology. Since studies of wetlands have played a crucial role in the development of the science of ecosystem ecology, this chapter will trace the historical development of the science as it applies specifically to wetlands. However, this chapter also serves as an introduction to the rest of the book. Examples of studies of wetland ecology will be used to demonstrate specific ecological principles; subsequent chapters will provide many additional examples.

"Ecology" is generally regarded to be the study of the relationships between organisms and their environment. However, the science of ecology has undergone radical changes in both philosophy and methodology over the past thirty years. One of the most significant changes is that the science has become less descriptive and more quantitative. Modern ecologists are usually as well versed in mathematics, statistics, and computer science as they are in traditional biology, chemistry and physics. A few simple mathematical and statistical procedures will be introduced in this chapter, but the reader requiring additional information should consult the references.

The study of ecology has been touched deeply by the many ideas arising from modern studies of evolutionary biology. In fact, many of us who received our professional training since the 1960's regard ourselves as "evolutionary ecologists" to call attention to our commitment to the new synthesis of ecology and evolutionary biology (and, perhaps, to distance ourselves from the purely descriptive methodologies of the past).

An "ecologist" is any individual who studies ecological relationships. The conventional wisdom has been that these people should be trained in traditional biological sciences. But in today's world of heightened environmental sensitivity and explosion in governmental regulation, any such limitation seems naive. Ecology has become an interdisciplinary profession, and "ecologists" today may be biologists, chemists, physicists, mathematicians, sociologists, attorneys, economists, planners, politicians, journalists, teachers, homemakers, or nearly any vocation (or avocation) in which individuals participate in the study of ecology in its broadest sense.

2.1 AN INTRODUCTION TO ECOSYSTEMS

2.1.1 Wetlands, and the Concept of an "Ecosystem"

During the past thirty years, it has become increasingly popular to analyze wetlands from the perspective of an ecological system, or an "ecosystem." The term "ecosystem" was apparently first used by the English ecologist A.G. Tansley (1935) although the general concept had been used much earlier (for example, in K. Möbius' use of the term "biocoenosis" in 1877, or in the S. A. Forbes' use of the term "microcosm" in 1887).

There is no universally accepted definition of an ecosystem, but most wetland ecologists accept the concept as described by E.P. Odum (1971) in the third edition of his classic ecology textbook: "Any unit that includes all of the organisms (i.e., the 'community') in a given area interacting with the physical environment so that a flow of energy leads to clearly defined trophic structure, biotic diversity, and material cycles (i.e., exchange of materials between living and nonliving parts) within the system is an ecological system or *ecosystem*."

Odum recognized the following components of an ecosystem: (1) inorganic substances (C, N, CO_2, H_2O, etc.) involved in material cycles; (2) organic compounds (proteins, carbohydrates, lipids, humic substances, etc.) that link biotic and abiotic components; (3) climate regime (temperature and other physical factors); (4) producers (or "autotrophs") which are primarily green plants capable of manufacturing food from inorganic substances and sunlight via photosynthesis; (5) macroconsumers ("phagotrophs"), which are animals or other heterotrophs that ingest other organisms or particulate organic matter; and (6) microconsumers ("saprotrophs" and "osmotrophs"), primarily heterotrophic bacteria and fungi, which are capable of decomposition of dead organismal matter, absorbing some of the decomposition products and making others available for use by producers. Components 1-3 comprise the *abiotic* (nonliving) components of an ecosystem, while 4-6 comprise the *biotic* (living) components. An ecosystem is, then, the basic unit of ecology because it includes both biotic and abiotic components, each influencing the properties of the other, and both necessary to sustain life.

In describing his concept, Odum also recognized that a functional analysis of an ecosystem would include considerations of (1) energy circuits, (2) food chains, (3) diversity patterns (spatial and temporal), (4) nutrient (biogeochemical) cycles, (5) development and evolution, and (6) control (cybernetics). Odum's concept of an ecosystem and how it could be analyzed most effectively has dominated analyses of wetlands and other ecosystems.

A potential problem in defining, classifying, or analyzing a particular ecosystem is the problem of "scope" (i.e., one observer may describe as a complete ecosystem what another describes merely as a subsystem of a larger ecosystem). How does one determine the limits of a particular ecosystem? In many instances, it is clear where the limits of an ecosystem lie. For example, the perimeter of a cypress swamp in the southeastern United States is often quite distinctive where it is completely surrounded by pine flatwoods that differ dramatically in many ecological parameters. On the other hand, a series of hardwood swamps in the floodplain of a major river may constitute a single ecosystem or a series of ecosystems depending on one's perspective.

The problem of scope (i.e., how inclusive or exclusive an ecosystem should be defined) is often treated by ecologists as a matter of convenience (in the words of H.T. Odum (1983), "one person's system is another person's subsystem"). Some ecologists refer to entire biomes (large regional systems characterized by vegetation, such as temperate forest, desert, etc.) as ecosystems, and many regard the entire biosphere (or "ecosphere," the entire area on the surface of the earth which contains all organisms and their environment) as one huge ecosystem (see Krebs, 1990).

Wetland ecosystems are often more clearly circumscribed than other ecosystem types. Nevertheless, wetland ecosystems *do* often grade into one another, so the scope of discussion of wetland ecosystems in this book will be narrow. It should be noted that the question of *legal* delineation of wetlands is not necessarily the same as the ecological boundaries. Problems and procedures for delineating wetlands will be described in detail in Chapters 5-6.

2.1.2 Communities

Whatever definitions ecologists have used to describe ecosystems, most include the concept of a biological "community" (sometimes called a "biotic community"), a concept that is particularly significant in the literature of wetland ecology. Applied initially to vegetation, a community can be defined as an aggregation of living organisms having mutual relationships among themselves and to the environment (Oosting, 1956). The concept of a "community" of organisms recognizes the fact that no organism lives in a vacuum, but is interdependent and interacts with other organisms and the environment.

From early theories about the structure of communities came the "continuum" concept (Whittaker, 1967). This concept recognizes that the ecological distribution (spatial or temporal) of any group of organisms is determined by responses to various environmental parameters. This response is often called "ecological succession," and will be discussed in section 2.2.3 below in the context of ecosystem succession.

2.1.3 Populations, Species, and Organisms.

Biological communities are composed of "populations," which can be defined as groups of organisms of a single species occupying a particular space at a particular time (Krebs, 1990). An entire discipline within ecology developed between the 1940s and 1970s known as "population ecology" (or "population biology"), largely as a result of the impact of the works of English biologists D. Lack (1946), and C.S. Elton (1946), Australian ecologists H.G. Andrewartha and L.C. Birch (1954), and Americans G. E. Hutchinson (1959) and R.H. MacArthur (1972), among many others. Population ecologists study interactions between and among populations of organisms such as competition (both intraspecific and interspecific), predation, growth and decline in numbers, and dispersal (see Emlen, 1984).

Refinements in population ecology have resulted in the development of many mathematical models of various relationships among species, populations, and communities. Mathematical models are often more objective than earlier observational techniques, and provide ecologists with a mechanism not only for describing interactions among the components of ecosystems, but also for making predictions of the effects of changes to ecosystems. Mathematical models have been of great utility for many years in wetland ecology and management. Some of the more important models will be discussed below, and elsewhere in this book.

The concept of a "species" has been defined in at least two ways. First, the notion of a "biological species" is used to indicate a population of organisms which, under natural conditions, are potentially capable of interbreeding with no significant decrease in fitness (i.e., survivability) of the resulting offspring. An alternative definition, the so-called "ecological species" concept, holds that a species is a number of related populations the members of which compete more with their own than with other species (Colinvaux, 1986). In most circumstances (and for the purposes of this book) the two definitions are the same.

The smallest living unit of any ecosystem is the "organism," or individual living thing. The concept of an organism is relatively simple in the case of most animals, but it may become complex when one considers plants, fungi, and certain other groups. For example, what appear to be isolated individual living things in many fungi (like toadstools in a wet prairie) are actually interconnected, genetically identical parts of a single organism.

Each organism within a community occupies a particular "habitat," which can be described as the total of environmental parameters necessary for that organism's survival. However, the term "habitat" is largely an arbitrary term, and is frequently used specifically for the environment of a population (or community, or ecosystem), while the term "microhabitat" is used for the environment of an individual organism (Ricklefs, 1973).

There have been many attempts to classify habitats, but none are uniformly accepted. Southwood (1977) attempted to classify habitats based on various structural and functional components. Begon (1985) proposed an alternative classification based on the effects of the size of an organism on residual reproductive value (survivorship and future fecundity).

Analysis of the various parameters of an organism's "habitat" have been included recently in the broader field of population ecology, particularly as applied to what has come to be known as "niche theory" (see Vandermeer, 1972). The "niche," or unique position in an ecosystem, of an organism (or population, or species) has been variously described as the place in a community of each organism (Elton, 1927), as the organism's specific adaptations for survival in its habitat (Colinvaux, 1982), or as a set of ecological conditions which allow a species to survive. The latter concept was originally proposed by MacFadyen (1957), but a more precise statement of the same conceptual model was that of Hutchinson (1965) who defined an ecological niche as a "multi-dimensional hypervolume of resource axes." This definition provided a useful tool for population ecologists to analyze ecological niches in many dimensions using modern analytical techniques. Hutchinson was also careful to distinguish between an organism's "fundamental niche" (niche hypervolume defined in the absence of competitors), and its "realized niche" (the typically smaller niche hypervolume resulting from shared resources).

2.2 COMMUNITY AND POPULATION ANALYSIS

Although studies of wetland ecosystems were among the first ecological studies, ecologists did not focus systematic attention on wetland ecosystems to any significant degree until the 1960s. The first comprehensive reviews of wetland ecology did not appear until the late 1970s (see Good et al., 1978; and Greeson et al., 1979). Since then, wetland ecosystems have become among the most intensely studied of any ecosystem. This increased interest probably resulted from several factors: (1) wetland ecosystems are relatively circumscribed and somewhat simpler to study than many other ecosystems; (2) wetlands have extremely important values as wildlife habitat, flood control, water quality improvement, and groundwater recharge systems; (3) the diversity of plants, animals, microorganisms, and abiotic components is relatively high; (4) wetlands are often highly productive; (5) as links between aquatic and terrestrial habitats, wetlands are interesting systems for scientific study; and (6) wetlands are among the most sensitive ecosystems to environmental degradation by humans.

In order to gain insight into the structure and function of wetland ecosystems, modern wetland ecologists have concentrated on several aspects of the ecology of wetland communities and populations: (1) classification of communities; (2) ordination (measurement of the ordering of individuals and populations within a community); (3) population structure; (4) competition (both interspecific and intraspecific) among members of wetland communities; (5) population changes (temporal and spatial); (6) ecological succession (changes in community structure over time or space); and (7) pollution and other forms of ecosystem degradation.

2.2.1 Community Classification and Ordination.

Classification. There are no universally accepted methods for classifying wetlands or other communities. Many wetland communities (and ecosystems) have been described simply by dominant plant species (e.g., "cypress swamp," "cattail marsh," or "mangrove swamp"). Because some wetlands (like other communities) are dominated by several plant species, or because dominant species may vary from one location to another, many communities cannot be classified effectively by dominant plant species alone. The classification of wetlands currently accepted by most wetland scientists and government regulators is that of the U.S. Fish and Wildlife Service (Cowardin et al., 1979), which is based on considerations of vegetation, soils, and hydrology, and will be discussed in detail in Chapters 5 and 6.

Early attempts to measure and classify communities were little more than formal methods for cataloging visible vegetation. Two techniques gained general acceptance by ecologists. The first was the product of the Zurich-Montpelier school of ecology. Often called the "Relevé method," this purely subjective technique requires the observer to follow a series of seven steps during which various notations are used to indicate qualities of the community (see Whittaker, 1962; Benninghoff, 1966). If the observer has followed the steps and made notations properly, the end result is called an "association." Because the Relevé method is subjective and not easily quantified, its popularity has declined since the 1970s. However, it still has value as a "quick and dirty" technique for characterizing communities, and retains some popularity among field ecologists. In fact, it is used on occasion in a preliminary fashion to determine the type and borders of jurisdictional wetlands.

A second descriptive technique, the "quadrat method," was developed by the Uppsala (Sweden) school of ecologists. With this method, a series of plots (called "quadrats") are set out at random within the community being sampled. Species of organisms are then listed, measured, or counted within each quadrat. The results from a series of quadrats may then be graphed to produce a "species-area curve" (Figure 2.1). Generally, as the area sampled increases, the likelihood of finding new species decreases, and the graph begins to approach horizontal. Rules exist for determining the point at which the graph reaches its "inflection point" (presumably, the point at which all but the rarest species are included). The area value lying directly

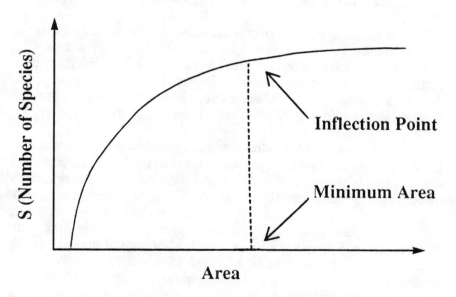

Figure 2.1: A species-area curve (see text; based on MacArthur and Wilson, 1967).

beneath the inflection point of the graph is the community's "minimum area" (i.e., the largest size of a single plot necessary to survey the community).

The primary advantage of the quadrat method over the Relevé method is that the results are much more quantifiable, such that data gathered from quadrats may now be used to analyze a wide variety of relationships between and among plant and animal communities for a variety of purposes. A review of modern uses of data from quadrat sampling can be found in Krebs (1989).

One recent use of species-area curves has been in the context of island biogeography (for example, see MacArthur and Wilson, 1967; MacArthur, 1972; and Simberloff, 1976). This theory holds that the smaller the area of an island, the fewer the species of organisms that will be supported on that island relative to a nearby mainland (or larger island). The concept of an ecological "island" has been shown to apply to virtually any isolated community, including wetlands. For example: (1) animals on isolated mangrove islands (Simberloff and Wilson, 1969, 1970); (2) arthropods on islands of *Spartina* plants in a salt marsh in Florida (Rey, 1981); and (3) mollusk species in lakes in New York state (Browne, 1981).

A problem often encountered when using the quadrat method is determining an appropriate size and shape for the quadrats. The simplest approach is to consult the literature in the field, but there is no guarantee that the correct size and shape have been used in the past. There have been several techniques suggested for picking an optimal quadrat size, based mostly on relative sampling costs (see Hendricks, 1956; and Wiegert, 1962).

The question of quadrat shape is problematic and unresolved. It has been argued that the optimal quadrat shape is round, because a round quadrat minimizes the "edge effect" (the tendency to make errors of inclusion or exclusion at the edge of a quadrat; see Krebs, 1989). Many ecologists continue to use square quadrats because they are easier to manage in the field, but a common approach is to use long, thin quadrats because they cross more "patches" of habitat than square or round ones. A popular form of quadrat sampling is called "line transects" (or "line intercepts"), in which a strip (actually a long, thin quadrat) of known area is searched and organisms located. Data collected from line transects can be used for a variety of purposes, including classifying and characterizing communities. For reviews of line transect procedures and applications, see Anderson et al. (1979), Gates (1979), Burnham and Anderson (1984), and Krebs (1989). For computer programs to calculate line transects, see Gates (1980), and Krebs (1989).

Classification of communities was largely a subjective process until the 1950s. Modern community classification relies an a variety of more objective mathematical and statistical procedures, such as linear regression and analysis of variance (ANOVA), and multivariate statistical analyses. There are several computerized numerical procedures used for classification of communities. Reviews of the various statistical techniques (and computer programs) can be found in Sneath and Sokal (1973), Clifford and Stephenson (1975), Sokal and Rohlf (1981), Legendre and Legendre (1983), Romesburg (1984), Ludwig and Reynolds (1988), and Krebs (1989).

The most commonly used multivariate classification techniques are known collectively as "cluster analyses," which group variables in various ways according to their matrix of correlation coefficients. The most frequently used techniques for cluster analysis for community classification (and taxonomy of organisms) has been the UPGMA (unweighted pair-group method using arithmetic averages) which produces a hierarchical classification "tree" showing relationships among variables (Figure 2.2). Reviews of UPGMA procedures are in Sneath and Sokal (1973), and Romesburg (1984). Several excellent computer programs are available for UPGMA analysis, including Rohlf et al. (1971), Orloci (1978), Rohlf (1985), Rohlf (1988), and Dixon (1990). For examples of the use of numerical clustering techniques to classify wetland communities, see Kortekaas et al. (1976), and Townsend et al. (1983). Cluster analyses are most often used for purposes of classification, but may be used for ordination studies as well (for example, see Williams, 1971; and Green, 1980).

Ordination. In addition to classifying communities, an important challenge for ecologists has been to determine how organisms are ordered along the various dimensions (biotic and abiotic) within and among communities. "Ordination techniques" permit analyses of the relationships between various environmental factors and community composition, as well as indicating species associations related to community structure. Unlike classification techniques, ordination does not assume that communities are discrete (although the results of

28 Wetlands

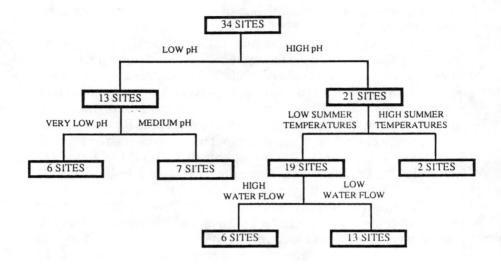

Figure 2.2: A classification tree of 34 invertebrate communities based on physicochemical factors in streams in southern England (based on Townsend, et al., 1983).

ordination analysis may be used to classify communities; see Gauch, 1982). Instead, ordination techniques allow the organization of communities along a graph in multiple dimensions using various community attributes as variables. Those communities that are most similar in community attributes will appear closer together on a graph. For an example of ordination used to classify wetland communities, see Townsend et al. (1983) which analyzed various physical and chemical characteristics of a series of stream invertebrate communities.

Like community classification, early ordination techniques were little more than observations of the relationships among the components of a community. Some elegant modern studies use analysis of variance (ANOVA) and multiple regression to separate out the effects of various environmental parameters on community structure.

For example, Pennings and Callaway (1992) used ANOVA to separate competitive and physical factors in determining community structure of a salt marsh in southern California. Cochran-Stafira and Anderson (1984) used multivariate analysis of variance (MANOVA), which considers interrelationships between many environmental variables simultaneously, to ordinate diatom communities in a northeastern Illinois bog.

However, most modern ordination techniques involve complex multivariate statistical procedures, and invariably require the use of computers. There is considerable confusion and disagreement as to which ordination (or classification) techniques are most appropriate for a particular purpose, and which computer programs should be used. A review of all of the various procedures and computer programs is beyond the scope of this book, but several excellent reviews of multivariate statistical procedures are Cooley and Lohnes (1971), Sneath and Sokal (1973), Sokal and Rohlf (1981), Gauch (1982), Pielou (1984), Digby and Kempton (1987), Krzanowski (1988), and Ludwig and Reynolds (1988).

Since the late 1960s, a variety of relatively simple but extremely useful mathematical procedures have been developed for examining relationships and classifying wetland communities, both in large and small-scale studies (see review in Grieg-Smith, 1983; Kershaw and Looney, 1985; Whittaker, 1991). Examples of these techniques are: (1) the "species-juxtaposition" technique, which determines whether a species occurs adjacent to another more or less than random expectation (Stowe and Wade, 1979); (2) the "species-region" technique, which determines an individual's likelihood of being one species whether or not it occurs in the same region as another species (Stowe and Wade, 1979); (3) the "joint-occurrence" technique, which uses chi-square analysis of primary and secondary occurrence data (Grieg-Smith, 1983; Whittaker, 1991); the "reciprocal averaging" procedure (Hill, 1973); and simple covariance/correlation analysis.

Frequently used multivariate ordination techniques include the various forms of "principal components analysis" (PCA, a form of factor analysis), which reduce clouds of data for N different variables into a series of vectors distributed in N dimensions. Axes that account for variability in the variables can then be extracted from these vectors, and graphed to demonstrate importance (each variable will contribute some proportion of the total variation for each axis). PCA can be used effectively to demonstrate relationships among variables within an ecosystem, and may even be used as a classification tool (Gauch, 1982). For an example of ordination used to classify wetland communities, see Townsend et al. (1983) which analyzed various physical and chemical characteristics of a series of stream invertebrate communities, resulting in a classification of 34 streams into five "classes" along environmental gradients. Figure 2.3 is a graph of the first two principal components resulting from PCA ordination of 16 physicochemical and biotic elements of a series of marshes adjacent to three Illinois rivers (X = Des Plaines River; Y = Fox River; Z = Rock River), based on the author's unpublished data.[1]

Another multivariate ordination technique that is gaining in popularity is "correspondence analysis" (COA) and its variant "detrended correspondence analysis" (DCA) (see Hill and Gauch, 1980;

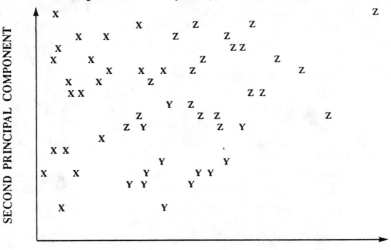

FIRST PRINCIPAL COMPONENT

Figure 2.3: Graph of first two principal components resulting from PCA ordination of marshes associated with three Illinois rivers (X = Des Plaines River, Y = Fox River, and Z = Rock River), based on biotic and physicochemical elements.

---

[1] The first principal component (eigenvalue = 0.77) accounted for 24.4% of variability in the data, while the second principal component (eigenvalue = 0.47) accounted for 17.4%. The two most important environmental variables were percent coverage by *Typha* (cattail), and percent dissolved oxygen.

and Ludwig and Reynolds, 1988). COA and DCA have the advantage over PCA of allowing examination of ecological relationships between species and environmental variables simultaneously (PCA can only perform one at a time). For example, DCA has been used effectively to elucidate patterns and processes in environmental and vegetational variables in Canadian riverine wetlands, and in freshwater tidal wetlands near Chesapeake Bay (Day et al. 1988; and Rheinhardt, 1992). Zampella et al. (1992) used DCA to demonstrate wetland-to-upland gradients in several soil and water characteristics in pitch pine (*Pinus rigida*) lowlands in the New Jersey Pinelands.

2.2.2 Population Structure

Species richness and diversity. Ecologists frequently concentrate their analyses on populations of organisms in order to understand the structure and dynamics of communities and ecosystems. One of the most important and frequently studied aspects of population structure within communities (and, therefore, ecosystems) is the number of species present and the relative abundance of each.

The number of species present in a community is known as "species richness" (McIntosh, 1967), which has become a valuable tool for population measurement in the study of wetlands. Unfortunately, it is often impossible to count every species in a community because some species may be extremely rare, or may be absent or dormant at the time of counting, or may be distributed unevenly. Furthermore, comparisons between communities will nearly always suffer from the statistical problem of differing sample sizes. As a consequence, direct measures of species richness are seldom practical or reliable. For examples of studies related to wetland ecology, see Simberloff (1976), Tilman (1977), Lubchenko (1978), Rey (1981), and Tonn and Magnuson (1982).

Several mathematical procedures have been developed to correct for various forms of bias in measurements of species richness (see review in Krebs, 1989). A technique known as "rarefaction" corrects for samples based on different sample sizes by standardizing all samples to one size (Sanders, 1968; Hurlbert, 1971; and Simberloff, 1972). A second technique, the "jackknife estimate," allows an estimate of species richness based on the frequency of unique (rare) species from quadrat samples. For a study analyzing the benthic fauna of a coastal creek community, see Heltshe and Forrester (1983). A variation on the jackknife estimate is the "bootstrap procedure," in which a parameter of interest is sampled, replaced, and resampled repeatedly (Smith and van Belle, 1984; Krebs, 1989). Meyer et al. (1986) used both jackknife and bootstrap procedures to measure population growth rates of the freshwater water flea, *Daphnea pulex*.

A more realistic alternative to species richness is the measurement of "species diversity," which takes into account not only species richness but also species "evenness" (abundance or importance relative to that of other species). There are many techniques for measuring species diversity (Peet, 1974; and Krebs, 1989), but the most common is the information theory based "Shannon-Weaver" measure:

$$H' = -\Sigma \, p_i \log p_i \quad (2.1)$$

where: H' = species diversity.
p_i = proportion of the total community or sample made up by species i.

Species diversity has been an extremely popular measure in population studies of wetlands, such as in analyses of competition, predation, community stability, and succession (discussed below).

Population size (i.e., the number of individuals in a population) must be known to estimate species diversity. There are many methods for estimating population size of plants, the most common being various forms of quadrats (including line transects). Animals are much more difficult to count. They often enter and leave an area with rapidity, and they are more difficult to observe in the field. Many species-specific methods for trapping, observing or otherwise counting animals are available. Among the more common methods used to estimate population size in animals are "mark-and-recapture" techniques. There are many variations on mark-recapture methods of estimating and following trends in population sizes and their uses (see reviews in Jolly, 1965; Krebs, 1989).

Competition. When similar species coexist in a community at the same time, it is often the case that they must compete for resources (food, space, mates, etc.). Competition may take the form of "interference" (competition for space) or "exploitation" (directly reducing resources) (see Miller, 1967). The consequences of interspecific competition (between members of different species) and intraspecific competition (between members of the same species) have been analyzed in detail for six decades if one recognizes the publication of the classic work of G.F. Gause in 1934 as a starting point.

Many approaches to measuring, analyzing, and interpreting competition and its effects are available. These approaches are discussed in many books and articles on competition, and the reader is referred to nearly any recent book on population biology for a review. For example, see MacArthur and Connell (1966), Hutchinson (1978), Emlen (1984), Hedrick (1984); and Pianka (1989).

The basic premise of interspecific competition theory is that increased numbers of competing species (or decreases in resources) under density-dependent conditions will result in one of two possible outcomes: (1) the various species will "partition" the resources in such a way that each narrows its niche (i.e., it uses a narrower range of resources) and they coexist; or (2) one (or more) species will become extinct. These results are natural outcomes of the "logistic growth equation" as proposed in the 1830s by Verhulst, but now known as the "Lotka-Volterra" competition models (Verhulst, 1838; Lotka, 1925; Volterra, 1926; and Hutchinson, 1978):

$$\frac{dN}{dT} = rN \frac{K-N}{K} \qquad (2.2)$$

where: r = the intrinsic rate of natural increase for that population.
N = number of organisms at time t.
K = number of organisms able to live in the environment, sometimes called the "carrying capacity."

In a population growing in the absence of competitors:

$$\frac{dN}{dT} = rN \qquad (2.3)$$

The major predictions of the models are shown graphically in Figure 2.4. A population growing under the influence of competition (intraspecific or interspecific) will increase in a sigmoid fashion called "logistic growth" until it reaches K, or the "carrying capacity" of the environment. Most populations will then reach some population size at or near K and maintain that population size with only minor fluctuations. There has been much discussion in the literature about the effects of altering K, which is nearly always affected by factors in addition to competition. For example, Elliott (1984) demonstrated changes in K that resulted from atmospheric changes during summer growing seasons in populations of trout (*Salmo trutta*) in English streams. It is likely that artificially altering energy flow by dredging or filling a wetland would have dramatic impacts on K for wetland species.

Of course, K for any species will be affected by the presence of competing species (the effect of species 1 on reducing K for species 2 is measured by the "competition coefficient," α_{21}). Incorporated into equation 2.3 above, this will then give the following model which demonstrates the effect of the presence of species 1 on growth of species 2:

$$\frac{dN_2}{dT} = r_2 N_2 \frac{(K_1 - N_1 - \alpha_{21} N_1)}{K_2} \qquad (2.4)$$

It should also be noted that different species may have very different relationships to K, even in the absence of competitors. Some species have very high birth and death rates, and their evolutionary strategy seems to be maximizing r in the above equations (called "r-selected" species) by investing their reproductive effort in producing many small eggs or seeds. Other species (called "K-selected" species) have much lower birth and death rates, investing reproductive effort in producing fewer, larger seeds or eggs (or investing in long-term parental

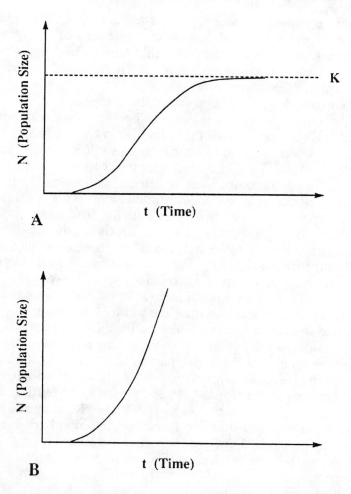

Figure 2.4: Population growth curves resulting from (A) growth in the presence of competitors ("logistic growth," equation 2.2), where K = carrying capacity; and (B) growth in the absence of competitors ("exponential growth," equation 2.3).

care of the offspring), and maintaining population size at or near K. For example, a freshwater sponge that releases millions of tiny eggs into the water every year (most of which will die) is r-selected, while a freshwater turtle that produces four to six relatively large eggs per year (many of which will survive) is more K-selected. Precisely because they can increase population size rapidly, r-selected species are generally much better colonists, and may take advantage of anthropogenic disturbances. For example, wetlands that have been drained or filled frequently undergo a period during which r-selected "weed" plant species rapidly colonize the area.

Interestingly, it has been shown that a species that is K-selected in one place may be more r-selected in another place. For example, Bidgood (1974) demonstrated that whitefish in one Canadian lake are r-selected (lower density, and lay more eggs), while in another lake they are more K-selected (higher density, and fewer eggs).

Interspecific competition is often a determinative factor in community structure. As demonstrated by Gause (1934) using laboratory populations of two species of the freshwater protozoan *Paramecium*, one species may have a competitive advantage over the other leading to local extinction of the inferior species (a phenomenon called "competitive exclusion"). This principle may explain why certain species do not occur in a particular habitat even though all environmental conditions appear favorable. On the other hand, if two or more species are able to decrease the width of the resource spectrum used by each such that overlap in niche utilization is minimized, then it may be possible for the species to coexist. This phenomenon is called "niche partitioning" (or "resource partitioning"), and is often used to explain the high species diversity in certain stable environments (e.g., many ecosystems in the tropics). An example of niche partitioning was demonstrated by Fenchel and Kolding (1976) and Kolding and Fenchel (1976) who determined that five species of amphipod crustaceans in Denmark's brackish Limfjord partition resources along salinity gradients, each species having its own very narrow salinity requirements.

Niche partitioning may be the result of specialization in resource gathering or related structures (called "character displacement" or "character divergence"), such that organisms are more similar where they occur apart and less similar where they occur together (Brown and Wilson, 1956; Grant, 1975). A case of character displacement in a wetlands community is for two species of mud snails (*Hydrobia*) which are the same size where they occur apart, but different sizes when they occur together (Fenchel, 1975).[2]

Actual measurement of competition under natural conditions is often difficult, and most researchers must rely on field manipulations or laboratory experiments to determine when competition actually occurs. Another approach is to use any of several measurements of overlap in resource usage (called "niche overlap") to imply competition (see Emlen, 1984; and Krebs, 1989). One of the most common (and simplest) measures is Morisita's index of similarity (Morisita, 1959; and Horn, 1966), which has been used to measure resource overlap in freshwater turtles (Berry, 1975). Another way to estimate competition is to measure the competition coefficient (α) from equation 2.4 above (see Jones, 1982).

Studies of competition in wetland communities abound. In a classic study of small-scale competitive exclusion, Connell (1961) studied two types of barnacles on a rocky shoreline in Scotland, and found that one genus (*Balanus*) outcompeted and excluded another (*Chthamalus*) from the lower zones of the beach, while the reverse was true in the higher (intertidal) zones. In another example, Pennings and Callaway (1992) examined competition and other factors affecting the distribution of salt marsh plants at Carpinteria Salt Marsh in southern California. The authors determined that: (1) pickleweed (*Salicornia virginica*) tolerated flooding better, and dominated in the low, regularly flooded portions of the marsh; (2) Parsh's glasswort (*Anthrocnemum subterminale*) tolerated high salinity better, and persisted in transitional and high marsh zones; while (3) both species grew best in intermediate zones, where interspecific competition was the most important factor in determining distribution (each species excluded the other from portions of the intermediate zones).

Many plants and animals have evolved adaptations which reduce competition, or provide a competitive advantage over competing species. These adaptations are reviewed in any book on population biology and/or theoretical ecology. Several adaptations in plants are of particular importance in determining community structure in wetland ecosystems. In plants for which light is a limiting factor (which includes

---

[2] Under certain conditions where resources are limited and generalized, character *convergence* may occur (MacArthur and Levins, 1964, 1967; and Cody, 1973). For an example of character convergence in a wetland species (freshwater turtles) see Berry (1975).

nearly all plants), there is often competition for light at the vegetational canopy favoring rapidly growing, tall trees. In addition, the leaves of many trees contain various allelochemics which retard the growth of competing species in a phenomenon called "allelopathy" (see Whittaker, 1970). The bare patches around the base of many coniferous trees and walnuts in hardwood swamps and bogs is probably partly the result of allelopathy.

Competitive relationships also explain the rapid increase in exotic plants in some wetland ecosystems. For example, the paperbark tree (*Melaleuca*) has been introduced from Australia into many cypress swamps in Florida with devastating results. It is both flood tolerant and fire resistant which allows it to coexist with natural cypress, but it recolonizes burned or drought affected swamps much more rapidly than native species. In southern Florida, *Melaleuca* has competitively excluded many native plants and replaced them as the dominant vegetation in many areas (Myers, 1983, 1984).

Predation. Predator-prey relationships have been shown to be additional factors affecting community structure. A "predator" is any organism that uses other living organisms (plant or animal "prey") as a source of energy. The predator population adversely affects the prey population, but is nevertheless dependent on it (Odum, 1971). This definition includes herbivores and parasites, but excludes "detritivores" that feed on the remains of dead organisms.

Predator-prey relationships can be modelled with the logistic equation in much the same fashion as was competition, using the Lotka-Volterra predation model (equations 2.5 - 2.6):

For the prey population, species 1:

$$\frac{dN_1}{dT} = r_1 N_1 - \tau_1 N_1 N_2 \qquad (2.5)$$

For the predator population, species 2:

$$\frac{dN_2}{dT} = \tau_2 N_1 N_2 - d_2 N_2 \qquad (2.6)$$

where: r_1 = intrinsic rate of increase of the prey species.
N_1 = number of prey individuals.
N_2 = number of predator individuals.
τ_1 = fraction of contacts fatal to prey individuals.
τ_2 = predator growth per prey contact.
d_2 = death rate of predators in the absence of prey.

This model predicts that prey numbers should fluctuate over time, increasing when predators are scarce and decreasing when they are abundant and predator pressure is more intense (sometimes called population "cycles"). By comparison, predator numbers should track prey numbers, always lagging slightly behind fluctuations in prey since predators are dependent on prey as an energy source. This model of fluctuations is called the "coupled oscillation hypothesis" (Chapman, 1931), and it has been tested successfully under laboratory conditions (Gause, 1934; Gause et al., 1936; and Utida, 1957).

Despite its intellectual appeal, the coupled oscillation hypothesis does not seem to fit natural population fluctuations very well, probably because: (1) prey populations often cannot find refugia to escape when predator numbers are high, so they are locally extirpated; and (2) emigration rates from other donor prey populations are too low to replenish the extirpated populations (May, 1975). Excellent models of predator-prey relationships are available, and the reader is directed to them (see Holling, 1973; and Emlen, 1984).

Several classical studies of predator-prey relationships have been conducted in intertidal marine wetlands. Paine (1966, 1969) removed all predatory starfish (*Pisaster*) from a rocky, intertidal community in coastal Washington state. After several years, the structure of the barnacle, mollusk, and algae community had changed completely, and species diversity had decreased. In a study of rocky, intertidal communities in New England, Lubchenko (1978) selectively removed herbivorous periwinkle snails (*Littorina*) and found that the snails' presence had determined the composition of the algal community.

Prey species (both plants and animals) have evolved many adaptations to reduce the effects of predation. A complete review of these strategies is beyond the scope of this chapter, but the reader is directed to any modern book on evolutionary biology for a more thorough review. Animals have many adaptations apparent even to the untrained observer, such as (1) escape mechanisms (a marsh fish which escapes a wading heron, or a grass frog which escapes a water snake), (2) armor (the shell on a freshwater turtle or mollusk), (3) schooling behavior (which reduces the risk to any individual prey organism, and may also serve to confuse the predator), (4) camouflage (a limpet is the same color and texture as the rock to which it is attached, or a cryptic moth has wing coloration which matches a cypress tree trunk), and (5) mimicry (many insects and their larvae resemble leaves and twigs, or the palatable viceroy butterfly mimics the coloration of the unpalatable monarch). Many plants have developed structural defenses such as thorns (as on a *Smilax* plant in a hardwood swamp), while others have developed chemical defenses which may be qualitative (e.g., tannins, lignins, and phytoliths which make many plants unpalatable) or quantitative (e.g., alkaloids which disrupt the physiology of the predator). A few animals have developed chemical defenses, such as the bufotoxins secreted by the parotoid glands of toads.

For their part, many predators have developed adaptations which counter those of the prey. For example, some insect larvae can sequester or detoxify alkaloid plant poisons; or an alligator snapping turtle can attract even the fastest fish into its open mouth by using its worm-shaped tongue as a lure. The evolution of interdependent structures and behaviors (as well as the evolution of symbiotic relationships) is known generally as "coevolution," and has drawn considerable recent attention from evolutionary biologists (see Nitecki, 1983; and Futuyma and Slatkin, 1983).

Measuring Population Dynamics. In order to study changes in community structure, it is necessary to determine the parameters of the life histories of the species of which it is composed, such as age structure of each population through time. The traditional method for measuring these parameters is through the construction of a survival-mortality schedule, often called a "life table." There is variation in the ways that ecologists represent life history parameters on life tables. Table 2.1 lists some standard life table parameters, which are used as headings of columns on a life table.

Other life history parameters can be calculated from those in the life table, such as "generation time" ($T = \Sigma\, xl_x m_x$); "net reproductive rate" (the average number of age class 0 offspring produced by an average newborn during its lifetime, $R_0 = \Sigma\, l_x m_x$); and "gross reproductive rate" (the total number of offspring that would be produced by an average organism in the absence of mortality, $GRR = \Sigma\, m_x$).

Once the life table values have been calculated, the results can be put to a variety of uses. For example, survivorship (l_x) can be plotted against age groups (x) to produce "survivorship curves." Such curves are important in characterizing populations as "Type I" (convex curve, high survivorship to old age then high mortality like affluent humans), "Type II" (linear, representing constant mortality), or "Type III" (concave, high mortality among juveniles but very low mortality in adults, like marine fishes with pelagic larvae) (Pearl, 1928). Reciprocal "mortality curves" can be created by plotting mortality (d_x) against age groups.

Of course, a life table is only as good as the data on which it is based. There are many methods available for collecting life history data, and several excellent guides are available. Reviews of methods for field data collection are Greig-Smith (1964), Andrewartha (1970), Mueller-Dombois and Ellenberg (1974), Brower and Zar (1977), Caughley (1977), Cochran (1977), Southwood (1978), Tanner (1978), Green (1979), Kershaw and Looney (1985), and Krebs (1989, 1990).

Life tables are particularly useful in the context of ecosystem management. Harvesting animals from one part of the age distribution often has the effect of changing the life table, some times with drastic consequences. For example, overfishing of adult fish shifts the age distribution to younger, smaller fish with lower fecundity. As a result, both the average catch and average size of fish caught decreases. Increasing the size of the mesh in fishermen's nets may allow more adult fish to escape, thus helping to reestablish the former age distribution

Table 2.1 Frequently measured parameters used in life table construction. The calculations of each value for each age interval are then arranged in columns below the parameter being measured.

x	=	age interval (this can be units of 1 year, 10 years, 1 month, etc.).
l_x	=	survivorship (number of individuals alive at the start of age interval x).
m_x	=	fecundity (number of offspring produced on average by a female of age x).
$l_x m_x$	=	realized fecundity (survivorship x fecundity).
$x l_x m_x$	=	age-weighted realized fecundity.
d_x	=	mortality (number of individuals dying between age intervals x and x + 1).
q_x	=	age-specific mortality rate (probability of l_x individuals dying before age x + 1, $q_x = d_x / l_x$).
p_x	=	age-specific survival rate (probability of l_x individuals surviving to age x + 1).
E_x	=	mean expectation of life for individuals alive at the start of age x: $E_x = \Sigma l_y / l_x$ (where l_y = next age group).
v_x	=	reproductive value (age-specific expectation of future offspring): $v_x = \Sigma l_y m_y / l_x$ (where l_y and m_y = next age group).

(Beverton and Holt, 1957). The goal of population managers should always be to maintain a stable age distribution.

2.2.3 Ecological Succession

The relative abundance of populations and species vary with time due to the dynamics of factors such as competition, predation, and changes in abiotic factors. Just as populations change over time, communities (and ecosystems) also experience temporal changes. The gradual change over time that takes place in an ecosystem is called "ecological succession," because populations of plants and animals are said to "succeed" or replace each other.[3] Succession is often regarded as the key process in ecosystem development. Few concepts have been more important to our understanding of wetland ecology.

E.P. Odum (1969, 1971) provided the conceptual framework for ecological succession accepted by most modern wetland ecologists. Ecological succession (ecosystem development) has three parameters: (1) It is an orderly process of community development involving changes in species structure and community processes with time, and is reasonably predictable; (2) it results from modification of the physical environment by the community; and (3) it culminates in a stabilized ecosystem in which maximum biomass and symbiotic relationships among organisms is maintained. Succession consists of a sequence of communities, each replacing another, in which each successive community is called a "sere" (the intermediate communities are "seral stages," with the first called the "pioneer stage"). The terminal, stabilized system is called the "climax." As explained in section 2.2.2 above, changes in community structure result from a variety of population interactions including predation and competition. Changes in biotic components of the ecosystem then result in changes in the abiotic components, modifying the physical environment to make it favorable for other species.

Most ecological studies of succession have focused on plant communities. Such studies are significant since it is the composition of autotrophic plant communities is usually the ultimate determining factor in heterotrophic plant and animal species composition.

The "Climax." According to the classical conception of succession as advocated by F.E. Clements and his followers, the terminal, stable ecosystem resulting from linear succession through a series of seral stages is called a "climax" (Clements, 1916, 1936). The climax was said to be at internal equilibrium and to be self-perpetuating, such that it behaved much like a "superorganism."

---

[3] The term "succession" was apparently first used by Tansley (1920), who based his concept only on plant communities. It was 15 years later in 1935 that he included animals and abiotic components and coined the word "ecosystem."

The climax that is characteristic for a particular ecosystem is determined by many factors, including weather, soil characteristics, hydrological cycles, and other characteristics. Under this conception of succession, wetland ("hydrarch") succession was once considered to be a series of intermediate, transient stages in the development of a terrestrial forested ecosystem (Mitsch and Gosselink, 1993). Certain characteristic organisms (known as "dominants") typify a particular climax. For example, the oak-hickory climax is common in the relatively dry soils of the midwestern U.S., while pine flatwoods dominate in the sandy soils of the southeastern states.

However, Clements' view of an orderly, self-perpetuating succession with a predictable climax has been challenged since the beginning. Most notably, H.A. Gleason (1926) advocated an "individualistic" model for succession which rejected the "superorganism" concept in favor of one which viewed the relationships among coexisting species as the result of similarities in requirements and tolerances (as well as chance). Under Gleason's view, the associations of species in any successional stage are far from predictable, and the structure of the "climax" (if it exists at all) is subject to change as would be the case for any seral stage.

The modern view of succession and climax formation is probably closer to Gleason's "individualistic" concept than to the "superorganism" view of Clements. Succession implies an orderly replacement of one community by another, leading to a predictable "climax" state. In reality, there are many variables affecting plant communities in successional stages, such that the "climax" may be anything but predictable (or stable). For example, Niering and Warren (1980) reported that the black grass (*Juncus gerardi*) and salt-meadow cordgrass (*Spartina patens*) communities, once the dominant communities along part of the Connecticut coast, had been nearly completely replaced by short saltwater cordgrass (*Spartina alterniflora*). Connecticut salt marsh vegetation, as well as many other wetland communities, are probably better characterized as a "mosaic" of climaxes that shift and change with time (Miller and Egler, 1950).

In addition, modern ecologists recognize that ecosystem development may be arrested at an intermediate stage if disturbances prevent succession from progressing to a true "climax." Such a condition is known as a "disclimax" (or "proclimax"). A disclimax may result from anthropogenic disturbance (an "anthropogenic subclimax";

Odum 1971), such as when dredging of canals near a highland marsh lowers groundwater levels creating a grassland, but many disclimaxes result from natural disturbances.

A form of disclimax of particular importance to wetland ecologists is the "fire disclimax." A variety of ecosystems require periodic burning which serves to clear underbrush, return nutrients to the soil, and otherwise to maintain the integrity of the ecosystem. The organisms within such a "fire disclimax" ecosystem are typically adapted to periodic burning. An example is the sawgrass marsh ecosystem of the Everglades (see Section 3.2.3). When the marshes dry, the sawgrass burns easily because its leaves are highly flammable. However, the growing bud of the plant lies buried in the soil, surrounded by overlapping leaves that protect it from the fire. Following a fire, the plants quickly regrow as a result of the nutrients released to the soil and decreased competition for space (Kushlan, 1990). Fire destroys many invading species, and may play a role in controlling the spread of invading exotic species such as *Melaleuca*. Prevention of fires by well-meaning forest managers has resulted in unnatural succession and large scale community changes in some ecosystems, while exposing others to massive destruction when fires eventually occur. Proper management of fire disclimax communities usually involves either permitting natural burns to take place, or to provide periodic controlled burns which maintain the fire disclimax community.

Autogenic and Allogenic Succession. Ecological succession is "autogenic" (Tansley, 1935) when it results from biotic changes within the ecosystem, as in the filling of a marsh by organic material from decomposing plant and animal remains. Autogenic succession may be "primary" (colonization of an uninhabited area, like sand dunes or lava flows), or "secondary" (replacement of a climax community after a disturbance, as in reestablishment of a hardwood forest after a period of agricultural use, or replacement of a mangrove community after a severe hurricane).

The rate at which primary succession of a newly formed ecosystem occurs often must be measured in centuries, as in the gradual colonization of a newly formed volcanic island. However, an analogous situation arises when intertidal boulders on the California coast are overturned by wave action. Under these conditions, it was demonstrated that five species of algae underwent successional stages to a climax monoculture within less than three years (Sousa, 1979).

Early successional (pioneer) species are typically rapid colonizers of the "r-selected" variety discussed in section 2.2.2 above. These colonizing species may alter physical conditions and availability of resources to such a degree that colonization by other species is made possible, a phenomenon known as "facilitation" (Connell and Slatyer, 1977). Alternatively, the colonizing species may have little or no impact on subsequent successional species (the "tolerance" model), or may make conditions even less favorable for new species (the "inhibition" model). The species that follow the colonists are typically more stable "K-selected" species.

A given species will usually have a characteristic position within the successional (seral) stages of an ecosystem. The relative position of any species will ordinarily be determined by such factors as resource utilization, competition and predator-prey relationships (see Harper, 1977). The characteristics of a species that determine its successional position have been called its "vital attributes" (Noble and Slatyer, 1979).

By contrast, "allogenic" succession results from geological, physical, or chemical forces outside the ecosystem, as in drying of a marsh due to lowering of regional ground water levels or decreased rainfall. The gradual seaward encroachment of salt marshes into estuarine habitats as a result of sediment deposition from rivers and streams is another example. Ranwell (1974) demonstrated the gradual seaward expansion of salt marsh (and woodland) into the estuary of the River Fal in England over a 95 year period. In New England, a dynamic interaction exists between salt marsh development resulting from sediment and peat deposition which tend to extend the marshes seaward, and gradually rising sea levels which tend to move the marshes landward (Redfield, 1972). It has been estimated that sea level rise along the northern Atlantic coast of the U.S. has been 1-3 mm/year in recent decades (Teal, 1986). For a discussion of the interaction of sedimentation, subsidence, and sea level rise related to salt marshes along the Gulf of Mexico, see Delaune et al. (1983).

Most wetland ecologists recognize the important interactions of both autogenic and allogenic processes in succession. In a useful application of these principles, van der Valk (1981) proposed a qualitative Gleasonian model to predict changes in the floristic composition of wetlands based on plant life histories. Van der Valk (1981) characterized this analysis as "allogenic," although it actually includes elements of competition and allelopathy, suggesting that the model actually parallels Gleason's, which combined elements of allogenic and autogenic succession (Gleason, 1917, 1927).

The particular composition of the vegetation of a particular wetland should usually result from (1) the destruction of all or part of existing vegetation (by herbivores, pathogens, or humans); (2) changes in the physical or chemical composition of the ecosystem that favor growth of some species over others; (3) interactions among the plant species (competition, allelopathy, etc.); or (4) invasion and establishment of new species. The van der Valk model suggests that it is then possible to predict the vegetational composition of a wetland if one knows the life history characteristics of all potential plant species (e.g., propagule dispersal, seed production and germination, growth rate under various environmental conditions, and susceptibility to pathogens, competitors, etc.). For example, a wetland experiencing "drawdown" (lack of standing water) will be composed of a particular set of plant species, but many of these will be locally extirpated and replaced by species with flood-tolerant life histories when the wetland becomes flooded.

The van der Valk model of succession examines only the plant community. However, heterotrophic plants and animals often exert great influence on community structure. For example, cyclical increases in herbivorous muskrat populations can result in removal of virtually all vegetation from midwestern prairie pothole wetlands, and new emergent vegetation cannot be re-established until a dry year exposes the seed bank (triggering another muskrat population explosion, and so on) (Weller, 1981).

It should be noted that succession may be characterized as "vertical" (i.e., temporal), or "horizontal" (i.e., spatial). Wetlands are among the best and most frequently used examples of horizontal succession, with open water replaced by submersed vegetation, then by emergent vegetation, and finally upland vegetation as one moves toward shore (Figure 2.5). Cochran-Stafira and Anderson (1984) studied horizontal, hydrarch succession of a northern Illinois bog-marsh-swamp system, and determined that vegetation and diatom seral stages were present, and maintained by factors such as temperature, Ph, color, and free calcium.

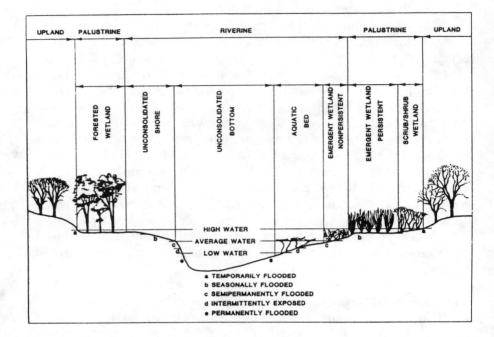

Figure 2.5: Diagrammatic representation of a riverine system with associated wetlands, including submersed and emergent wetland plants. Note the use of the U.S. Fish and Wildlife Service classification (from Cowardin et al., 1979; see Figure 1.1).

2.2.4 Food Webs, Nutrient Cycling, and Energy Flow

The process of succession in wetlands includes a series of often complex cycles among the biotic and abiotic components of the ecosystem. Charles Elton (1927) recognized a phenomenon that he called a "food cycle," where organisms within an ecosystem transfer nutrients from one trophic level to another with numbers of organisms (or biomass) decreasing from plants (autotrophs) through progressive steps to herbivores and carnivores (heterotrophs). Elton demonstrated these relationships with a "food pyramid," showing in simplistic fashion that the number of organisms at each level decreases from bottom to top. Many of Elton's (1927) observations were based on data collected from ponds and other wetlands. His work formed the basis for many subsequent studies of wetland food pyramids (for example, see Juday, 1940; Lindeman, 1941a, 1941b; among others).

Ecologists (including Elton) recognized, however, that relationships among the organisms within an ecosystem are more complex than can be represented by Elton's simple food pyramid (Figure 2.6). A more realistic representation is seen in a "food web," which demonstrates more effectively the non-linear passage of

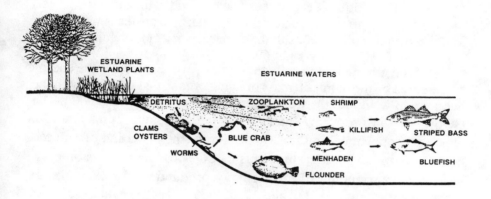

Figure 2.6: Simplified food pathways from estuarine wetland vegetation to invertebrate and fish species of commercial and recreational importance (from Tiner, 1984).

nutrients among various organisms within the ecosystem. One of the earliest food webs to be described was that of Hardy (1924), who demonstrated the transfers of nutrients within the North Sea from phytoplankton, through various planktonic crustaceans and other animals, eventually to the various size classes of herring. Adult herring were the top carnivores in the food web, but each size class fed on a variety of different species of planktonic organisms. Food web models have been developed for a variety of wetland ecosystems, including salt marshes (Montague and Wiegert, 1990), freshwater marshes (Patten et al., 1989), western riparian systems (Faber et al., 1989), and tidal freshwater wetlands (Simpson et al., 1983; Odum et al., 1984).

Productivity. The importance of transfer of nutrients among trophic levels within an ecosystem is that it represents the transfer of energy among individual organisms and trophic levels that drives wetland ecosystem dynamics. But while the study of energy production and transfer is of paramount importance in ecosystem ecology, it is usually difficult (or impossible) to measure. Consequently, ecologists generally must rely on indirect inferences from measurements of biomass, gas exchange, or nutrient transfers.

Beginning with ecologists such as Elton (1927) and Lindeman (1942), studies of aquatic ecosystems have provided the theoretical basis on which much of our understanding of energy transfer is based. The interested reader should consult the classical work of Hutchinson (1949, 1967), Teal (1957), and Wright (1959).

The gross energy production of any trophic level is the energy represented by the biomass (mass of living organisms) together with the energy that went into the work of producing it (Colinvaux, 1986). The energy used in producing the biomass is represented by respiration, which is proportional to the CO_2 produced (the oxidation of reduced carbon compounds). Measurements of CO_2 production plus biomass give an approximation of gross energy production (for examples in aquatic ecosystems, see Hall and Moll, 1975; Pomeroy et al., 1981). As an alternative, direct measurements of isotopic carbon (as C^{14}) can be measured (Steeman-Nielsen, 1952).

Lindeman (1942) defined the "productivity" (rate of production) of a trophic level as the total rate of energy that flows into that trophic level.[4] The efficiency of energy transfer between trophic levels is called "ecological efficiency" (or "Lindeman efficiency"). Productivity has often been measured indirectly by measuring the "standing crop" (biomass/unit time/unit area, often as $g/m^2/yr$) of plant and animal communities, because it is relatively easy to measure (see reviews in

---

[4] Lindeman's hypothesis has been challenged as contradictory, however, because he argued that both efficiency of energy use, and percent of energy lost due to respiration increase with trophic levels. A more realistic model would show energy gains from other than the previous trophic level, such as from bacteria, detritivores, and use by some organisms of multiple trophic levels (see reviews in Burns, 1989; and Higashi et al., 1991).

Odum, 1971; and Whittaker, 1975). Dickerman et al. (1986) have discussed and compared methods of comparing biomass data in wetlands. Productivity of specific wetland types will be discussed in Chapters 3 and 4.

The "primary productivity" of an ecosystem is the rate of biomass production by plants (primary producers) per unit area. It can be expressed as units of energy, dry organic matter, etc. The total energy fixed by photosynthesis is the gross primary productivity (GPP). The proportion lost to the ecosystem in respiration is respiratory heat (R), and the difference is net primary productivity (NPP = GPP - R). NPP represents the rate of production of biomass that is available to consumers (heterotrophic animals, bacteria, and fungi). The rate of consumer biomass production is "secondary productivity." Reviews of these concepts can be found in most recent ecology texts (see Whittaker, 1975).

Energy is made available to organisms in an ecosystem through organic matter. When this matter is generated within the ecosystem, it is called "autochthonous" matter. In aquatic ecosystems, autochthonous matter is generated by photosynthesis by submergent or emergent plants in shallow water, or by planktonic algae in open, deeper water. However, a substantial amount of organic matter enters aquatic ecosystems from outside the ecosystem ("allochthonous" matter), such as when leaves from riparian plants enter a wetland. The relative importance of autochthonous and allochthonous contributions to the total input of organic matter depends on the size and depth of an aquatic ecosystem, and the nature of surrounding communities (see review in Mitsch and Gosselink, 1993).

The relative importance of autochthonous and allochthonous contributions to the total input of organic matter depends on the size and depth of an aquatic ecosystem, and the nature of surrounding communities. In wetland ecosystems, it has generally been found that "detritus" (decomposing plant and animal material) forms a primary source of carbon compounds (for example, see Odum et al., 1984; Smock et al., 1985; Twilley et al., 1986; Moran et al., 1988; Jordan et al., 1989; and Findlay et al., 1990). The role of organic input into specific wetland types will be discussed in Chapters 3 and 4.

A simple but useful model often used to demonstrate energy and nutrient transfer relationships within an ecosystem is the "hydraulic" model, which uses the analogy of a series of interconnected pipes (or lines) to demonstrate energy cascades in a linear fashion (see reviews in H.T. Odum, 1983; Colinvaux, 1986; Mitsch and Gosselink, 1993). Various forms of this conceptual model have been used to demonstrate energy relationships in wetlands (see Mitsch and Gosselink, 1993; Teal, 1962, 1986 (salt marsh); Valiela, 1984 (carbon transfer); Twilley, et al., 1986 (carbon transfer in mangrove forests)).

Biogeochemical Cycles. Chemical materials exist within wetlands and other ecosystems as gasses in the *atmosphere* (e.g., as O_2, or CO_2), as solids in the *lithosphere* (e.g., as calcium in limestone), or as liquids and dissolved materials in the *hydrosphere*. The movement of these various chemical materials within the ecosystem compose the so-called "biogeochemical cycles," so named because they are transferred between the biotic and abiotic components of an ecosystem. A wetland ecosystem may be considered "open" or "closed," depending on the degree to which these materials are brought in from outside the particular ecosystem, or cycled within it.

Biogeochemical cycles have their greatest impact within the soils of an ecosystem. Wetland soils are both unusual and critically important to maintaining wetland organisms (see Mitsch and Gosselink, 1993). Wetland soils are called "hydric soils," because their frequent saturation by water creates anaerobic conditions which favor the growth of wetland plants, called "hydrophytes" (see Tiner, 1991a). Wetland soils and hydrophytes will be discussed in detail in Chapters 3-6. Detailed discussions of biogeochemical cycles in wetlands are in Pomeroy and Wiegert (1981), and Mitsch and Gosselink (1993).

Among the most important biogeochemical cycles within wetlands is the "nitrogen" cycle (see reviews in Whitney et al., 1981; Mitsch and Gosselink, 1993). Nitrogen (N) is an extremely important element for all organisms, since it is necessary for the construction of proteins and other materials necessary for life. However, nitrogen is often the most limiting nutrient in hydric soils, particularly in salt marsh wetlands (see Section 4.2; see also Simpson et al., 1978; Marinucci et al., 1985; Jordan et al., 1989). Nitrogen typically becomes available to wetland plants in the form of the ammonium ion (NH_4^+) or as organic nitrogen, both of which result primarily from the presence in the soil of decomposing plant or animal material. Ammonium ions are then oxidized to NO_3^- in a process called "nitrification," which depends on aerobic soil and plant bacteria (*Nitrosomonas* and *Nitrobacter*). If the NO_3^- is not assimilated by wetland plants or microbes, it may be converted to nitrous oxide (N_2O) or molecular nitrogen (N_2) in an anaerobic process called "denitrification," and be lost to the atmosphere. However, N_2 may be converted back into organic nitrogen by certain bacteria located in the soil or on plant roots, in a process called "nitrogen fixation." Nitrogen fixing and nitrifying bacteria are abundant in salt marsh soils, but are nearly absent from low Ph peat soils such as that found in northern peat bogs (see Section 3.3; see also Moore and Bellamy, 1974; Whitney et al., 1981).

Another important biogeochemical cycle is the "phosphorus" cycle. It is a major limiting element in many freshwater wetlands, but is probably less important in salt marshes where it is relatively abundant (see Mitsch et al., 1979; Whitney et al., 1981; Brown et al., 1984; Jordan et al., 1989; Mitsch and Gosselink, 1993). Phosphorus occurs in wetland soils in several forms, both as soluble and as insoluble molecules, which may be organic or inorganic. It does not cycle through the atmosphere as does nitrogen, but it may be unavailable to wetland plants under aerobic conditions if it is bound as insoluble phosphates, adsorbed into soils, or bound into living, organic compounds. However, phosphorus may be released in soluble form under the relatively anaerobic conditions that often exist in flooded wetland soils.

Still another important cycle is the "carbon" cycle. Gasseous carbon is cycled between soils and the atmosphere as carbon dioxide (CO_2), or organic gasses such as methane (CH_4). Carbon is also made available to soils by the disintegration of lime rock ($CaCO_3$), and the decomposition of organic material. A large percentage of available carbon in wetlands is in the form of "lignocellulose" within the cell walls of vascular plants (Moran et al., 1988). Because lignocellulose is largely indigestible by grazing animals, most primary production in wetland ecosystems enters the detrital, non-living carbon pool. Under anaerobic conditions such as those in waterlogged soils, important pathways by which organic matter in the detrital pool can decompose and re-enter the cycle are by "fermentation" (organic molecules, typically carbohydrates, are reduced into simpler forms, such as alcohols; Wiebe et al., 1981), or by "methanogenesis" (conversion to methane; Valiela, 1984). Other biologically important biogeochemical cycles such as the sulfur, iron, and manganese often interact with the carbon cycle as well as with each other (see review in Mitsch and Gosselink, 1993).

The Hydrologic Cycle. Perhaps the most important biogeochemical cycle in wetland ecosystems is the hydrologic, or water cycle. Wetlands depend on the cyclical entry and exit of water for the production of the wetland soils, hydrophytic vegetation, and hydrology by which "wetlands" are defined (see Section 1.2.1, and Chapters 5 and 6). Because wetlands are typically transitional between upland (terrestrial) ecosystems and open water (aquatic) ecosystems, their biotic and abiotic stability can be dramatically altered by relatively small changes in water cycles.

Water may enter wetland ecosystems as precipitation, surface water flow or runoff, or as ground water recharge (if the wetland is below the groundwater level). Water exits from a wetland by "evapotranspiration" (evaporation from the surface of open water, soils, or the surface of plants), by surface runoff, or by groundwater recharge (see reviews in Novitzki, 1979; Winter, 1980; Mitsch and Gosselink, 1993; and Winter and Llamas, 1993). Most wetlands experience seasonal and yearly patterns of fluctuation in both surface and ground water, which are known as the "hydroperiod" of that wetland. The balance between inflows and outflows of water define the "water budget" for the wetland (see reviews in Winter and Carr, 1980; Brown, 1990; Winter and Llamas, 1993; Mitsch and Gosselink, 1993).

2.3 WETLAND FUNCTIONS AND VALUES

Discussions of wetland conservation, management, or development nearly always include considerations of the various "functions" of wetlands and their "values" to people. For many years environmental philosophers have argued that the environment has "intrinsic" value which alone should justify its preservation, independent of any usefulness to humans. Whether wetlands have an intrinsic value is a philosophical question, but there is little argument that the presence of anthropocentric values of wetlands are the justification for laws and regulations that protect wetlands from degradation and destruction at the hands of humans.[5]

---

[5] In recent years, issues of environmental valuation have come under scrutiny in the context of the protection of "biodiversity" in a variety of ecosystems, including wetlands

The terms "functions" and "values" as applied to wetlands are used more-or-less interchangeably. The U.S. Department of the Interior (1989) has adopted the following definition, based on § 301 of the Emergency Wetlands Resources Act of 1986 (16 USC § 3921):

> "WETLAND FUNCTIONS AND VALUES - The various products, services, functions and values which wetlands provide to society, including fish and wildlife habitat, water supply, improvement of water quality, flood control, erosion and shoreline protection, outdoor recreation opportunities and education and research."

This definition demonstrates the anthropocentric nature of discussions of wetland functions and values, but also provides the framework on which most discussions have centered. These provide a useful starting point, even though wetlands undoubtedly provide many additional functions and values (for examples see Kusler, 1983; Tiner, 1984, 1985a, 1985b; Burke, et al., 1988; Leslie and Clark, 1990; Metzler and Tiner, 1992).

2.3.1 Fish and Wildlife Habitat

A huge variety of plant and animal species utilize wetlands during all or part of their life cycles, many of which are of economic interest to humans. Indeed, it is estimated that wetlands contribute between 60% and 90% (over $10 billion per year) of the U.S. commercial fish catch (Feierabend, 1992).

A complete list of all wetland plants and animals would be far too extensive for this book, but many species typically associated with various freshwater and saltwater wetland types are discussed in Chapters 3 and 4 (see also Niering, 1985, 1991; Ewel and Odum, 1984; Reed, 1988; Mitsch and Gosselink, 1993).

Animal Reproduction and Development. Wetlands are often regarded as biological "nurseries" because many animal species depend on wetlands for reproduction and early developmental stages. Many

(see Zedler, 1988). Several excellent reviews are available, for example Norton (1987, 1988), Ehrenfeld (1988), and the papers in Decker and Goff (1987).

species are dependent on wetlands during early developmental stages. For example, many species of saltwater fish require the low salinity, energy-rich conditions of brackish tidal wetlands for early developmental stages. It has been estimated that two-thirds of commercially important fishes require estuaries and salt marshes as nurseries or spawning grounds, particularly along the Atlantic and Gulf coasts. Familiar species include menhaden, sea trout, croaker, striped bass, and drum (Tiner, 1984). Likewise, coastal wetlands adjacent to spawning streams are vital for many salmon species along the northwest Pacific coast.

Most freshwater fishes are dependent on wetlands for nurseries and spawning, as well as for food production (Niering and Warren, 1980; Tiner, 1984). For example, marshes along the Great Lakes are spawning and feeding grounds for species such as northern pike, muskellunge, walleye, yellow perch, small and largemouth bass, catfish, and many "minnow" species (Jaworski and Raphael, 1978). Freshwater marshes, and swamps adjacent to rivers and lakes are important to most freshwater fish species in the southeastern U.S. (Ewel, 1990; Kushlan, 1990). River swamps and bottomlands adjacent to rivers are also valuable as feeding grounds for adult fishes during times of flood. Spawning of certain fishes is tied to these floods, which allows the juveniles to take advantage of available food and shelter that is provided by the wetlands (see Day et al., 1989; Kushlan, 1990).

Wetlands not associated with rivers and lakes may be particularly important to the early life stages of many arthropods and amphibians. Nonadjacent wetlands, particularly those which experience periodic drawdowns, are less likely to support populations of predatory fishes, thus providing a necessary condition for many larval stages. For example, terrestrial salamanders (*Ambystoma tigrinum* and *A. cingulatum*) of Florida's pine flatwoods or mixed hardwood forests require nonadjacent wetlands (cypress swamps or hardwood swamps) without predatory fishes for reproduction and larval development (Moler, 1992). Similarly, many species of insects and other arthropods require wetlands for their larval stages (e.g., *Anopheles* mosquitoes).

Wetland Animals. A variety of other amphibians, reptiles, birds, and mammals require wetlands during substantial parts of their life history. Amphibians are more dependent on wetlands than most groups of animals due to their aquatic larvae. Familiar amphibian species include a variety of frogs and toads (e.g., *Rana, Bufo, Acris, Pseudacris, Hyla*), and many salamanders (e.g., *Ambystoma, Amphiuma, Eurycea, Taricha*). Familiar reptile species which typically inhabit wetlands include the American Alligator, *Alligator mississippiensis* (and the endangered American Crocodile, *Crocodylus acutus*), freshwater turtles, such as painted turtles (*Chrysemys*), pond turtles (*Pseudemys, Clemmys*), snapping turtles (*Chelydra*), water snakes (*Nerodia*), and the venomous water moccasin (*Agkistrodon*). Typical wetland mammals include the musk rat (*Ondatra*), nutria (*Myocastor*), beaver (*Castor*), and the moose (*Alces*).

Many birds are closely associated with wetlands. In fact the economic importance of many wetland bird species has been the driving force behind many wetland protection actions. A huge variety of aquatic birds depends on wetlands for breeding and feeding, including many species of ducks (e.g., *Anas, Aythya, Bucephala, Mergus*), geese (*Anser, Branta, Chen*); herons, egrets, and cranes (*Egretta, Ardea, Nycticorax, Grus*); coots and gallinules (*Fulica, Gallinula, Pophyrula*); kites, hawks, eagles, and ospreys (*Rostrhamus, Elanoides, Buteo, Haliaetus, Pandion*); and a staggering variety of gulls, terns, avocets, flamingoes, and other wading and shore birds.

Wetland species comprise a disproportionately large number of endangered species. As of 1991, the U.S. Fish and Wildlife Service had listed 595 plant and animal species as endangered or threatened. Of this number, 256 (43%) are wetland-dependent species, while wetlands provide essential habitat to 60% of all threatened and 40% of all endangered species (Feierabend, 1992).

Wetland Plants. Many species of plants are associated with wetlands. In fact, the presence of characteristic wetland plants ("hydrophytes") is used to delineate wetlands for regulatory purposes, and to delineate and classify wetlands. Wetland plants may be characterized as "submersed" (i.e., completely submerged, such as *Chara* and *Potemogeton*), "emergent" (i.e., those plants with a root system and stem below the water, but which reach to or above the surface, such as *Nymphaea* and *Typha*), or "terrestrial" (Figure 2.7). Plants may be "obligate" wetland species that only occur in wetlands (and are useful indicators of wetlands even when there is no water present), or "facultative" species which occur in wetlands as well as in certain upland habitats (Tiner, 1991a). Additional discussion of wetland

Figure 2.7: Emergent wetland (marsh) adjacent to the Crystal River, near Glen Arbor, MI. Note both submersed and emergent vegetation. (J.F. Berry)

plants is provided in the Chapters 3-6.

The importance of wetland plants to wetland ecosystems should not be underestimated. Wetland plants are the primary food source for most wetland animals, as well as for many non-wetland species. Plants provide cover from predators and assist in thermoregulation in many wetland animals.

Plant productivity is extremely high in wetlands, particularly in tidal salt marshes (Figure 2.8). Productivity has been estimated at 650-950 $g/m^2/yr$ in New England salt marshes (Niering and Warren, 1980), 444-810 $g/m^2/yr$ in Florida mangrove swamps (Twilley, et al., 1986), and may exceed 1200 $g/m^2/yr$ in some southern low-salinity marshes (Cahoon and Stevenson, 1986).

The decomposition of wetland plants by fungi and bacteria forms the "litter" layer on the soil surface, which is the source of nutrients for many wetland plants and animals (Lugo and Snedaker, 1974; Smock et al., 1985; Moran et. al, 1988; Jordan et al., 1989), and which forms the rich, organic soils that accumulate as wetlands mature (Winkler, 1988; Foster and Wright, 1990). These rich soils then form a nutrient source for other plants and animals.

Seed-producing plants play a crucial role in the ability of wetlands to regenerate following periods of drawdown and drying. Van der Valk and Davis (1978) recognized three categories of plants that form the "seed bank" of periodic prairie marshes: (1) emergent species that germinate in exposed mud or shallow water; (2) submersed and free-floating species with seeds that can withstand a
year in exposed mud before germinating in shallow water; and (3) mud-flat ephemerals which require dry mud for germination. Gerritsen and Greening (1989) identified three categories of seed bank plants in the Okefenokee Swamp of Georgia whose success fluctuated with water drawdown and flooding: drought-adapted species (e.g., beak-rush, *Rhynchospora*); inundation-adapted species (e.g., water lily, *Nymphaea*); and generalists (e.g., pipewort, *Eriocaulon*). The role of seed banks has been studied for several other wetland types as well, including marshes (Smith and Kadlec, 1983; Leck and Simpson, 1987) and river swamps (Schneider and Sharitz, 1986).

Other groups of organisms receiving increased attention are the algae, fungi, and bacteria that inhabit wetlands. For example, a group of anaerobic bacteria known as "methanogens," which break down cellulose and release methane, sulfur, and other organic gasses, are

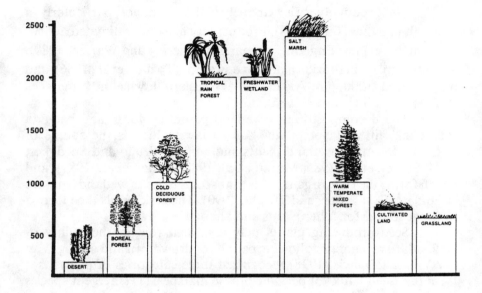

Figure 2.8: Net primary productivity of selected ecosystems in g/m^2/year (from Tiner, 1984). Note the high primary productivity of salt marsh and freshwater wetlands.

responsible for much of the characteristic smell of freshwater marshes, swamps, and bogs. A number of species of anaerobic (and some aerobic) bacteria as well as fungi are the primary detritivores in wetlands which break down dead plant and animal material and make nutrients available to other organisms. In addition, algae, especially blue-green algae, may form a "periphyton" mat on the surface of the substrate of wetlands which may survive periods of prolonged drying. The presence of a periphyton layer may be used to determine if an area is a wetland or not (see Tiner, 1993b; and Chapter 6).

Recreation. The recreation opportunities associated with fishing and waterfowl hunting are among the strongest anthropocentric justifications for wetland protection currently recognized. Tiner (1984) reported that in 1980 alone, 5.3 million people spent $638 million on hunting waterfowl and other migratory birds. Recreational fishing is increasing in popularity, particularly in marine ecosystems. In 1975, sport fishermen spent $13.1 billion to catch wetland-dependent species (Peters et al., 1979). Collecting of shellfish of various kinds, either as food or decorations, remains an important wetland-dependent industry in many states, and is often of commercial importance. Other recreation associated with wetlands is generally non-consumptive. Hiking, boating, swimming, nature observation, and photography are apparently increasing in popularity.

2.3.2 Groundwater Recharge and Water Quality

Wetlands are crucial in maintaining the quality and volume of both surface and ground water. They have the ability to intercept sediments, nutrients, and a variety of organic and inorganic chemicals and heavy metals, which are often in the form of pollutants. By removing these materials, they serve to provide clean surface water and to cleanse water percolating beneath the surface as groundwater. The materials trapped in the subsurface soils provide necessary nutrients for plant and animal life. In addition, evaporation from wetlands plays a crucial role in providing atmospheric water for the natural hydrological cycle.

Surface water in wetlands is generally of high quality. The presence of aquatic plants, fungi, and bacteria is important in removing organic contaminants and inorganic materials (particularly nitrogen, phosphorous, and sulfur-containing molecules) from the water (Metzler and Tiner, 1992). In addition, wetlands play an important role in reducing turbidity, particularly during flooding. Wetlands are also important in sediment retention, serving to remove many natural sediments before they can enter adjacent flowing rivers, streams, and ponds (Boto and Patrick, 1979; Novitzky, 1979.; Tiner, 1984).

An important function of wetlands is the removal of nutrients, which prevents eutrophication or enrichment of open water (Klopatek, 1978; Richardson et al., 1978). Tiner (1984) noted, however, that each wetland has a limited capacity to absorb nutrients, such that it is possible to overload a wetland and reduce its ability to perform this function. Destruction of many coastal wetlands along the Great Lakes contributed to the eutrophication that has seriously degraded water quality and the commercial fisheries in several midwestern and eastern states.

The ability of wetlands to function in removal of nutrients and other materials has been put to a practical use in sewage treatment (see Wilen and Tiner, 1993; Zachritz and Fuller, 1993). For example, the Brillion Marsh in Wisconsin and the Tinicum Marsh near Philadelphia receive sewage, and both dramatically reduce biological oxygen demand (BOD), phosphorous, nitrates and ammonia, and sediments, while increasing oxygen content of the water (Grant and Patrick, 1970; Odum et al., 1975; Boto and Patrick, 1979; Salvesen, 1990).

Several firms around the U.S. have experimented with the use of created wetlands as an alternative to traditional sewage treatment techniques. With these systems, partially treated sewage ("gray water") is introduced into a constructed wetland planted with wetland plants (rushes and bulrushes seem to be favored). Gray water is allowed to move gradually through subsurface plant root systems. The bacteria and the plants themselves, and adhesion to soil particles, remove most of the nutrients and pollutants, such that the water at the end is reasonably clean. The advantage to these systems is that they avoid the use of the traditional sewage treatment infrastructure. However, they are relatively land-intensive, such that they are not practical for large populations (especially urban areas) where land is restricted or extremely valuable.

Once surface water has passed through a wetland and seeped into the ground, it becomes groundwater. Most commercial and domestic water used in the U.S. is derived from ground water, either from wells or directly from subsurface aquifers. Therefore, the ability of many wetlands to cleanse water is of great economic and public health importance.

The recharge potential of a wetland depends on many factors, including the type of wetland, geographical location, season of the year, soil type, position of the water table, and local weather (Tiner, 1984). Studies have indicated that the most effective wetlands for groundwater recharge are prairie potholes (Stewart and Kantrud, 1972; Winter and Carr, 1980) and cypress swamps (Odum et al., 1975), although floodplain swamps and marshes may also contribute (Klopatek, 1978).

It is likely that other wetlands, particularly those associated with limestone karst areas, also serve this important function. However, not all wetlands are recharge sites.

2.3.3 Flood and Erosion Control

One of the most visible and anthropocentrically valuable functions of wetlands is their ability to ameliorate the effects of stormwater runoff and reduce floodwater damage. This results from the capacity of many wetlands to store runoff for a period of time and gradually return it as surface flow. Surface runoff gradually fills a wetland, which then slowly releases it to streams, lakes or into ground water. The gradual return has the effect of dampening the "pulse" of runoff normally associated with large storms, and to mitigate greatly the effects of flooding (Novitzky, 1979; Tiner, 1984; see Figure 2.9).

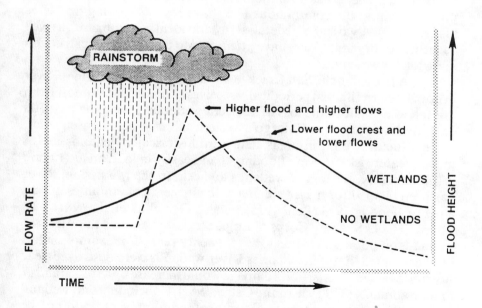

Figure 2.9: Diagrammatic representation of the value of wetlands in dampening the effects of storms by reducing flood crests and flow rates after storms (Tiner, 1984).

Destruction of natural wetlands often results in increased property damage from flooding. For example, in the Devil's Lake basin in North Dakota it has been demonstrated that prairie pothole wetlands store almost 75 percent of entering stormwater runoff (Ludden et al., 1983).

Depending on factors such as surface area, depth, connections to streams and ponds, the type of wetland itself, and degree of anthropogenic disturbance, different wetlands have varying capacities to store excess runoff, and the dampening effects will vary accordingly. It is also possible to exceed the capacity of a wetland to store water, such that the storage function is lost.

In many urban and suburban areas around the U.S., residences and other structures have been built in low-lying areas that were historically wetlands adjacent to rivers, streams, or lakes. Unfortunately, it is not uncommon for these areas to become severely flooded when the wetlands are temporarily reestablished during times of high water. Too often the result is loss of human life and extensive property damage. It has been estimated that about 134 million acres of property within the coterminous U.S. have severe flooding problems (Tiner, 1984). A variety of federal, state, and local governments (as well as insurance organizations) have attempted to discourage construction in historical wetlands.

As an example, Salt Creek (a tributary of the Des Plaines River in northeastern Illinois) periodically receives excess surface runoff from rainfall as a result of the loss of adjacent wetlands during the past 100 years of development in Chicago's western suburbs. As a consequence, those structures built in areas that were historically wetlands are at much greater risk of severe flooding than nearby upland areas (Figure 2.10). Similarly, the loss of wetlands adjacent to the Mississippi River has been blamed for massive damage to adjacent communities during a major flood in 1973 (Belt, 1975; see also Gosselink et al., 1981).

The U.S. Army Corps of Engineers has examined the effects of wetlands on flood prevention for over twenty years. For example, it has been estimated that if the Charles River wetlands were destroyed near Boston, Massachusetts, flood damage would increase by as much as $17 million annually (Thibodeau and Ostro, 1981). Concluding that wetland protection was the most cost efficient method to prevent flooding, the Corps acquired nearly 8,500 acres of wetlands in the Charles River basin in 1983 for the purpose of flood protection (Metzler and Tiner,

Figure 2.10: During floods in 1987, homes built in historical wetland areas adjacent to Salt Creek, DuPage County, IL, experienced severe damage from flooding. (P.F. Ries)

Yet another important wetland function is the ability to decrease runoff velocity (Tiner, 1984). Many wetlands serve as buffers between sources of surface runoff and property. In particular, coastal wetlands serve to dampen the effects of strong storm surges and inhibit erosion (Mitsch and Gosselink, 1993). Wetlands reduce erosion of shorelines by increasing durability of the shoreline by binding it with roots, by dampening wave energy through friction, and by reducing current velocity through friction (Dean, 1979; Metzler and Tiner, 1992). Other related benefits include reduced turbidity of water, and reduced siltation (Hammer, 1992).

In coastal areas where wetlands have been removed, severe erosion is often a direct result. This is particularly evident in southeastern Florida where systematic removal of mangroves has left coastal areas exposed to wave action (particularly during storms with strong offshore winds), and severe erosion has resulted. In a few coastal areas, planting of wetland vegetation has helped reduce erosional effects (see Metzler and Tiner, 1992).

2.3.4 Wetland Valuation

Several writers have struggled with the issue of wetland valuation. If wetlands indeed have value, what are they "worth" in terms of the marketplace? Issues of valuation often become critical when, for example, a private landowner negotiates sale of a wetland with a government agency, or for mitigation agreements (see Chapter 8).

Unfortunately, placing a precise monetary "value" on wetlands is difficult. If one assumes that wetlands have an "intrinsic" value (see Norton, 1987, 1988), then it may be argued that wetland values are priceless (or extremely expensive). Even if wetland values are perceived as "extrinsic," it remains difficult to quantify them.

There have been numerous attempts at valuing wetlands for a variety of purposes (see reviews in E.P. Odum, 1979; Lonard et al., 1981; Scodari, 1990; Anderson and Rockel, 1991; and Mitsch and Gosselink, 1993). Most attempts to value wetlands have used market-based economic models, in which wetland values are assessed based on some kind of scaling of wetlands against more easily valued commodities, reduction of wetland values to some common denominator that is more easily valued (Odum, 1979; Mitsch and Gosselink, 1993), or calculating the replacement value of the wetland functions that are lost or degraded. Most of these models make assumptions that wetlands can be valued much as other commodities in the marketplace.

Scodari (1990) provided a thoughtful economic analysis of wetland valuation, in which he recognized the inherent difficulties of valuing wetlands or any natural resource. Current methods are clearly inadequate, and future attempts at valuing wetlands will require: (1) improved biological and ecological data bases; (2) improvement of economic data bases; (3) improvement of communication between scientists, economists, and decisionmakers; and (4) legislative and administrative reforms to include non-market guidelines for valuation.

Most studies recognize that there are serious shortcomings in most methods for wetlands valuation. For example, it is unclear how wetlands should be valued for economic purposes when their values in terms of wetland functions differ depending on the area where they are located. It has also been argued that there are no credible methods for evaluating most wetland functions (see Stavens and Jaffe, 1990). It is not surprising that those wetland functions for which little scientific data are available have received the least attention for economic valuation purposes. Valuation models in the future must consider with much more precision the exact benefits to be derived from wetland functions which will allow a favorable balance against the costs resulting from wetland destruction and degradation (Anderson and Rockel, 1991).

2.4 WETLAND CONSERVATION

The foregoing discussion would seem to suggest that wetlands have many important values, and that they are worth preserving. But important questions remain, namely (1) how severe are losses of wetlands and wetland degradation due to anthropogenic factors, (2) are there prophylactic measures to prevent further losses, and (3) what are the prospects for recovery of wetlands that are degraded, or creating new wetlands to replace those that are lost? The remainder of this chapter will attempt to answer some of these questions, while others will be considered in the remaining chapters in this book.

2.4.1 Wetland Losses

Although estimates vary over a considerable range, it appears that nearly half of the wetlands in the contiguous United States have been lost, and many others have been seriously degraded (Figure 2.11), since settlement by American colonists (Dahl, 1990; Berry, 1992a). During the period of intensive development and wetland conversion between 1954 and 1974, 95 million acres of wetlands were lost (Frayer et al., 1983). During that period, annual wetland losses averaged about 458,000 acres (400,000 acres of inland and 18,000 acres of coastal wetlands), with about 396,000 acres per year (87%) resulting from conversion to agricultural uses (U.S. Fish and Wildlife Service, 1989). Losses of forested wetlands were estimated at 2.74 million ha (1.1 million acres) from 1940-1980 (Abernethy and Turner, 1987). Between the mid-1970s and mid-1980s, 97.8 million acres of freshwater wetlands and 5.5 million acres of coastal estuarine wetlands were lost (Dahl et al., 1991).

Wetland losses among the various types of wetlands have not

been equal. Losses of forested wetlands and emergent wetlands have been 5.8 and 2.7 million acres respectively from the mid-1950s to the mid-1970s, with agricultural conversion accounting for most losses (Tiner, 1984). Shrub wetlands and pocosins in the mid-Atlantic states are declining due to conversion for agricultural uses, including pine plantations. Prairie pothole wetlands have declined considerably in North and South Dakota, and Minnesota due to draining for agricultural purposes (Frayer et al., 1983). Losses of estuarine wetlands and salt marshes has been greatest along the Gulf coast (Frayer et al., 1983; Frayer and Hefner, 1991).

There have been slight increases in certain deepwater habitats and ponds due to construction of manmade lakes, reservoirs, etc. (although this construction was sometimes at the expense of existing natural wetlands). Scrub-shrub wetlands have made small gains, with the creation of 200,000 acres (often unvegetated) (Tiner, 1984; Burke et al., 1988). Clearly, losses of wetlands far exceed gains.

The rate and intensity of wetland losses has not been uniform in all states within the U.S. (Dahl, 1990; Dahl et al., 1991). For example, Illinois, with a history of intensive agricultural conversion and urban development, contained approximately 8.3 million acres of wetlands (23% of land surface area) prior to settlement, but only 918,000 acres (2.5% of land surface area) now remain. Between 6,000 and 9,000 acres per year are lost to development and agricultural conversion in Illinois, and only about 6,000 acres of Illinois wetlands are still of high quality (Simon et al., 1993). Under pressures similar to those in Illinois, Iowa has lost over 95% of its natural wetlands (Bishop, 1981).

Wetland losses in Florida are perhaps more noticeable than those in other states, due to the large number of highly visible and sensitive wetlands (e.g., the Everglades). Between the mid-1970s and mid-1980s, Florida lost a net 260,300 acres (2.3%) of its nearly 12 million acres of wetlands, but this estimate does not include degraded wetlands (see Frayer and Hefner, 1991). From the 1950s to the 1970s, Florida wetlands were lost at a much higher annual rate of 72,000 acres (Hefner, 1986), during which time over 2,500 acres of mangroves were lost from Collier County (southwestern coastal Florida) alone (Patterson, 1986). Among the hardest hit Florida wetlands has been the Everglades, in which ditching and draining has reduced its 3,600 square miles by 67% (Kushlan, 1986). For examples of wetland losses in other

Figure 2.11: A degraded wetland in a heavily industrialized area near Lake Calumet, in northeastern Illinois. In degraded wetlands such as this, some wetland values are inevitably lost. (J.F. Berry)

states, see Tiner (1984), and Burke et al. (1988).

The *causes* of wetland losses are as varied as are anthropogenic activities. Burke, et al. (1988) divided threats to wetlands from humans into two categories: (1) *Direct threats*. Includes activities such as drainage for crop or timber production, or mosquito control; dredging for stream channelization; construction of dikes, dams, levees and seawalls; discharge of pollutants; and mining of wetlands for peat, coal, sand, gravel, phosphate, and other materials. (2) *Indirect threats*. Includes sediment diversion by dams, channels and other structures; hydrologic alterations by canals, spoil banks, roads, and other structures (as well as groundwater withdrawal); and subsidence due to extraction of groundwater, oil, gas, or other materials. Using a different approach, Leslie and Clark (1990) distinguished between two types of factors that cause wetland alterations: (1) "Discrete events or individually distinct actions," such as fill or excavation, drainage, spills, fires, and clearing; and (2) "Gradual, subtle, on-going processes," such as long-term changes in water level or sediment loads, concentration of nutrients or contaminants, or changes in the biological community.

Tiner (1984) estimated that 87% of recent (post-1950s) wetland losses resulted from drainage for agricultural uses, 8% from urban development, and 5% from other development. Other causes of wetland losses may be more subtle, such as gradual inundation of coastal freshwater wetlands resulting from rising sea levels (see Delaune et al., 1983; Teal, 1986), or increases in salinity from deicing salts used on roads in northern states (see Wilcox, 1986). In some cases, anthropogenic activities seemingly unrelated to wetlands have profound effects, such as when marshes in the northeastern U.S. become acidified by acid rain which originates as air pollution in the Mississippi River valley (NAPAP, 1990; Charles, 1990).

Among the more insidious factors that have led to historical wetland losses are the various governmental programs which have actually encouraged wetland destruction (Baldwin et al. 1990). Admittedly, these programs were created mostly during a time when the values of wetlands were misunderstood, but many persist. For example, the Swamp Lands Acts of 1849, 1850, and 1860 encouraged drainage of wetlands, which were considered to be threats to public health or public nuisances.

Many modern programs may indirectly result in wetland losses. For example, when the Federal Highway Administration provides funds to state or local governments for highway construction, wetlands may be directly impacted by construction activities, or indirectly affected by pollutants that enter the wetlands as runoff. In addition, certain activities which damage wetlands may be exempted from the federal § 404 permit process for policy reasons (see Section 7.3.3). Even the U.S. Environmental Protection Agency may encourage wetland losses by providing funds for the construction of sewage treatment plants that damage wetlands.

2.4.2 Wetland Protection

Various forms of wetland protection will be discussed throughout this book. Nevertheless, it is useful to categorize the various forms that wetland protection might take. Many federal laws and regulations protect wetlands in one way or another, and some of these will be discussed in detail in Chapters 7 and 8 (see also Kusler, 1983; Want, 1993; Bingham et al., 1990; Silverberg and Dennison, 1993).

Perhaps the most important such law is §404 of the Clean Water Act (33 USC § 1344). This law requires a permit from the U.S. Army Corps of Engineers for the deposit of dredged or fill materials into the "waters of the United States," which regulation includes wetlands. Unfortunately, § 404 falls far short of regulating all types of wetland destruction or degradation (Berry, 1992a). Section 10 of the Rivers and Harbors Act of 1899 (33 USC § 403) prohibits obstruction of navigable waters (including some wetlands), but it is also limited in its scope and applicability to wetland destruction.

The National Environmental Policy Act (NEPA, 42 USC § 4321-47) requires federal agencies to prepare an environmental impact statement (EIS, a detailed analysis of environmental impacts) for all major federal activities significantly affecting the quality of the human environment. Issuance of § 404 dredge and fill permits by the Corps often have the potential to cause significant environmental damage, such that an EIS is occasionally required. Other federal laws which have the effect of protecting wetlands include the Endangered Species Act (16 USC § 1531 *et seq.*, prevents the taking of any species listed as endangered or threatened), the Marine Mammal Protection Act (16 USC § 1361 *et seq.*), and the Coastal Zone Management Act (16 USC § 1451 *et seq.*; see Section 7.6).

Approximately 27 states have laws which protect wetlands in some fashion (some are called "wetland" protection laws, but many are not). Some states have laws which provide greater protection for wetlands than federal laws, but most do not. As an example, Florida's Wetlands Protection Act (Fla. Stat. § 403.91 *et seq.*) regulates deposit of dredge and fill material through the Florida Department of Environmental Regulation in much the same way as the Corps regulates the federal § 404 permit program. Illinois requires a joint permit from the Corps and the Illinois Department of Transportation before dredge or fill material may be placed in a wetland. New Jersey's Freshwater Wetlands Act (N.J. Stat. § 13:B-9(a)) requires any person who proposes any activity in a freshwater wetland to obtain a permit from the state Department of Environmental Protection. Delaware has recently announced a plan to rehabilitate 10,000 acres of privately-owned urban wetlands to restore tidal influence for water quality improvement, mosquito abatement, and flood control (Northern Delaware Wetlands

Rehabilitation Program; see Barrette, 1993; and see Section 7.9).

In addition, many local governments (regions, counties, cities, etc.) have ordinances which protect wetlands by restricting construction in wetlands, or preventing runoff or pollutants from entering. These ordinances are typically either zoning ordinances, development regulations, or building codes. For examples of such ordinances and related activities, see Kusler (1983), Burke et al. (1988), and Callahan et al. (1992), and see Section 7.9.

It is often the case that the best mechanism for protecting a particular wetland is to purchase the property outright from the owner, or to obtain the property as an exchange in mitigation proceedings (see Chapter 8). The Federal Government has a variety of acquisition and funding programs that deal with wetlands. These programs may involve outright purchases of land from governmental or private individuals, conservation easements that prevent certain uses of protected land, or various contractual agreements which serve to protect the wetlands. A few examples include programs such as the National Wildlife Refuge System, by which 90 million acres are set aside for wildlife conservation (mostly waterfowl) and administered by the FWS. Under the National Estuarine Research Reserves program, the National Oceanographic and Atmospheric Administration (NOAA), along with state agencies, designates a series of estuaries for conservation purposes, such as the valuable Rookery Bay National Estuarine Sanctuary in Collier County, Florida. The Bureau of Land Management (BLM) acquires critical riparian habitats in the West primarily by exchanges of land with states and private individuals (nearly 170,000 acres have been acquired since 1988). Other federal acquisition programs are discussed in Goldsmith and Clark (1990), and U.S. Fish and Wildlife Service (1990).

One of the most promising new federal programs is the National Wetlands Priority Conservation Plan, which was authorized under the Emergency Wetlands Resources Act of 1986 (16 USC § 3901 *et seq.*). Under the Plan, the FWS has designated a 10 year acquisition plan to protect over 2.5 million acres of waterfowl habitat, at an estimated cost of $563 million (U.S. Fish and Wildlife Service, 1989; Goldsmith and Clark, 1990).

In addition to federal acquisition programs, many states have similar programs (reviewed in Goldsmith and Clark, 1990). Florida has (or will) purchase hundreds of thousands of acres of wetlands through programs such as the Conservation and Recreational Lands (CARL) program (Myers and Ewel, 1990). New Jersey's Green Acres program provides grants and low-interest loans to municipalities for acquisition for open space and recreation (over 650,000 acres, mostly wetlands, have been purchased).

Many private organizations and individuals have played key roles in acquiring or otherwise protecting wetlands. Among the most visible of such organizations are The Nature Conservancy with its National Wetlands Conservation Project; Ducks Unlimited, which has acquired over 3 million acres of wetlands; and the National Audubon Society, which has acquired nearly 250,000 acres of wetlands. These and other organizations (reviewed in Goldsmith and Clark, 1990) have also participated in negotiations that have led to the placement of hundreds of thousands of acres of wetlands in conservation easements, conservation land trusts, and other mechanisms that protect wetlands over long periods of time.

A major obstacle for wetlands protection is education of the public. A number of public and private organizations make information about wetland values and wetland protection available to the public in a variety of formats. The U.S. Fish and Wildlife Service has assumed a leadership role in public education, having produced hundreds of publications and delivered thousands of talks and classes on various aspects of wetland science and conservation. Many states have similar programs, as do local governments, park and recreation districts, and schools.

Many private organizations have strong public education programs, such as the National Wildlife Federation, the Sierra Club, the National Audubon Society (and its various state and local organizations), the Izaak Walton League. Many state and local environmental groups are of particular value because they provide education directly to local communities.

The list of programs, initiatives, etc. promoting wetland protection may seem huge at first glance. Nevertheless, wetlands losses in the U.S. continue at an alarming rate. Programs which serve to educate the general public of the values of wetlands are an important first step, but it is only through aggressive enforcement of laws and regulations, along with federal, state, and local governmental support that the tragedy of wetland losses can be reversed.

3

Freshwater Wetlands

James A. Kushlan

INTRODUCTION

The purpose of this chapter is to introduce the major types of freshwater and inland wetlands with emphasis on their salient characteristics. As discussed more fully in Chapters 2 and 5, wetlands are ecosystems having shallow water standing above the soil surface or having a soil saturated with water for periods of time that are sufficient to produce characteristic soils or vegetation. Thus, freshwater wetlands are identified by specific water conditions including their shallowness, soils, and plants. Specifically, inland freshwater wetlands are those having consistently low concentrations of ocean-derived salt (less than about 0.5 parts per thousand) in the water and in the hydric soils, plants adapted to fresh water, and limited influence of diurnal tides. Estuarine and marine wetlands, which lack this suite of characteristics, are described in Chapter 4.

There are many types of freshwater wetlands. Approaches to describing, identifying, and delineating wetlands for legal purposes are discussed in Chapters 5 -7. As noted there, these can be controversial, reflecting both the high stakes in the outcome of classification and delineation exercises, and the inherent difficulty one has in achieving a universally acceptable nomenclature for any set of natural ecosystems that vary in a continuous rather than discrete manner. It is important to regulators to have a consistently applied way to delineate and name wetlands, so that they are able to distinguish wetlands and delineate their boundaries for permitting purposes. But such delineation is far less important to scientists, who are more interested in ecosystem function within a wetland than in drawing a boundary or insisting on a specific name. In the United States, the scheme of classification and nomenclature created by the U.S. Fish and Wildlife Service has been adopted and used with some success by some regulators (Cowardin et al., 1979). However, such universal classification schemes may or may not reflect vernacular names or reflect all the differences and similarities among wetlands that have ecological importance in a specific study.

Most wetlands are, in fact, readily identifiable, irrespective of the workings of universal classification schemes and controversial details of wetland delineation. For the purposes of this chapter, freshwater wetlands need only be categorized into vernacularly familiar types that are as easily recognizable by the nonscientist as the scientist. In this chapter several such readily identifiable types are reviewed that together encompass much of the diversity of freshwater wetland ecosystems in the United States. The divisions used here can be readily correlated with the more finely divided regulatory definitions of Cowardin et al. (1979).

Shallow freshwater ecosystems may be most conveniently divided into swamps, marshes, and bogs, as further described in Sections 3.1, 3.2, and 3.3, respectively. These may be subdivided in many ways, such as by dominant vegetation, geography, basin morphology, nutrient sources, or water supply. Recognition of a wetland by dominant plant type is one of the more common approaches to classification. By convention, "dominance" is the coverage of more than 30% of the surface area of a wetland by a plant type. Another important approach to classification is by basin morphology and water source. In this scheme, wetlands can be classified into palustrine wetlands or floodplain wetlands. Palustrine wetlands generally occupy basins. They are shallow with limited influence of flowing water and so often have an extensive cover of aquatic emergent or submersed vegetation. Such palustrine ecosystems are the ones most commonly thought of as wetlands by both scientists and the public. Floodplain wetlands are influenced by and associated with the flowing water of rivers and streams.

An important characteristic of wetlands is the shallowness of their water depths. However, freshwater wetlands and deep water habitats are not always ecologically separable. In many cases, emergent or submersed wetlands form a fringing habitat adjacent to deep water ecosystems, which are likely to have considerable interchange of water and materials. In general, wetlands are aquatic sites shallower than 2 m deep. Although somewhat arbitrary, an ecological basis for this demarcation that 2 m is about the depth to which erect, self-supporting aquatic plants could be expected to grow. Beyond the depth of rooted plants, the patterns and pathways of energy and nutrient flows differ from those in emergent wetlands. Of course, the 2 m limit is not always the case, because some shallower wetlands lack rooted plants and deep, clear aquatic systems support luxurious plant growth. Ponds may be wetlands in some classifications and not in others. They often are depressions within extensive emergent wetlands. For completeness and contrast, the major deep water aquatic systems, streams, and lakes, are briefly noted in Section 3.4.

3.1 SWAMPS

In different parts of the world, the term "swamp" has different meanings, an important point if one wishes to access the literature. In Europe, the term generally refers to herbaceous or forested wetlands with mineral-rich nonorganic soil. In other parts of the world, swamp refers to dense and extensive herbaceous wetlands, such as the reed swamps of Africa. In the North America, however, the term "swamp" typically means a wetland dominated by woody plants.

Swamps are of two types, swamp forests (or forested wetlands) and shrub swamps (or shrub wetlands). Swamp forests are wetlands dominated by trees, whereas shrub swamps are dominated by shorter woody plants. Since most swamps contain both tall and short woody plants, as well as herbaceous vegetation, a delimiting rule is useful. A forested wetland would be more than 30% covered by woody plants taller than 6 m; a shrub wetland would lack this dominance by tall trees and would be more than 30% covered by shorter woody plants (Cowardin et al., 1979).

Both swamp forests and scrub swamps may be dominated by broad-leaved trees or narrow-leaved trees, by deciduous trees or evergreen trees. Broad-leaved swamp trees include alder (*Alnus*), ash (*Fraxinus*), bay (*Persea*), or gum (*Nyssa*); narrow-leaved swamp trees include wetland conifers such as spruce (*Picea*), cedar (*Chamaecyparis*), cypress (*Taxodium*), or tamarack (*Larix*). Shrub wetlands may be dominated by broad-leaved plants, such as alder, ericaceous shrubs (*Vaccinium, Chamaedaphne, Andromeda, Ledum, Kalmia*), or birch (*Betula*), or by narrow-leaved plants such as dwarf cypress, spruce, or pine (*Pinus*). It is worth noting explicitly that in wetlands, narrow-leaved trees are not necessarily evergreen, nor broad-leaved plants deciduous. Two of the most important North American swamp trees, the cypress of the south and the tamarack of the north, are deciduous conifers.

These and other tree and shrub species can combine into a large variety of dominant plant communities, depending on how finely one wishes to subdivide vegetative descriptors. Some of the major and readily identifiable sorts of swamps in the United States include the cypress swamps of the southeastern United States, palustrine hardwood swamps found over much of the continent, shrub wetlands, and floodplain forests associated with river valleys. These categories, of course, overlap to a great extent.

3.1.1 Cypress Swamps

Cypress swamps are characterized by a peculiar, ancient conifer, the bald cypress (*Taxodium distichum*), which gets its common name from its deciduous habit. The best development of this swamp association occurs typically in relatively deep moving water of the humid southeastern USA (Figure 3.1; Plate IV-V) (Ewel and Odum, 1984). Those occurring in floodplains are a type of southern bottomland forest (divided in this chapter between cypress and hardwood forests). They are often associated with hardwoods and after disturbance may be succeeded by hardwood swamps. Cypress swamps also develop as stillwater, palustrine swamps, and as shrub swamps. They have many colloquial names, some of which have a high degree of technical specificity, such as domes, heads, strands, sloughs, scrub, and brakes.

Geophysical Characteristics. Cypress swamps occur in a wide variety of geomorphic situations ranging from broad, flat floodplains to isolated basins. These swamps are found throughout much of the lowlands of the southeastern and south-central United States. The most northern example of this type of swamp is in southern Illinois (Anderson and White, 1970). Several sorts of cypress swamps are easily recognizable.

Cypress domes are small, more or less circular swamps up to about 10 hectares, having taller trees in the center and smaller trees

Figure 3.1: Cypress swamp, showing cypress trees in the Okefenokee Swamp, Georgia. (J.A. Kushlan)

along the margin, giving the canopy a domed appearance. Also called cypress heads, they are stillwater, palustrine systems, occupying topographic depressions in clay or sandy soils, topped by a layer of peat in the center, and underlain by an impermeable clay or hardpan layer. Dwarf cypress covers much of southern Florida both east and west of the Everglades. Trees are generally of shrub stature and are locally called scrub cypress, with interspersed domes or strands of taller cypress trees. These swamps occur in sandy soils, sometimes thinly veneering, intermittently outcropping limestone bedrock. Although the surface water does flow, it does so slowly that these swamps are more like basin wetlands.

Cypress strands are similar to cypress domes but elongated in the direction of surface water flow. Also like domes, taller trees occur in the center of the strand with shorter trees toward the periphery. Cypress strands, often called sloughs, are riverine systems, essentially shallow, slow-moving, intermittent streams in a very gently sloping topography. Cypress riverine swamps are floodplain systems that occur in periodically to permanently flooded depressions within a river flood plain. Fringing cypress swamps occur along the sloping margins of lakes and ponds.

Wetland Criteria Characteristics. *Vegetation.* The cypress swamp takes its name from its dominant tree, the bald cypress, a deciduous conifer (Ewel and Odum, 1984). The wide distribution of these swamp types is due to the broad tolerances of the cypress with respect to water levels, anaerobic sediments, and nutrient conditions. Historically, two species have been recognized taxonomically, with the smaller pond cypress being distinguished by leaf morphology. Whether the pond cypress (called *Taxodium ascendens*) is taxonomically different from the bald cypress at the species, subspecies, or varietal level remains unresolved. Useful references on cypress swamp vegetation include Monk (1966), Musselman et al. (1977), Schlesinger (1978), Clark and Benforado (1981), Wharton et al. (1981, 1982), Conner and Day (1982, 1987), Schomer and Drew (1982), Drew and Schomer (1984), Dorge, et al. (1984), Duever et al. (1984), Conner et al. (1986), Tiner (1988), Hook and Lea (1989), Ewel (1990), Mitsch and Gosselink (1993), and Wilen and Tiner (1993).

Bald cypress can be very large trees, not atypically up to 40 m tall with a trunk diameter in excess of 1.5 m. Trunk diameters in excess of 5 m and heights in excess of 50 m have been reported. Many of the oldest and largest bald cypress trees were lumbered early in the present century, since their soft wood is rot resistant. It is likely that some of these lumbered trees were 400-600 years old. Trees over 200 years old remain in some swamps.

A characteristic adaptation of the tree is its growth of trunk buttresses and root protuberances called knees. Buttressing occurs especially in deeply flooded water, such as occur in fringing swamps and deep strands. The buttresses are reminiscent of those of tropical rain forest trees, which function in providing support for a shallow rooted large tree. It is almost certain that this is their function in bald cypress as well. The height of buttresses on old trees can be an indication of water levels prior to drainage. The knees also occur primarily under flooded conditions. They are elongated, tapered structures that emerge from the water to as much as 1 m or more above the soil. At present, the function of the knees is unclear, but they are similar to the pneumatophores of many swamp trees, and so most likely function in root respiration.

Despite the wide tolerances of adult trees, conditions for establishment of cypress seedlings are more restricted (Conner et al., 1986, Huenneke and Sharitz, 1986). An important requirement of cypress seeds is that they need nonflooded soil to germinate. As a result, seasonal or periodic drydowns may be required to establish a forest. Drying is seasonal in forests, such as dwarf cypress or domes, but is less frequent in fringing cypress swamps. After seed germination, shoot elongation is relatively rapid to keep pace with rising water levels. Since cypress can persist for decades or even centuries, germination events need not be frequent for a forest to maintain itself without logging or other human disturbance. Natural regeneration in existing cypress swamps appears to be low.

The stature of a cypress depends primarily on the quality of its growing conditions. The taller stature of trees in the center of a cypress dome or strand reflects deeper soils and longer hydroperiods. The trees become smaller towards the periphery where the organic soil is shallower and hydroperiods shorter. Dwarf "pond" cypress growth form occurs under even poorer growing conditions, particularly in shallow, nutrient poor, sandy soils. Dwarf cypress swamps also experience shallow water depths, shorter hydroperiods, limited surface water flow and, therefore, nutrients, and recurrent fires.

A co-dominant in deepwater cypress swamps is the water tupelo (*Nyssa aquatica*). The tupelo is sometimes found in pure stands within a cypress swamp, and dominance of this and other broad-leaved trees in cypress swamps appears to have increased as a result of logging that opened the canopy in previous decades. Other typical and co-dominant trees include the black gum (*Nyssa sylvatica* var. *biflora*), pumpkin ask (*Fraxinus profunda*), and slash pine (*Pinus elliottii*) in shallower sites.

Many of the cypress swamps in southern Florida are undergoing invasion by a exotic swamp tree, the cajeput or paperbark (*Melaleuca quinquenervia*, see discussion in Section 2.2.2). In Australia this tree dominates swamps that are subject to wide extremes of flooding, drying, and fire. As a result it is highly predisposed for success in the less stringent cypress swamps of semitropical North America. Melaleuca strands and domes are replacing cypress strands and domes. Those swamps that have experienced invasions by melaleuca appear to have suffered reduced biodiversity among both understory plants and associated animals.

The understory of cypress swamps varies greatly, depending on such factors as geography, light penetration, history of logging, water depth, and availability of hummocks, knees, or fallen timber, where plants can gain a foothold. In domes and strands, in addition to the gums, and depending on location, the understory may include swamp bay (*Persea palustris*), sweet bay (*Magnolia virginiana*), holly (*Ilex cassine, Ilex glabra*), lyonia or fetterbush (*Lyonia lucida*) willow (*Salix caroliniana*), red maple (*Acer rubrum*), wax myrtle (*Myrica cerifera*), buttonbush (*Cephalanthus occidentalis*), and virginia willow (*Itea virginica*). In lake-edge swamps, tupelo, ash (*Fraxinus caroliniana*), and maple (*Acer* spp.) are associated with the cypress. In floodplain swamps the understory includes buttonbush, hackberry (*Celtis laevigata*), willow (*Salix nigra*), and swamp rose (*Rosa palustris*). In dwarf cypress, the understory is sparse but includes wax myrtle, holly, smaller cypress and pines.

The herbaceous layer includes aquatic species, although more terrestrial annuals occur during the dry season. The cypress dome herbaceous layer includes ferns (*Woodwardia virginica, Osmunda cinnamomea*), lizard tail (*Saururus cernuus*), red root (*Lachnanthes caroliniana*), pipewort (*Eriocaulon compressum*), various vines, and panic grasses (*Panicum* spp.). The cypress strand has the most diverse herbaceous layer, owing to the abundance of fallen logs and knees that provide a substrate for colonization. However, the density and coverage of herbaceous plants may be restricted where light penetration is low due to the dense canopy cover. The herbaceous layer in cypress strands includes ferns (*Blechnum serrulatum, Nephrolepsis exaltata, Thelypteris kunthii*), and grasses (*Andropogon* spp.). The deeper water pools are characterized by water fern (*Azolla* sp.), duckweed (*Spirodela polyrhiza*), and water lettuce (*Pistia*), and in recent years, water hyacinth (*Eichornia crassipes*). The dwarf cypress has a diverse understory of grasses and flowering herbs. Where light penetration is sufficient in cypress swamps, epiphytic algae can be important contributors to production (Atchue et al., 1983).

Cypress has rough bark, which along with the moderate climate in the closed canopy of a cypress forest makes the tree an excellent substrate for vines and epiphytic plants. These especially include ferns, such as resurrection fern (*Polypodium polypodies*), spanish moss (*Tillandsia usneoides*), and other bromeliads (*Tillandsia* spp.), and orchids in the extreme south.

Soils. The base substrate of cypress swamps may be sand, clay or organic. As a result, soil chemistry varies from acidic to neutral Ph, low to high ion exchange capacity, and low to high organic content. Peat deposition is characteristic of deeper water sites, such as the center of a cypress dome or strand. The thickness of the deposit decreases toward the edge. Strands tend to deposit peat in the deeper pools and adjacent to the central drainage way. Peat does not usually accumulate in shallow water, short hydroperiod sites, or particularly at sites subject to periodic burning.

Hydrology. Cypress swamps are wet for much of the year but most experience a seasonally fluctuating water regime. Because water conditions vary from year to year, intermittent droughts may last several months, during which the swamp can be without surface water. Higher productivity appears to occur in cypress swamps with greater water flow probably because flowing water aerates the soil and imports nutrients. Decomposition and solubilization products, such as tannin, may stain the water brown.

Water flow patterns vary among cypress swamps. Cypress domes are stillwater systems, flooded by rainfall and surface flow. Cypress strands occur in slowly flowing waters, following the downstream trend of the watercourse. Water flow in dwarf cypress is even slower. Fringing cypress swamps are connected hydrologically with their adjacent lake during high water seasons and subject to wind and density driven exchanges of water. Floodplain cypress swamps experience river flows during flood and are more isolated from the river during other parts of the year.

Hydroperiod also differs among swamps. Fringing and floodplain cypress swamps have the longest hydroperiod since both connect to the deeper water habitats in the wet season. They may not dry in some years. Domes and strands have the next longest hydroperiod, usually experiencing a seasonal dry period, although the deeper central ponds of the dome or strand may retain water through a normal dry season. Dwarf cypress swamps have the shortest hydroperiods and may be dry for several months per year.

Cypress swamps can be affected by environmental alterations. Drainage decreases the hydroperiod, whereas damming water courses increases the hydroperiod. The basic forest that developed in response to the naturally fluctuating hydrology may persist for some time under altered conditions, but recruitment patterns may be altered leading to succession to other forest or wetland types.

Because most cypress swamps are underlain by impermeable soil layers and the wetter swamps lay down a peat layer on top of the substrate, there is little vertical discharge or recharge. Ground water recharge does occur at times of high water along the edges of cypress domes and presumably in other cypress swamps that occupy distinct basins.

Relations between cypress and hydrology are reciprocal. The needle-leaved cypress has lower evapotranspiration than the nearby upland forest, and its deciduous habit further reduces evapotranspiration during the dry, winter season in Florida. Invasion of the evergreen malaleuca may increase annual water loss in these swamps.

Characteristic Fauna. The cypress swamp fauna is characteristic of southeastern swamps. The types of animals comprising the invertebrate community depend on water depth and duration of flooding, current, substrate, food availability, and oxygen levels (Sklar, 1985). Characteristic groups include oligochaetes, fly larvae, amphipods, isopods, crustaceans (especially crayfish), and mollusks. The large amount of detritus and deep water conditions in the deeper cypress swamps produce a substantial invertebrate community contrasted with those in shorter hydroperiod sites.

Fish communities respond to fluctuating water levels, with larger individuals being found in deeper ponds during the dry season (Kushlan, 1976). Species adapted to endure low oxygen levels have higher survival during dry periods (Kushlan, 1974a). These include fish that can breath atmospheric air, such as bowfin (*Amia calva*) and gar (*Lepisosteus* spp.), those that can endure low oxygen, such as bullheads (*Ictalurus* spp.), and those that can use the oxygenated surface layer, such as mosquitofish (*Gambusia affinis*). The larger species are, therefore, common in the deeper swamps and those with deeper ponds that serve as dry season refugia, whereas mosquitofish are common in shallow systems, such as dwarf cypress. Fringing and floodplain cypress swamps serve as spawning sites during flooding.

Various aquatic vertebrates of the southern United States are also found in cypress swamps. Water snakes (*Nerodea* spp.) and the cottonmouth (*Agkistrodon piscivorus*) can be exceptionally abundant, especially around ponds in the dry season. The American alligator (*Alligator mississippiensis*) is common throughout its range, and is often responsible for maintaining deep water ponds (Kushlan, 1974b).

Birds are abundant in cypress swamps, especially in winter. Among those frequently found there are the pileated woodpecker (*Dryocopus lineatus*), prothonotary warbler (*Protonotaria citrea*), and blue-gray gnat-catcher (*Polioptila caerulea*). Wood ducks (*Aix sponsa*) nest in tree cavities. Limpkins (*Aramus guarauna*), with their piercing call, are especially typical in Florida. Wading birds feed in shallow water, with green-backed herons (*Butorides virescens*) occurring throughout the swamps. Larger species of herons, ibises, and the wood stork (*Mycteria americana*) nest in colonies in cypress trees and feed in shallow water. Two extinct birds species were found in cypress swamps, the Carolina parakeet (*Conuropsis carolinensis*) and the ivory-billed woodpecker (*Campephilus principalis*).

Mammals are less abundant than in other nearby habitats. Those that are found typically include the white-tailed deer (*Odocoilus virginiana*), the black bear (*Ursus americanus*), panther (*Felis concolor*), fox squirrels (*Sciurus niger*), mink (*Mustella vison*), and raccoon (*Procyon lotor*). The introduced wild boar (*Sus scrofa*) occurs over much of the area.

Wetland Values. The hydrologic values associated with cypress swamps are many. Floodplain swamps serve to retard runoff during floods, decreasing the severity of downstream flooding. Cypress domes probably recharge ground water at their periphery. Because of their leaf morphology and deciduous habit, cypress trees transpire less water annually than an equivalent evergreen or broad leaved tree.

Cypress domes have been shown to be capable of serving as sinks for nutrients, and probably serve this function in some locations (Atchue et al., 1983; DuBusk and Reddy, 1987; Dierberg and Brezonik, 1983). Cypress swamps are highly productive in supporting wildlife, including hunted species, such as squirrels, deer, and hogs, and nongame wildlife. Some species, such as the limpkin, common in cypress swamps, have a limited distribution in the United States. The wood stork, which nests and feeds in cypress throughout its North American range, is an endangered species. Floodplain cypress swamps may be important as nursery grounds for game fish, but this is not well understood. Cypress has been harvested for many decades for its wood. Even today, lumbering of cypress knees and trees to make curiosities and of dwarf cypress for poles continues in some areas. Given the slow growth rates, sustainable harvest of timber may not be practical. Cypress swamps create some of the more spectacular landscapes in North America. They epitomize southern swamplands.

Unusual Characteristics. Cypress itself is an unusual species, with its knees, buttresses, deciduous habit and ancient lineage. Fire is an important element of cypress swamps. Short-hydroperiod dwarf cypress swamps burn relatively frequently, often with little harm to the tress as the fire moves quickly across the sparse herbaceous layer and does not ignite the mineral soil. Longer-hydroperiod swamps burn infrequently. A dry season fire in a cypress dome or slough can kill large trees by burning out their roots, toppling the trees, and burning the accumulated peat, in some cases resulting in the formation of deeper pools, which inhibits recolonization of the site by cypress.

3.1.2 Palustrine Hardwood Swamps

Palustrine hardwood swamps are stillwater wetlands that occupy basins (Figure 3.2; Plate I) (Monk, 1965, 1966; Wharton et al., 1982; Vince et al., 1989). Hardwood swamps are dominated by various species of broad-leaved hardwoods, which may be deciduous or evergreen. (In this context, hardwoods are distinguished from softwoods, such as cypress and pine.) These woody plants are of tree stature (6 m or greater), or would reach that stature given sufficient developmental time. These forests are often associated with cypress swamps. In some cases, they have developed in former cypress swamps following logging. Their rapid growth rates and the intolerance of cypress for shading probably account for the shift in dominance. Palustrine swamps are in some places called heads or hammocks. Some of North America's finest examples of wetlands fall in this category, including the Okefenokee Swamp and Great Dismal Swamp. Additional useful references on palustrine hardwood swamps are Dabel and Day (1977), McCormick and Somes (1982), Cohen et al. (1984), Tiner (1985a, 1985b, 1988, 1991c), Jahn and Anderson (1986), Erickson and Leslie (1988), Faber et al. (1989), Veneman and Tiner (1990), Metzler and Tiner (1992), and the references in Section 3.1.1.

Figure 3.2: Palustrine hardwood swamp in Virginia. (R.W. Tiner)

Geophysical Characteristics. Palustrine swamps develop in topographic depressions where interrupted drainage and sufficient rainfall or groundwater seepage flood the soil. They may occur in relatively isolated depressions surrounded by upland, in low spots along floodplains, or in various depressional landscapes. Suitable conditions are found broadly over the eastern, southern, and central states. Wet conditions may lead to the development of highly organic, enriched surface layers and organic deposition, which further enhances wetland development. Streams may enter or leave palustrine swamps, but their hydrology is not dominated by moving water.

Wetland Criteria Characteristics. *Vegetation.* The vegetative character of palustrine swamps depends on environmental factors, including rainfall, soil, drainage, drought, frost, fire and storms. The vegetation is more diverse in mild southern states than in the more climatically harsh north. Where water is relatively deep and hydroperiods relatively long, palustrine swamp forests are generally dominated by a few species. The diversity of plants tends to increase towards the edges of the swamp as hydroperiods and peat accumulation decrease.

Swamp black gum and red maple (*Acer rubrum*) are dominant trees of the flooded palustrine hardwood swamps in the southeast. The gum is flooding tolerant and develops adventitious pneumatophore roots when deeply flooded. Like cypress, with which it is often found, it requires dry conditions for seed germination. Loblolly pine (*Pinus taeda*) can co-dominate in many forested wetlands in the South. Along the Atlantic coast, white cedar (*Chamaecyparis thyoides*) dominates sites having shorter hydroperiods, forming evergreen cedar swamps. Other trees that may become prominent in places depending on hydroperiod include water tupelo, sweet gum (*Liquidambar styraciflua*), swamp bay (*Persea palustris*), ash (*Fraxinus caroliniana*), sweet bay, beech (*Fagus grandifolia*), laurel oak (*Quercus laurifolia*), swamp white oak (*Quercus bicolor*), overcup oak (*Quercus lyrata*), fetterbush (*Lyonia lucida*) and loblolly bay (*Gordonia lasianthus*), the latter being particularly invasive after disturbance. Willows (*Salix* spp.) are common either as trees or in the understory. Various hollies (*Ilex* spp.), arrowwood (*Viburnum dentatum*), and buttonbush are also common in the understory in the southeast.

A subtype of the southern palustrine swamp is the bay swamp. Also called bayheads, these found at the headwaters of streams and on peat filled basin. They are dominated by broadleaved, evergreen "bay" trees that make up the understory of other swamps. These include sweet bay, swamp bay, loblolly bay, and sometimes slash pine.

Further north and inland, northern conifers become more important components of palustrine swamps. These include tamarack (*Larix laricina*), fir (*Abies*), and northern white cedar (*Thuja occidentalis*). Hardwood trees include ash (*Fraxinus nigra*), red maple, poplar (*Populus balsamifera*), northern pin oak (*Quercus palustris*) and formerly, American elm (*Ulmus americana*). The understory is better developed further north, due to the sparser canopy. It includes many shrubs such as red-osier dogwood (*Cornus stolonifera*), speckled alder (*Alnus rugosa*), winterberry (*Ilex verticillata*), and spicebush (*Lindera benzoin*), as well as herbs and grasses. The understory is less developed in cedar shrub swamps.

These species combine into a number of recognizable swamp types. In the northern midwest, hardwood swamps are dominated by red maple, black ash, yellow birch, and formerly by American elm. Further north, tamarack is a pioneer tree, and its swamps occur on glacial outwash and in depressions. Northern white cedar swamps include tamarack, white spruce, balsam fir, and yellow birch. At each site, the diversity of plant species depends on hydrology and nutrients, with greater diversity in swamps having higher nutrients levels.

Fires can significantly impact the vegetation of these wetlands. Fires are more common in short hydroperiod wetlands, or in those having been drained, and can result in loss of organic soil. Pond pine (*Pinus serotina*), pitch pine (*Pinus rigida*), black titi (*Cliftonia monophylla*), and titi (*Cyrilla racemiflora*) require fire for regeneration.

Soils. The underlying base soil can be important in the colonization of certain plants into a wetland. Many cedar swamps, for example, occur on calcareous soils. Most basin swamps have mineral soils, since they occur in situations where relatively short hydroperiods lead to the oxidation of organic matter that might otherwise accumulate. Some tend to accumulate organic matter leading to the creation of an organic or peat soil. Greater organic matter accumulation occurs where decomposition rates are reduced due to prolonged flooding, acidic water and soil, low dissolved oxygen concentrations, and low nutrient content of leaves and stems.

Hydrology. The water source for hardwood swamps is primarily rainfall or ground water. Surface water flow is less important and, if present, is derived from a relatively small catchment basin and perhaps delivered via small streams. Any surface flow is therefore very sluggish, seasonal, and usually broadly spread over the landscape. This contrasts with the strong, directional, and more channelized flow affecting floodplain swamps or the water level influenced flow in fringing swamps. Surface flow ceases early in the dry season, and the hardwood swamps become isolated and the water stagnant.

The locally driven, dynamically restricted hydrology is a critical factor in determining the vegetation and soils of palustrine swamp forests. Water saturation of the soil creates longstanding anaerobic conditions, such that only a few tolerant tree species can survive where hydroperiods are long. Additionally, the surface water may be depleted of oxygen for extended periods of time, affecting plants, animal survival, and water and soil chemistry. The pattern of water level fluctuation, resulting in periodic dry seasons, affects seed germination and survival of plants permitting establishment of species requiring aerobic germination and requiring surviving plants to be capable of enduring both flooded and dry conditions.

The source of water to a palustrine swamp forest and its nutrient content are important determinants of water quality and productivity. Generally, low overland flow means that these wetlands are exposed to a limited nutrient regime, contrasted with wetlands having greater water flow. Export of nutrients and organic material is also low.

Characteristic Fauna. In the anaerobic sediment and, sometimes, waters of palustrine swamps, the benthic invertebrate fauna is dominated by groups well adapted to such conditions, especially midge larvae (chironomids). Crayfish (*Cambarus* spp., *Procambarus* spp.) are characteristic of this swamp.

The isolated location of many palustrine swamps combined with periodic dry downs and deoxygenation events means that fish populations are usually small unless the swamp periodically has surface water connections to other wetlands or to a river. In part because of low fish numbers, amphibians such as frogs and salamanders are common in these wetlands. Snakes and turtles may occur or not. Song birds nest and feed in the trees. Larger species include the barred owl (*Strix varia*) and pileated woodpecker. The presence of aquatic birds, such as herons, depends on food availability. Among mammals, deer, black bear, squirrels, raccoons, and beaver (*Castor canadensis*) are typical over much of North America, the latter creating conditions conducive for the development of still water swamps and marshes.

Wetland Values. Basin hardwood swamps serve as flood retention features in their relatively small catchments and may serve as recharge zones at their periphery. It is likely that the organic peat can serve as sinks for materials. Swamp trees have produced important forest products, especially high value hardwood lumber. They are productive for birds and mammals, including those traditionally hunted.

Unusual Characteristics. Because of their dependence on rainfall and localized water flows, palustrine swamp forests are susceptible to external forces, including human modification of their watershed. Drainage, diversion, impoundment, and extraction of water can all affect the sensitive annual pattern of water level fluctuation. Within the wetlands, forest product harvest and pollution load can substantially alter its character. Changes include reduced tree growth, shift in canopy dominance, loss of peat, and reductions in animal populations. The addition of nutrients can increase production and carbon accumulation, and also shift vegetation patterns.

Fire can be an important factor in the long-term development of these swamps. As palustrine swamps often occur under fluctuating water regimes and deposit organic soil, there will be times when the highly flammable peat will dry and be susceptible to fire. Such fires, if they burn deeply into the peat, can persist for weeks burning ever deeper and changing the hydrological character of the site. In this way fire also sets back any successional trajectory (see Section 2.2.3). In southeastern North America, periodic burning generally favors cypress over hardwoods.

3.1.3 Shrub Swamps

Shrub wetlands are separable from forested wetlands by the stature of their dominant plants (Figure 3.3) (Richardson, 1981; Lugo et al., 1989; Schalles and Shure, 1989). In many cases, the dominant plant species are among those also found in forested swamps. For the most part, their dwarf stature is due to stunting in response to edaphic conditions. These shrub swamps or shrub carrs are treated separately here to call attention to those naturally limiting conditions and to reinforce the understanding that shrub swamps are fully functioning natural wetlands. Evergreen shrub swamps of the southeast United States are called pocosins. In situations where the catchment basin is small and water is derived mostly from rainfall, pocosins may be considered to be southern shrub bogs.

Geophysical Characteristics. Shrub wetlands develop in similar situations as swamp forests, but usually under more nutrient limiting conditions, caused principally by limited surface water flow. Scrub swamps range from a few to thousands of hectares. Most form in

Figure 3.3: Willow shrub swamp in Maine. (R.W. Tiner)

depressions and at ground water seeps, such as spring heads. They also form between streams on the southeastern coastal plain of the U.S., with their base growing over time as peat accumulation overtops the surrounding land. Carolina bays are elliptical basins of uncertain origin that generally support shrub swamps. Shrub swamps are common in the upper midwest (Minnesota, Wisconsin, Michigan). These systems tend to be transitional from marsh to upland.

Wetland Criteria Characteristics. *Vegetation.* Shrub wetlands are characterized by their short stature, usually due to nutrient deficiencies and also, in some cases, recurring fires. As a result of the short stature and lack of shading by a tall canopy, these wetlands usually develop into impenetrable thickets of shrubs and vines. Shrub may be successional features, succeeding from marshes following a reduction in fire frequency.

Many shrub swamps in North America are characterized by

buttonbush, alders, willows, and dogwoods (*Cornus* spp.). Tree saplings are also common, especially red maple and poplars. Useful references on shrub swamps are Moore and Bellamy (1974), Richardson (1981), Larsen (1982), Schomer and Drew (1982), Drew and Schomer (1984), Tiner (1985a, 1985b, 1988, 1991c, 1993c), Windell et al. (1986), and Crum (1988).

Pocosins are shrub swamps identifiable more from their setting and hydrological situation than by their vegetation, which they share with other wetlands (Figure 3.4). These may be shrub swamps or swamp forests, depending on the tree height, although shrub stature is more typical. The shrubs are evergreen, usually with scerophylous characteristics, although this is certainly not universal. Dominants include red maple, swamp cyrilla or titi (*Cyrilla racemiflora*), loblolly bay, lyonia or fetterbush (*Lyonia lucida*), sweet bay, and redbay (*Persea borbonia*), Atlantic white cedar, cypress, and pond pine, which forms the canopy of most forest pocosins. Other shrubs and vines include holly (*Ilex glabra, Ilex coriacea*), laurels, blueberries, huckleberries (*Gaylussacia* spp.), and greenbriar (*Smilax laurifolia*). Peat moss (*Sphagnum* spp.) occurs where the canopy is sufficiently broken to

Figure 3.4: Ocean Bay, a Carolina Bay pocosin in Francis Marion National Forest, Berkeley County, South Carolina. (R.W. Tiner)

allow sunlight penetration to the floor. Pocosins support several unique plants including white wicky (*Kalmia cuneata*), arrowleaf shieldwort (*Peltandra sagittaefolia*), spring flower goldenrod (*Solidago verna*), and rough-leaf loosestrife (*Lysimachia asperulaefolia*). Useful references on pocosins are Richardson (1981), and Sharitz and Gibbons (1982).

Also in the South are shrub swamps associated with standing or fluctuating surface water, which may or may not be peat depositing. The most widespread is the dwarf cypress swamp (see Section 3.1.1). Shrubby swamps are also dominated by swamp privet (*Forestiera acuminata*), elm, willow, alder, and birch. In the upper midwest, shrub swamps are dominated by willow, elder, red-osier dogwood, and buttonbush. Along the gravel flood plains of intermittent streams, shrub swamps are dominated by buttonbush and willow. In the north, shrub swamps are dominated by buttonbush, willows, alders, silky dogwood, or red-osier dogwood. In the northern peatlands, stunted trees and shrubs invade bogs, creating a shrub bog (see Section 3.3).

Soils. Soils differ ranging from mineral soils (sandy humus) to organic soils (muck, peat). Usually, due to limited outflow, they are more generally organic. Peats are most common in basins but are also laid down over alluvial clays in flood plains. In the East, shrub peatlands have expanded by paludification over thousands of years, much as have northern bogs. Peats deposited in basins, such as in Carolina bays of the southeast, may be over 4 m thick. Peat accumulation is inhibited by periodic fires. Highly organic soils are generally nutrient and cation deficient. This, combined with the limited surface and ground water inflow, accounts for the low nutrient conditions in most shrub wetlands.

Hydrology. Hydrologic conditions also differ among shrub wetlands, so generalizations are difficult. As peat accumulates, the substrate tends to hold water and sustain wetland conditions divorced from ground water, which may create a shrub wetland perched above the water table. These perched shrub wetlands discharge water to the adjacent lower landscape, especially in winter when the soil is saturated and evapotranspiration is low.

In most shrub wetlands, being nutrient limited, overland flow is limited and confined to that resulting from high rainfall events. Except for perched wetlands, surface drainage is usually minimal, and evapotranspiration may be the most important water output. Evapotransporation is seasonal and also dependent on characteristics of the plants. In pocosins many plants appear to be adapted to conserve water. Rainfall is generally seasonal, which leads to water depth fluctuations. Even in perched wetlands, ground water varies seasonally, decreasing during summer periods of high evapotranspiration.

Characteristic Fauna. The fauna of shrub wetlands is little studied. The fish fauna depends on the shrub wetlands' connections to more permanent water bodies. Seasonal drying inhibits the persistence of many fish species. Shrub wetlands are important to amphibians, including the endangered pine barrens tree frog (*Hyla andersoni*). They appear to be less important for reptiles because of their seasonality, although rattlesnakes (*Crotalus adamanteus*) and alligators occur in the South. Mammals and birds characteristic of nearby uplands frequent shrub wetlands during low water conditions. These include bear, deer, bobcat (*Lynx rufus*), gray squirrel (*Sciurus carolinensis*), and marsh rabbit (*Sylvilagus floridanus*).

Wetland Values. The peat development of peat-forming shrub swamps controls aspects of local hydrology by maintaining higher soil water levels than would otherwise be the case. Shrub wetlands also affect runoff and transpiration.

Although of small stature, some plants such as cedars have forestry importance. Shrub wetlands also are important wildlife habitat, especially as refuges for upland game species during the dry seasons.

Unusual Characteristics. Shrub wetlands are typically maintained in this form primarily by low nutrients. It is important to note that these are fully functioning wetlands and are not necessarily less valuable because of the small stature of their trees. Many shrub swamps have been logged, especially for their cedar and cypress. Others have been drained for farming or mined for peat, thereby altering their wetland character.

Recurring fire has been an important factor in the development of pocosins, as evidenced by the root sprouting capabilities of pocosin plants and the serotinous cones of the pond pine. All of the typical pocosin plants recover quickly after a fire and may rapidly become dominant. Some herbs and grasses sprout only after fire passage.

Pocosins are considered fire subclimax associations.

3.1.4 Floodplain Forests

Wetland forests occurring in floodplains take their character from the influence of the nearby river, which controls ground water levels and the surface water flow regime, and therefore material transport. Floodplain wetlands are dynamic, productive, and intimately constrained by the hydrology of the river valley and its watershed (Figure 3.5) (Clark and Benforado, 1981; Wharton et al., 1982; Conner and Day, 1982; Lugo et al., 1989; Ewel, 1990). In the humid southeast, they are in some places known as bottomland hardwood swamps, bottoms, or backswamps.

Geophysical Characteristics. These forests occur within a river floodplain, delimited at one margin by a natural levee created by coarse sediment deposition at the river's edge and at its other margin by terrestrial conditions toward higher elevations. Between these borders, poorly drained soil and a high water table controlled by the river

Figure 3.5: Floodplain forest swamp, showing the swamp adjacent to the Myakka River, Florida. (J.A. Kushlan)

elevation often result in wetland conditions. The topography may be one of gradually rising land surface or a more complex arrangement of alternating highs and lows.

The river valley is a dynamic landscape. The river channel changes as it downcuts or as it meanders across a steady state floodplain, altering the location of wetlands, including forests. Often topography can be complex, with ridges representing old depositional events, swales representing old erosional and flow features, and oxbows representing former river channels. When surface flows are high during exceptional flood events, the waters can uproot, smother or damage standing trees, deposit materials from the river, and alter basin topography itself. In northern climates, ice floes similarly can destroy trees and alter the topography.

Wetland Criteria Characteristics. *Vegetation.* Well developed floodplain forests, especially in warm, humid climates, generally are dominated by densely spaced plants of tree stature. Their diversity and structure are less well developed in drier and in cooler climates. In most cases, the floodplain forest is more diverse and better developed than forests of the adjacent upland. In the best developed systems, the understory is reduced due to limited light penetration combined with the effects of flooding depth and frequency.

Forest development and the wetland character of the vegetation depend on the ground water, flooding regime, and ice conditions. In most cases, riverine control of the water table leads to saturated soil conditions. Where a hydraulic gradient exists, the various forest communities (following the individual adaptations of the constituent plants) may follow the gradient from hydric to mesic forests, or they may respond to the more complex pattern of topographic highs and lows with wetlands occupying the low sites (Larson et al., 1981). In some cases, a river edge forest may seldom flood and be fully terrestrial. In the moist floodplain, nutrients and water are highly available, but adaptation to periodic soil anaerobiosis is required of plants. Wetland adaptations required depend on the timing, duration, and depth of flooding. Where hydroperiods are long, trees may give way to herbaceous marsh. Ice floes may limit vegetation development by periodic scouring. Useful references on floodplain forests are Brinson et al. (1981), Clark and Benforado (1981), Larsen et al. (1981), Wharton et al. (1981, 1982), Ewel (1990), Faber et al. (1989), Hook and Lea

(1989), Minshall et al. (1989), Gosselink et al. (1990), and Sharitz and Mitsch (1993).

Floodplain forests occur under a wide variety of conditions throughout the U.S. from the broad floodplains of the lower Mississippi to the narrow riparian fringes of the arid southwest. As would be expected, they show substantial differences in their vegetation.

The river floodplains of southern North America are often dominated by cypress swamps, as described in Section 3.1.1. In the south, non-cypress floodplain forests tend to occur in drier and less deeply flooded sites than the cypress. The vegetation of these bottomland hardwood swamps is variable, depending especially on hydroperiod. In deeper sites, along with cypress, water tupelo occurs with planertree (*Planera aquatica*), and Carolina ash (*Fraxinus caroliniana*), with a sparse understory including buttonbush. Slightly higher sites support overcup oak (*Quercus lyrata*) and water hickory (*Carya aquatica*). On shallower sites, a more complex forest develops of red maple, green ash (*Fraxinus pennsylvanica*), American elm (which has been eliminated by disease over much of its range), and laurel oak (*Quercus laurifolia*). Shorter hydroperiod conditions, such as on ridges, support sweetgum, and black gum.

In the floodplains of the eastern and midwestern states, silver maple (*Acer saccharinum*) is often dominant, with red maple, elm, eastern cottonwood (*Populus deltoides*), black cherry (*Prunus serotina*), and white ash (*Fraxinus americana*). The understory includes American hornbeam (*Carpinus caroliniana*), hackberries (*Celtis* spp.), and willows.

To the north on mineral soil in floodplains, the swamps are dominated by conifers, especially northern white cedar (*Thuja occidentalis*), tamarack (*Larix laricina*), and white spruce (*Picea glauca*), along with the broad leaved black ash (*Fraxinus nigra*) with speckled alder (*Alnus rugosa*) in the understory. Even further north in Canada and Alaska, floodplains are dominated by willows and alders and at higher elevations by balsam poplar (*Populus balsamifera*).

In drier climates, development of the floodplain forest depends on rainfall. In the eastern grasslands, the fringing forest is composed of river birch (*Betula nigra*), sweetgum, black gum, American sycamore (*Platanus occidentalis*), and cypress. In relatively dry areas, willow and cottonwood dominate the river edge forest, occurring in an open canopy with boxelder (*Acer negundo*) and bur oak (*Quercus macrocarpa*). In the deserts, floodplain vegetation is naturally dominated by screwbean mesquite (*Prosopis pubescens*), cottonwood (*Populus fremontii*), and willows. Tamarisk (*Tamarix pentandra*), introduced within this century, has invaded many riparian wetlands and become dominant in places.

Soils. The base soil of the floodplain develops from the mineral rich alluvium transported by the river. In some locations, colluvium from the adjacent upland forms or contributes to the soil base. These soils are high in nutrients, although nutrient availability is affected by soil moisture in that anaerobic conditions usually persist for much of the year. Under the influence of riverine conditions, organic matter tends not to accumulate as it is oxidized by overflow water or by air during the dry season, or is physically removed by surface water flow. Fires, also not uncommon, also remove organic accumulation. Thus, these swamps tend not to deposit peat except in topographic depressions having a long hydroperiod, usually associated with an elevated water table. Such situations become basin wetlands.

Hydrology. Floodplain forests are characterized by the influence of the flowing water of the river itself. Groundwater levels in the floodplain are often controlled by river elevation. Relatively flat floodplains have a high water table and waterlogged soils. Periodic surface flooding leads to rapid and sometimes prolonged periods of inundation, often seasonal. In the South, periodicity of flow depends primarily on seasonal variation in rainfall, whereas in the midwest and northern areas, flow seasonality tends to depend on the timing, duration, and amount of the spring thaw.

The surface hydrology may therefore fluctuate seasonally from deep water during flooding to a lack of surface water at other times. In addition to rain and snowfall, the topography and slope of the valley and the river, distance from the river channel, frequency of flooding, icefloes, soil characteristics, and bedrock conditions all affect the hydrology of the floodplain. The hydrologic gradient from the river, in turn, determines vegetation development.

With fluctuating water levels and hydroperiods shorter than in basin wetlands, floodplain swamps are dynamic. They have relatively high dissolved oxygen concentrations during flow conditions and a nutrient store replenished by river water materials. With both groundwater and surface water inputs, floodplain forests have access to a substantial nutrient pool, resulting in high productivity.

Characteristic Fauna. Floodplain forests are among the most productive wildlife habitats (Wharton et al., 1981). Fish populations are high seasonally, as they enter the floodplain during high water and retreat to the river when water levels fall (Guillory, 1979; Finger and Stewart, 1987). In the wetland, they can access this high productivity. It is likely that floodplain forests of the United States provide important nursery grounds for stream fishes, as has been so well demonstrated in tropical floodplain forests (Goulding, 1980).

Beavers are characteristic of this environment and have the ability to change floodplain forests to palustrine forests or marshes by impounding water. Otter, swamp rabbits, and raccoon are other typical aquatic mammals. Upland game animals enter the forest during low water conditions. Populations densities may be highest at the swamp-upland ecotone.

Birds are locally quite abundant. The floodplain forests of the lower Mississippi provide a primary wintering habitat for waterfowl from the central North American flyway. They also serve as nesting and feeding areas for wood duck, ibises, wild turkey (*Meleagris gallopavo*), barred owl (*Strix varia*), downy woodpecker (*Picoides pubescens*), red-bellied woodpecker (*Melanerpes carolinus*), pine warbler (*Dendroica pinus*), common yellowthroat (*Geothlypis trichas*), Canada warbler (*Wilsonia canadensis*), ovenbird (*Seiurus aurocapillus*), northern waterthrush (*Seiurus noveboracensis*), wood peewees (*Contopus virens*), veery (*Catharus fuscens*), wood thrushes (*Hylocichla mustelina*), and grey catbird (*Dumetella carolinensis*).

Floodplain forests are particularly important in the arid West where they serve as nesting habitat for many birds, including mourning dove, verdin (*Auriparus flaviceps*), and orioles (*Icterus galbula*) (Wauer, 1977). They also are important habitat for migratory birds.

Wetland Values. Floodplain wetlands have proven to be crucial flood retention features of the landscape. As a river floods its valley, the downstream movement of water is slowed. Although the total amount of discharge is the same per event, its release is delayed, decreasing flood stages downstream over what would have occurred without the floodplain forests. Floodplain forests are intimately tied to the ground water table, and serve as water sources to the river during lowering water conditions. These forests have superior value as wildlife habitat, both for hunting and for nonhunted wildlife. In the south, these forests serve as a primary wintering area for migratory waterfowl. Deer are characteristic and heavily hunted in most areas. It is likely that these forests have important fishery values, as nursery areas for river sport and food fishes. Forest products are an important value of these wetlands. They also have important aesthetic values in the floodplain.

Unusual Characteristics. Not only are floodplain forests determined by hydrological conditions, they significantly influence the overall water regime of the river. Floodplain forests have been recognized as important flood control features. Floodplain forests are highly productive, owing to the capture of nutrients and materials from the river. This productivity is translated into forest products and both hunting and fishing. Floodplain forests also provide corridors for the dispersal of plants and animals between remnant patches of upland habitat, a function that may become increasingly important as forested habitats are altered.

3.2 MARSHES

Marshes are wetlands dominated by herbaceous, usually emergent, plants (Good et al., 1978; Weller, 1981; Gore, 1983; Kushlan, 1990). Various marshes are similar in structure, and presumably function, although the species of dominant plants differ from one site to the next. Freshwater marshes are found in many situations, including those in palustrine and floodplain environments, discussed in Sections 3.2.1 and 3.2.2. The largest contiguous marsh systems in North America are the Everglades and the prairie potholes, which for this reason are treated separately in Sections 3.2.3 and 3.2.4, respectively.

3.2.1 Palustrine Marshes

Palustrine marshes are emergent herbaceous wetlands that form in basins (Figure 3.6; Plate I-II). These may be tiny and isolated or may cover thousands of square kilometers. The largest single marsh in the U.S. is the Everglades, and the largest marsh complex is the prairie pothole region. Other marshes include those along the Gulf and

Figure 3.6: Palustrine marsh in central Maryland, with red maple swamp in the background. (R.W. Tiner)

Atlantic coast, marshes inland of the tidal freshwater marshes, the playas of the high plains of Texas and New Mexico, northern fens, the tule marshes of the west, and the many small to large marshes found elsewhere where water pools at the soil's surface. Useful references on palustrine marshes are Weller (1981), Schomer and Drew (1982), Nelson et al. (1983), Simpson et al. (1983), Drew and Schomer (1984), Odum et al. (1984), Tiner (1985a, 1985b, 1988, 1991c, 1993c), Herdendorf et al. (1986), Windell et al. (1986), Herdendorf (1987), Conner and Day (1987), Nachlinger (1988), Kushlan (1990), and Mitsch and Gosselink (1993).

The water sources of basin marshes may be rainfall, surface water, and groundwater in any combination, with the latter two dominating overall. Outflow typically is restricted. As a result of minerals and nutrients transported into the wetland by these water sources, palustrine wetlands can be very productive and can accumulate peat.

Geophysical Characteristics. Palustrine marshes typically develop in topographic depressions. They also develop on flatter ground where drainage is sufficiently retarded and in river valleys (see Section 3.2.2) behind drainage restrictions. They may occur on any soil, even over highly permeable karst topography when rainfall is high and drainage restricted by peat deposition. In some cases, marshes represent a fire subclimax, with periodic burning inhibiting shrub invasion. Appropriate marsh-forming conditions occur over most of the non-arid United States, resulting in the wide distribution of these wetlands.

Other marshes develop on floating mats suspended above the basin bottom. Freshwater floating marshes are particularly common in coastal areas, such as in Louisiana, and along the margins of deeper water habitats.

Wetland Criteria Characteristics. *Vegetation.* Marsh vegetation is rather uniform in its overall aspect, consisting of emergent and submersed plants growing in shallow water to damp soil and submersed or floating leaved plants in deeper areas. Many of the same plants are found throughout North America, although in differing relative abundance and stature in different regions and sites. Herbaceous wetland plants are adapted to flooded conditions in several ways. Annuals sprout during periodic dry periods, to be succeeded by emerging perennials over the following wet periods. Some species have long-lived seeds that can spend many years in the soil, which becomes a seed bank, before sprouting when conditions turn appropriate. Perennials may allocate a substantial proportion of their energy to root production, especially storage structures, which are used for overwintering or dry season survival. After establishment, vegetative spreading is a typical way for marsh plants to reproduce and dominate a site. Often, the history of colonization determines which plants become dominant. Because of seasonal drying, fires can be recurring. Wetland plants, both annuals and perennials, resprout soon after a light, fast burn. In Winter, above soil portions of the plants die back.

The sedges, grasses, and other herbaceous plants that can dominate a marsh site are relatively few, although many more species are found peripherally. Typical marsh species include common reed (*Phragmites australis*), wild rice (*Zizania aquatica*), cattail (*Typha latifolia, Typha angustifolia*), maidencane (*Panicum hemitomon*), cutgrass (*Leersia oryzoides*), reed canary grass (*Phalaris arundinacea*), bluejoint (*Calamagrostis canadensis*), sawgrass (*Cladium jamaicense*), bulrush (*Scirpus* spp.), spikerush (*Eleocharis* spp.), arrowhead (*Sagittaria* spp.), pickerelweed (*Pontedaria cordata*), smartweeds (*Polygonium* spp.) arrow arum (*Peltandra virginicum*), sedges (*Carex* spp.), rush (*Juncus* spp.), and marsh fern (*Thelypteris palustris*). These species can form a variety of recognizable marsh types, depending on their relative abundance and cover.

Common reed is a predominant marsh plant at many sites, especially in disturbed areas. Cattail marsh occurs similarly, with best development under conditions of high nutrient loading and deep sediments. Reed and cattail dominate a large proportion of small North American basin marshes. Both of these species have superior production efficiencies.

Floating marshes often have maidencane, arrowhead and various vines and herbs growing on a floating mat of roots, tubers, and detritus. In more hydrologically complex marshes, rooted plants sort themselves out by water depth and hydroperiod. At the higher margins of a wetland, various grasses and sedges predominate. Toward deeper water, tall, herbaceous emergents occur. Floating leaved and submersed plants occur next, adjacent to open water if present. To what extent this zonation recapitulates successional patterns differs among situations.

Palustrine marshes occur in the far north, especially where permafrost underlies the tundra. These marshes are dominated by a few species such as bulrush (*Scirpus cespitosus*), cottongrasses (*Eriophorum angustifolium, Eriophorum spissum*), beakrush (*Rhynchospora alba*), sedges (*Carex* spp.), and pendant grass (*Arctophila fulva*). These northern marshes contain mosses, such as sphagnum, but it does not dominate the water quality of the site as it does in true bogs (see Section 3.3). Northern marshes that deposit peat can develop into sphagnum bogs under appropriate conditions.

In temperate latitudes, a myriad of identifiable marshes occur. Among the more notable are tule marshes, beaver dams, and savannahs. The tule marshes of the western U.S. occur (or occurred) along the lower reaches of rivers. These are dominated by *Scirpus* of several species, along with cattail, common reed, and sedges. In the lower prairies of North America, there are many marshes, dominated by cattail, bulrush, and sedges (*Carex* spp.). Also present are common reed, cutgrass, arrowhead, spikerush, and burreed (*Sparganium* spp.). These marshes undergo a multiyear cycling, based on hydrological conditions (Weller, 1981). During periods of high water level, emergent plants (cattail, bulrush, arrowhead) flourish, covering much of the open area of the marsh. As wet conditions continue, the emergents begin to die, especially due to muskrat cutting of the emergent stems, opening up the marsh. During droughts, the emergent plants reestablish themselves, beginning with annuals such as *Bidens* spp. and *Rumex* spp. and the "marsh cycle" begins again.

In arid regions, palustrine marshes can be quite saline (Chapman, 1974). Vegetation includes saltworts (*Salicornia* spp.), sea blites (*Suaeda* spp.), iodine bush (*Allenrolfea occidentalis*), sacaton (*Sporobolus airoides*), and grasses (*Distichlis spicata*).

Beaver ponds are common throughout most of North America. Beavers, in damming drainage features and cutting down trees, create a palustrine environment that supports herbaceous vegetation (Figure 3.7). Along the southeast coast, marshes called savannahs occur under short hydroperiod conditions. These frequently burning marshes are dominated by grasses (*Andropogon virginicus, Ctenium aromaticum, Aristida stricta*).

Marshes known as "fens" occur in the north (Figure 3.8). These are peat accumulating wetlands that receive drainage from surrounding mineral soils, such that water and wetland soils are rich in minerals. Fens generally support marsh vegetation such as reeds, sedges, and grasses (see Glaser, 1983, 1987), and are occasionally considered to be transitional stages in the development of a bog (Mitsch and Gosselink, 1993).

Further south, on the southeast coastal plain and in Florida, marsh vegetation becomes more complex, and species change dominance over surprisingly small increments of elevation and hydroperiod (Kushlan, 1990). In short hydroperiod marshes, beakrush, maidencane, redroot *(Lachnanthes caroliniana*), or sawgrass may predominate in a sparse stand. In slightly deeper water sawgrass, pickerelweed, arrowhead, fire flag (*Thalia genticulata*), maidencane, bulrush, or spikerush predominate, alone or together. In locations of high nutrient or deep soil conditions, cattail often flourishes. Submersed and floating leaved plants may occur in even deeper water. These include bladderwort (*Utricularia* spp.), water lily (*Nymphaea* spp.), fanwort (*Cabomba caroliniana*), water hyssop (*Bacopa* spp.), primrose

Figure 3.7: A beaver pond in Minnesota. (R.W. Tiner)

willow (*Ludwigia repens*), and chara (*Chara* spp.).

Soils. The base substrate of a marsh may be varied, including sand, clay, organic soil, or even bedrock. The important circumstance is that a combination of substrate permeability and water supply are together sufficient to initiate marsh development. Retarded drainage may be due to the impervious character of the soil, the existence of a hard pan, decreased percolation in organic soil, or a high water table even in a substrate having high porosity.

Organic content of the soil is usually relatively high in palustrine marshes. In many, perhaps most, palustrine marshes peat deposition occurs, even in warm, circumneutral waters. Under such conditions, accumulation is the result of high production despite relatively high decomposition rates. Deposition increases in deep, deoxygenated water, under permanently flooded conditions, and in cool climates. Oxidation of the soil is higher in shallow, oxygenated water and during dry periods. Deep reaching fires can burn out peat deposits, lowering soil elevation, and changing the water regime at a site.

Figure 3.8: A fen in Minnesota. (R.W. Tiner)

Hydrology. Water in inland, freshwater marshes comes from three sources: rainfall, groundwater and surface water. Marshes in well watered areas gain most of their water supplies from overland and ground water flow. Marshes in arid areas depend more on periodic rainfall and short-term runoff for their water supply. The western playas occupy small basins with a relatively restricted watershed and are dry for considerable periods (Barclay and White, 1981).

In freshwater marshes situated on the coastal plain (inland of tidal freshwater marshes), water levels are controlled by sea level elevation. However, in most inland marshes, water levels fluctuate with variations in rainfall and runoff. Palustrine marshes may lack standing surface water seasonally or infrequently in dry years. This characteristic of inland marshes is important relative to wetland delineation, in that surface water might not be present in a marsh at all times.

The importance of fluctuating water levels in most palustrine marshes cannot be overstated. Much of the plant and animal life found in these systems either depend on or can acclimate to appropriately fluctuating water depths (Kushlan, 1989a). The permanence of flooding, hydroperiod, and hydropattern decisively affects the character of the palustrine marsh ecosystem.

Some marshes discharge to ground water, particularly at their margins, and by surface flow when the marsh edge is overtopped. However, most basin marshes have limited surface outflow.

Palustrine marshes tend to develop where surface and ground water effects predominate. Because these inputs carry loads of dissolved and suspended materials, marshes can have high levels of nutrients and minerals, including carbonates. As a result, marsh waters are well buffered, generally circumneutral, and have high specific conductivity.

Characteristic Fauna. Among invertebrates, flies are the most uniformly present group in palustrine marshes. These include such characteristic groups as midges, blackflies, and mosquitoes. Macroinvertebrates such as prawns, crayfish, and snails are often common. Fish are abundant where drydowns are not severe or if recruitment sources are available during higher water periods. Those that are most abundant are often small stature species better adapted for survival during the low oxygenation conditions of the dry period. Aquatic snakes, turtles, and the alligator may be typical, depending on location. Marshes are highly productive for aquatic mammals and birds. Muskrat (*Ondotra*), raccoon, otter, and rice rats (*Oryzmyzes*) are typical mammals. Beavers create and sustain many palustrine, herbaceous wetlands. By their dam building activities, they turn bottom land, especially swamp forests, into herbaceous marshes.

Birds sort themselves according to vegetative habitat, and populations of marsh birds change as the emergent vegetation goes through multiyear cycles. Characteristic are: redwinged blackbird (*Agelaius phoeniceus*), yellow-headed blackbird (*Xanthocephalus xanthocephalus*), long-billed marsh wren (*Telmatodytes palustris*), swamp sparrow (*Melospiza georgiana*), sang sparrow (*Melospiza melodia*), black tern (*Chlidonius niger*), spotted sandpiper (*Actitis macularia*), killdeer (*Charadrius vociferus*), mallard (*Anas platyrhynchos*), pintail (*Anas acuta*), blue-winged teal (*Anas discors*), ring-necked duck (*Aythya collaris*), ruddy duck (*Oxyura jamaicenis*), least bittern (*Ixobrychus exilis*), American bittern (*Botaurus lentiginosus*), Virginia rail (*Rallus limicola*), sora (*Porzana carolina*), sandhill cranes (*Grus canadensis*), whooping cranes (*Grus americana*), American coot (*Fulica americana*), common gallinule (*Gallinula chloropus*), and pied-billed grebe (*Podilymbus podiceps*).

Some marshes are particularly important for animal life. Fifteen species of waterfowl bred on marshes in the upper midwest (Smith et al., 1964). Morning doves (*Zenaida macroura*) nest in the playa marshes of Texas (Gutherie, 1981). The marshes of the southeastern and southern United States support many thousands of colonial waterbirds (herons, ibese, storks). The tundra marshes of the Arctic are the primary nesting areas for shorebirds, geese (*Branta canadensis, Anser albifrons*), and certain ducks, such as the pintail (*Anas acuta*).

Wetland Values. Freshwater marshes have important hydrological functions. They absorb floods and runoff, delaying its down gradient movement, thereby reducing flooding conditions elsewhere. They often serve as groundwater recharge sites, especially in porous soil and along their margins. It is now well established that these marshes can serve as a sink for contaminants and fertilizers. These materials are sequestered in the sediments, taken up by plants, or transformed by microorganisms. Of the various wetlands, it appears that marshes dominated by herbaceous plants appear to have the greatest ability to sequester and transform environmental contaminants. Marshes support both wildlife and fishes, particularly waterfowl. Rare and endangered species of birds and amphibians depend on palustrine marshes.

Unusual Characteristics. Palustrine herbaceous wetlands are widespread and, as such, may be the most important wetlands in many areas. Water recharge, discharge, nutrient sinks, wildlife habitat, and fish production are some of the important functions of these marshes. The fluctuating water conditions means that these sites are not always flooded, yet fulfill important wetland functions.

3.2.2 Floodplain Marshes

Floodplain marshes occur in river valleys. Although similar to palustrine marshes, they are influenced by river flows and are exposed to periodic flooding by river water, with attendant stresses and nutrient pulses (Figure 3.9). These may occur in basins in floodplains where

Figure 3.9: Floodplain marsh along the Rio Verde in Arizona. (R.W. Tiner)

periodic flow conditions are conducive to herbaceous growth. They often occur interspersed with floodplain swamps and shrub swamps (see Sections 3.1.2 and 3.1.3). They also occur in situations where water flows over the floodplain for long periods during the year.

Geophysical Characteristics. These marshes occur in many kinds of situations within the floodplain. The most common site is in basins, which may be oxbows or shallow drainageways (sometimes called sloughs). They also occur along the river edge in situations where trees cannot survive, such as where flooding is great or icefloes occur. These are particularly extensive within the floodplain forests along the large rivers of North America, such as the Mississippi and the lesser rivers along the Gulf and Atlantic coastal plain. They also occupy large areas of floodplain where slightly sloping topography and seasonal rainfall lead to extensive flooding over many months of the year. This is the case in Florida along the St. Johns and Kissimmee Rivers. The largest of the flood plain marshes is the Everglades, which is distinguished by lacking a core river, except as it approaches the coast (see Section 3.4)

Wetland Criteria Characteristics. *Vegetation.* The vegetation of floodplain marshes is similar to palustrine marshes, consisting of the same suite of species. Most are the same as was described for palustrine marshes (Section 3.2.1). Some floodplain marshes remain deeply flooded and support dense stands of emergents. Others flood less frequently or for shorter duration and support a sparser vegetation that includes annuals and upland species.

In the south, shallowly sloping floodplains of coastal plain rivers support extensive marshes (called prairies or wet prairies). These are dominated by water lily, pickerelweed, sawgrass, maidencane, torpedo grass (*Panicum repens*), or smartweed, depending on the hydroperiod. More terrestrial species are added toward the gently sloping margins of more upland sites.

Soils. Floodplain soils are either mineral or organic depending on the intensity of the scouring action of the flow. However, most have an organic soil. In basins and other sites protected from erosive water flows, peat can accumulate. The type and degree of accumulation depends on elevation, which determines the length of the dry period.

Hydrology. These marshes differ from palustrine marshes by their hydraulic connection to riverine conditions. High water tables are maintained by the head of the river. Periodic flooding events carry nutrients and materials into the marsh. Depending on the region, seasonal flooding is related to a wet-dry rainfall cycle, or to the Spring snow melt. In northern regions, the scraping and gauging of icefloes can benefit floodplain marshes over woody swamps. Some floodplain marshes are flowthrough systems where water moves across the surface in a sheet flow for many months of the year. Hydroperiod and flood frequency, of course, depend on elevation relative to the river.

The combination of herbaceous growth, flowing water sources, and warm climate leads to high productivity in these wetlands.

Characteristic Fauna. These marshes contain all the characteristic species of palustrine marshes, but in addition support more mobile species seasonally. During the high water period, fishes can move from the river into the floodplain marsh and may spawn there. They retreat back to the river as the marshes dry. Similarly, migratory and regionally mobile birds use the marsh, determined by the water depths. Snakes and turtles are similarly common. In the south the American Alligator may be abundant in this habitat. Wading birds (herons, ibis) feed on fish being concentrated by decreasing water level in the dry season. Water fowl use flooded areas in the winter. Sandhill cranes (*Grus canadensis*) often feed and nest when the marshes are flooded. These are highly productive habitats for fish and game.

Wetland Values. Like floodplain forests, floodplain marshes serve to retard flooding. They also recharge ground water, and can discharge to the river under low water conditions. Like palustrine marshes they can serve as sinks and transformers of materials. They are important wildlife habitat, especially for waterfowl. Also like palustrine forests and in conjunction with connecting riparian habitats, these marshes serve as wildlife corridors.

Unusual Characteristics. Floodplain marshes vary from small potholes to large expanses covering most of a shallow river valley. They are deeper and more permanent near the river than on the higher land farther from the river. They usually occur where fires are rather frequent, and most plants are well adapted to rapid burning.

3.2.3 The Everglades

The Everglades is the largest continuous freshwater marsh in the U.S. (Gleason, 1974; Kushlan, 1989b, 1991). Located in southern Florida, it is a flowthrough, floodplain wetland 100 km long by as much as 65 km wide. The Everglades is dominated by herbaceous, emergent vegetation but also contains extensive stands of submersed marsh and shrub swamp (Figure 3.10).

Geophysical Characteristics. The Everglades overlies a trough in the limestone bedrock of southern Florida (Hoffmeister, 1974). The trough has filled with peat over the Everglades' 10,000 year history. The result of this depositing is that the Everglades occupies a relatively uniform, shallow basin, deeper along the center of its basically north-south axis, shallower to both sides rising to slightly higher ridges.

Along this north-south axis, a ridge and swale topography is elongated in the direction of water flow, with higher ridges showing hydrodynamic sculpting. Deeper pools occur, as do patches of higher

Figure 3.10: The Everglades, showing a ground view of a sawgrass dominated marsh. (J.A. Kushlan)

ground, on peat mounds or limestone highs. Because of the extensiveness of the wet landscape, these high ground patches are called islands. The limestone and sand deposits underlying the Everglades are highly permeable.

Wetland Criteria Characteristics. *Vegetation.* The vegetation of the Everglades is dominated by herbaceous emergents, especially sawgrass, with more complex communities occurring on progressively higher ground. The presence of a particular dominant plant species depends on water depth and duration of surface flooding, nutrients, fire, and recruitment history.

The deepest sites are open water ponds that support submersed (*Najas* spp.) and floating vegetation, such as water lily and spatterdock (*Nuphar luteum*), or submersed plants especially bladderwort

(*Utricularia*). Bladderwort, as characteristic of the southern Everglades as sawgrass, supports a thickly growing community of epiphytic algae. These deepest marshes are elongated in the direction of the surface current and serve as the primary flow ways.

On slightly higher ground and slightly shorter hydroperiods, emergent plants take over. Depending on the water depth and hydroperiod, these may be dominated by spikerush, cattail, maidencane, pickerelweed, arrowleaf, or sawgrass, which is the most widespread plant cover of the Everglades. On higher sites, shrubs such as willow (*Salix caroliniana*), holly, and buttonbush may form shrub swamps on the ridges and on islands. At the periphery of the Everglades, the shallow, short hydroperiod marsh supports foxtail (*Erianthus giganteus*), black rush (*Schoenus nigricans*, and muhly (*Muhlenbergia fillipes*). Ground that is emerged for most of the year supports a more complex forest of tropical hardwoods.

Soils. The soil of the core of the Everglades is an endogenous peaty muck, unconsolidated at the surface and compacted below. The deepest deposits, near Lake Okeechobee, were historically up to 6 m deep. This thins to less than a few cm in the southern portion of the marsh. Along the periphery, and mixed with the organic soil in the interior, are marl substrates. These have formed in situ from calcium carbonate precipitation by epibiotic algae that cover the plants and bottom. The peat is deficient in trace nutrients but is highly organic. Dry season fires can burn the peat and the organic marl, either superficially or deeply depending on water conditions.

Hydrology. The water source of the Everglades is rainfall and overland flow. Rainfall is seasonal with 85% of the 130 cm of annual rain falling in six months. Water moves from topographic highs near Lake Okeechobee toward the coast over a gradient of about 0.3%. Because of the slowness of the flow, the Everglades has characteristics of both palustrine and floodplain marshes. The flow is sufficient to create the ridge and swale lineations that further direct the flow patterns.

In the dry season, flow decreases and finally stops, and water levels drop. The marsh surface becomes exposed first along the higher periphery, moving inland with the drying season toward the central trough of the marsh. In some years, all but the deepest sloughs and ponds will be without surface water. The rainy season begins in late May, on average, and the Everglades fills up rather quickly.

There appears to be no definitive aquiclude between the groundwater moving through the porous limestone and the surface water. However, the peat and marl deposits no doubt inhibit rapid movement of water between the two. Recharge probably occurs along the periphery and through canals that have been cut through the peat into the limestone baserock.

Characteristic Fauna. The Everglades is well known for its wildlife (Kushlan, 1976, 1989b; Sykes, 1983; Frohring et al., 1988; Kushlan and Jacobsen, 1990). The small fish community is dominated in numbers by mosquitofish (*Gambusia affinis*), but includes a number of minnows (Cyprinids) and sunfish (*Lepomis* spp.), which survive the dry season in alligator ponds and deeper marshes and have fast population growth rates, allowing them to reinvade the marsh soon after rising water levels. Small invertebrate larvae are an important link in the food chain. These include dragonflies, mayflies, mosquitos, gnats, and deer flies. Macro-invertebrates include prawns (*Palaemonetes* spp.), crayfish (*Procambarus alleni*), and apple snails (*Pomacea paludosa*).

Amphibians and reptiles are common, including water snakes (*Nerodea* spp., *Agkistrodon piscivorus*), and turtles (*Pseudemys* spp.) as well as the American alligator. Mammals include the white-tail deer, Florida panther, roundtail muskrat (*Neofiber alleni*), and mink (*Mustella vison*).

At one time populations of wading birds nested in large numbers in this system. Prominent species include wood storks, white ibis (*Eudocimus albus*), great blue herons (*Ardea herodias*), great egrets (*Egretta alba*), tricolored herons (*Egretta tricolor*), snowy egrets (*Egretta thula*), and little blue herons (*Egretta caerulea*). Populations of these species have declined significantly in the past several decades due to hydrological changes that have taken place in the Everglades. Other characteristic birds include the snail kite (*Rostrhamus sociabilis*), which specializes on catching apple snails. Mottled ducks (*Anas fulvigula*) are found year round, with a few other species wintering in the marsh.

Wetland Values. The Everglades is the primary means of flood control and the primary water supply for the large human population in southern Florida. It has tremendous value for recreation, including hunting, fishing, and nature study. Everglades National Park attracts hundreds of thousands of visitors each year. The Everglades is a nutrient-poor system which can absorb nutrients only with changes in vegetation community. This system is recognized by all major international programs as a wetland of worldwide importance.

Unusual Characteristics. The Everglades is the largest expanse of wetlands in the United States and the only basically tropical wetland system located there. Its historical importance as a wildlife habitat led to the creation of national parks and wildlife refuges to preserve a portion of its landscape. Unfortunately its hydrology has been altered and degradation has occurred despite this concern (Blake, 1980; Kushlan, 1987). Considerable effort is being made to understand the system and reverse its degradation.

3.2.4 Prairie Potholes

Prairie potholes are small, palustrine marshes scattered over the northern prairies from South Dakota and Minnesota through Manitoba, Saskatchewan and Alberta (Figure 3.11) (Stewart and

Figure 3.11: Prairie pothole marsh in North Dakota. (R.W. Tiner)

Kantrud, 1972; van der Valk and Davis (1978); Leitch et al., 1979; Weller, 1981; van der Valk, 1985, 1989; Hubbard et al., 1988; Kantrud et al., 1989). The complex of marshes and prairie covers 780,000 square kilometers.

Geophysical Characteristics. The pothole marshes occur in basins formed by moraine and meltwater landforms left by retreating glaciers. The basins occupy depressions in till deposits formed by scouring, iceblocks, drainage dams, and riverine deposits. The accumulation of glacial silt sealed the basins. The irregularities of the landscape provided a variety of situations where marshes were able to develop within the prairie. Most marshes are less than 0.4 hectares (Cowardin et al., 1981).

Wetland Criteria Characteristics. *Vegetation.* The prairie pothole marshes are characterized by emergent plants interspersed with open water. The dominant plants include cattail (both species), bulrush (*Scirpus validus, Scirpus fluviatilis, Scirpus acutus*), burreed (*Sparganium eurycarpum*), sedges (*Carex* spp.), and arrowhead. Floating leaf and submersed plants include lilies (*Nymphaea* spp.), pondweed, stonewort, and bladderwort (van der Valk, 1985, 1989; Hubbard et al., 1988; Kantrud et al., 1989).

The vegetation follows a zonation based on water depth. Lilies, pondweed, and bladderwort occur in the deeper, more open areas. Bulrush occurs next to open areas, followed by cattail, then sedges and arrowheads, and finally the terrestrial prairie grasses.

Prairie potholes undergo a 5-20 year marsh cycle alternating periods of drying, regenerating marsh, and degrading marsh (Weller, 1981). The marsh cycle begins during drought with plants sprouting from the seed bed of the marsh. These include both annuals and the dominant perennials. After flooding, the annuals die out leaving the emergent perennials. Submersed and floating leaf plants next appear. These reach their maximum stands and then decline.

Soils. The base soil of prairie potholes is glacial till with a high percentage of clay and silt with low permeability. Higher percentages of sand and gravel occur on outwash areas. Soils underlying the marshes have a well developed soil profile.

Hydrology. The hydrology is driven by rainfall, snowfall, runoff, and ground water. Ground and surface water hydrology is further constrained by glacial topography and soil permeability. Annual precipitation is moderate (30 -60 cm) and evapotranspiration is seasonally rather high. In fact, evapotranspiration generally exceeds rainfall, so runoff and snowmelt are essential to maintain water in these marshes. Because of variation in watershed contribution and groundwater relationships among marsh basins, they differ in their water depths and permanence. Some prairie pothole marshes are only temporarily flooded, some are flooded seasonally, some semi-permanently, and some permanently.

Water level in the marshes tends to reflect the ground water, especially at high water stages. At lower stages, some marshes discharge to the ground water at their edges contributing importantly to the aquifer. Other marshes receive discharge from ground water and can be interconnected by subsurface water flow. Semipermanent marshes tend to be flowthrough systems, receiving ground water and discharging to groundwater.

Rainfall varies among years and this variation can substantially alter the hydrological situation on a marsh. A cycle of plant growth is initiated by periodic drying during low water conditions.

Water chemistry varies as these marshes have a wide range of salinity, due to variation in materials carried from the glacial materials.

Many of the potholes have been drained for farming. Others are plowed to the edge, disrupting the quantity and quality of their water source.

Characteristic Fauna. Fish are few to nonexistent since prairie potholes dry periodically and are not confluent with refugia that might allow fish to survive droughts. As a result of the limited ichthyofauna, the food chain of prairie potholes is based on the invertebrates found there. These include isopods, chironomids, amphipods, snails, cladocerans, and copepods. Populations of these groups vary seasonally and according to the marsh cycle. Isopods are most abundant early in the cycle, snails and amphipods in midcycle, and cladocerans and copepods of open water systems characterize late stages.

Fathead minnows (*Pimephales promelas*) and brook sticklebacks (*Culaea inconstans*) are among the few native fish that can survive the difficult environment of the potholes, mostly in marshes with connections to deepwater refugia. A number of reptiles and amphibians use prairie potholes. Notable are the tiger salamander (*Ambystoma tigrinum*), the chorus frog (*Pseudacris woodhousei*), and the leopard frog (*Rana pipiens*).

The importance of these wetlands lies in the fauna they support (van der Valk, 1989). It is estimated that as much as 75% of the waterfowl produced in North America come from this region. The amount of recruitment is correlated with the number of potholes with water in them during the beginning of the breeding season. The most important species is the mallard (*Anas platyrhynchos*). Others include pintail (*A. acuta*), blue-winged teal (*A. discors*), gadwall (*A. strepera*), green-winged teal (*A. crecca*), canvasback (*Aythya valisineria*), redhead (*A. americana*), ring-necked duck (*A. collaris*), lesser scaup (*A. affinis*), hooded merganser (*Lopodytes cucullatus*), ruddy duck (*Oxyura jamaicenis*), wood duck, and Canada goose (*Branta canadensis*). Various species use marshes with specific hydrological conditions. Temporary marshes are used only by dabbling ducks (*Anas* spp.). Seasonal marshes and semipermanent are also used by diving (*Aythya* spp.) and the ruddy duck. Waterfowl appear to make best use of the potholes when the emergent vegetation and open water areas are interspersed, leaving both shelter and open feeding areas.

Many other birds use these marshes including, grebes, rails, coot, American bittern, killdeer, willet (*Cataptrophorus semipalmatus*), marbled godwit (*Limosa fedoa*), Wilson's phalarope (*Phalaropus tricolor*), black tern (*Chilonius niger*), marsh wren, yellow-headed blackbird, red-winged blackbird, and savannah sparrow. These species use specific vegetation stands, which change with the marsh cycle.

Mammals are exceptionally important in the ecology of prairie potholes. During the regeneration stage of wetland development, muskrat (*Ondatra zibethicus*) populations increase, and their activities are thought to hasten the decline of the marsh and its succession from emergent marsh to open aquatic habitat. Other aquatic mammals include beaver and mink. Shrews, mice, hares, weasels, fox, coyote, and deer use the marsh periodically.

Wetland Values. The prairie potholes serve a number of important functions, the most critical being flood retention and ground water recharge. They also function to maintain water quality, which is of importance given the heavy agricultural use surrounding them. The

Freshwater Wetlands   119

most important international value, of course, is the production of a significant portion of the continent's migratory waterfowl.

Unusual Characteristics. The importance of the prairie pothole ecosystem is the combination of intensive agriculture and waterfowl production. The combination makes this one of the most valuable systems in the world. With much of the marsh system having been drained, the future depends on an optimization of the several important values they serve.

3.3 BOGS

Bogs are wetland communities typified by the dense surface cover of aquatic moss, *Sphagnum* spp., which creates a highly acid water chemistry. These wetlands are found across the north temperate zone of North America and Eurasia and are similar in appearance and functioning (Figure 3.12, and Plate III) (Heinselman, 1970; Moore and Bellamy, 1974; Larsen, 1982; Gore, 1983; Larsen, 1982; Johnson, 1985;

Figure 3.12: A bog in New York. (R.W. Tiner)

Tiner, 1985a, 1985b, 1988; Damman and French, 1987; Glaser, 1987; Crum, 1988; Mitsch and Gosselink, 1993).

Bogs characteristically derive their nutrients primarily from rainfall. The low nutrient availability and high acidity results in peat deposition, which in turn influences the development of plant and animal communities. These nutrient poor peatlands may be herbaceous bogs or shrub bogs, with a successional relationship often existing between the two. Bogs are called muskegs in Canada and moors in Europe. They also are a type of mire, in European terminology.

Geophysical Characteristics. Bogs have been well studied and are known to occur in several situations. Basin bogs are those that occupy depressions including former lakes, which become filled with peat after which the bog begins to emerge above the landscape. As a result basin bogs are called raised bogs. Bogs also cover flat landscapes, often extensively. These blanket bogs have generally grown out from a foci such as a depression. Another type of bog occurs on slopes, where an undulating topography develops a series of peat bog ridges tending across slope separated by intervening pools. These are called string bogs. In all these bogs, the peat deposition process itself creates the final land form.

Peat grows not only up but out. As peat accumulates, new growth can also move sideways encroaching upon and overtopping the soil and plants adjacent to its margins. It is in this way that blanket bogs spread from basin loci through woodlands. Along the edges of lakes and pools, peat can grow out over the water, forming a floating shelf. These wetland situations are called quaking bogs.

Wetland Criteria Characteristics. *Vegetation.* Bogs are characteristically dominated by a few species of peat moss, a bryophyte. These included *Sphagnum cuspidatum, S. major, S. papillosum, S. magellanicum,* and *S. capillifilium.* Sphagnum is well adapted to survive and prosper under conditions of low nutrients and acid water. Acid water is not required for its survival, but it is competitive over other plants under such conditions. *Sphagnum* grows in densely packed mats. It has a high concentration of air in its cells, which keeps it floating toward the surface. Because decomposition is slowed by cool temperatures, acid water, deoxygenated conditions, and low nutrient concentrations, the peat located below the actively growing moss accumulates, creating the bog conditions.

Bogs have a low plant diversity relative to other wetlands because of the lack of mineral soil, lack of nutrient availability, and acidity. However, some are capable of invading acidic peat substrate. Cotton grass (*Eriophorum vaginatum, Eriophorum angustifolium*) is a common associate. It is a deciduous perennial, which conserves nutrients by translocation prior to leaf death. Sedges also are common associates. These include *Carex oligosperma, C. limosa, C. pauciflora*, and *Rhynchospora alba*. Other herbaceous plants include liverwort (*Cladopodiella fruitans*), the moss *Drepanocladus*, and orchids (*Calopogon tuberosus, Habenaria clavellata, Habenaria blephariglottis, Pogonia ophioglossides*, and *Arethusa bulbosa*). Carnivorous plants are also typical. These include the bladderwort (*Utricularia cornuta*), *Sarracenia purpurea, Drosera intermedia*, and *Drosera rotundifolia*.

Woody plants can also invade, although their growth is slow and stature usually slight; species that are trees elsewhere may only achieve shrub stature in bogs. Shrubs include cranberry and blueberry, heather (*Calluna vulgaris*), crowberry (*Empetrum* spp.), black spruce (*Picea mariana*), scotch pine (*Pinus sylvestris*), leatherleaf (*Chamaedaphne calyculata*), laurel (*Kalmia politifolia*), heaths (*Ledum groenlandicum*), Labrador tea (*Ledum palustre*), willow, bog birch (*Betula pumila*) and tamarack. Trees, such as black spruce, can occur on higher islands in the bog.

Most of these species are not restricted to acid conditions, but are competitive there. Among the adaptations for survival are: reduced oxygen consumption, aerenchyma and lacunae, root aeration, low nutrient requirements, sequestering of nutrients during the Winter, deep penetration of roots, symbiotic nitrogen fixation and carnivory, adventitious roots and shoot elongation to ovoid overgrowth by the moss (Crum, 1988).

As succession occurs, typical plants change. Bogs typically develop from mineral, groundwater marshes. First, sphagnum invades and eventually alters the water and soil chemistry. Then a mixed bog develops as other herbaceous plants and small woody plants invade. Finally a shrub bog develops.

Soils. The functional substrate of bogs is the sphagnum peat itself, which is waterlogged, acidic, and anaerobic at depth. Sphagnum also dominates water chemistry, particularly through its control of pH and the ability of living sphagnum to take up dilute nutrients.

Bogs deposit peat. Conditions leading to peat deposition and bog formation include cool climate, ample precipitation, high humidity, poor drainage and waterlogged conditions, low oxygen levels, and usually low pH.

Bog peat is deposited over various soil and geologic bases but is soon divorced from the influence of the underlying mineral soil. Bogs are generally successional ecosystems, having developed over mineral soils through the autogenic activities of the bog plants. In succession, the transition from soil dominated processes (owing to mineral soil or mineralized peat) to sphagnum dominated processes occurs rather sharply.

Hydrology. Bog vegetation occurs typically where ground and surface water inflow is minimal, thus the wetland depends primarily on rainfall. Ground and surface water may influence bogs more in the early stages of bog formation. Bogs develop along the edges of lakes that have a flow through hydrology. However, as bog development occurs, the bog is increasingly isolated from other water sources.

The water holding capacity of peat is many times its own weight, and as a result, affects such characteristics as effective water table, temperature, nutrients, and gases. Water is contained not only among the peat particles but also in water holding cells in the sphagnum. As the accumulating peat raises the bog, it carries with it its water table. The bog water supply may then become perched above the ground water and higher than flows of surface water, which bypass it. The bog may become a water source.

The limited inflow of surface and ground water results in the bog's low fertility and therefore explains many of its characteristics. The water is highly acidic with low concentrations of minerals and high cation exchange capacity in the peat. The acidity comes from the oxidation of organic compounds, especially those containing sulfur, by the sphagnum's extraction of cations from the water and exchange with hydrogen ions, the production of galacturonic acid at the growing tips of sphagnum, and the leaking of humic acid from decomposing peat (Clymo, 1964).

Characteristic Fauna. Animals are generally scarce in bogs, owing to the low production, low nutrient concentrations, acidity, and an inability to access and consume the dominant biomass - sphagnum. Insects are the dominant consumers, especially flies (psyllids, tipulids). Snails are nonexistent.

Among vertebrates, those that are mobile among habitats are the most common. Herbivores and carnivores enter the bog from the surrounding habitats or higher areas of the peat lands. Mammals include hares, lemming, and weasels. Large mammals, such as moose, occur along the margins. Birds include sora, sandhill crane, great gray owl (*Strix nebulosa*), short eared owl (*Asio flammeus*), palm warbler (*Dendroica palmarum*), northern waterthrush (*Seiurus noveboracensis*), and swamp sparrow (*Melospiza georgiana*).

Wetland Values. Bogs determine and create the water table of their basin. Since prehistoric times, bogs have been mined for their peat, which is used for fuel, and more recently for horticulture. Bogs cover much of the Northern Hemisphere and are part of the traditional live of that huge region. Berries are farmed in bogs and harvested commercially. It is likely that bogs influence atmospheric gasses, considering their wide distribution.

Unusual Characteristics. Peat creates its own wetland. Peat bogs have played an important role in human history for centuries. For example, they provide fuel. As a result, peat has been mined as a source of fuel in situations where fuel is limited. This practice continues in many parts of the world.

3.4 DEEP WATER AND OTHER INLAND AQUATIC ECOSYSTEMS

Wetlands are often found in association with deep water habitats, especially rivers and lakes. The environmental conditions in these ecosystems differ in important aspects from those in wetlands. Nonetheless they interact as water and organisms move between them.

3.4.1 Rivers and Streams

A river or stream is composed of water and its load that moves along a defined and repeated track, following or cutting a channel. The predominating physical and biological feature is the role of moving water in structuring this ecosystem.

Geophysical Characteristics. The stream channel is cut through the substrate, soil, sediment, or rock by the erosive force of water moving down gradient. The velocity and volume of water moving down the stream depend on the catchment area, the width, depth, and gradient of the stream bed, and the amount of rainfall, snowmelt, and upstream discharge. Typically, streams originate in upland areas, such a mountains or hills. There they are relatively swift, shallow, and narrow. Velocity decreases as the stream falls in elevation. The size of the stream increases with distance from its source, joining other streams and becoming wider, deeper and carrying a great load of sediment and other materials. When it reaches flatter land, such as in the coastal plain, it becomes more sluggish and meanders as it cuts along out along the outer bends and deposits materials along the inner bend. As meanders come closer together, the stream can break through the neck of land separating the channels, taking a more direct course and isolating oxbows from the main channel. Flow varies seasonally and from year to year, from no flow in intermittent streams to flood stage. Repeated flood stages, annually or less frequently, cut a valley larger than the stream itself. During floods, materials are both transported to and removed from the flood plain.

Ecological Characteristics. *Vegetation.* Vegetation in the channel itself it usually limited by the flowing water and lack of substrate in upper reaches, and by high flows and lack of light in lower reaches. Where the water is clear, algae grows on rocks. In slower streams, or in protected sites, submersed plants can form plant beds. Typical plants include *Potomogeton, Myriophyllum, Wolffia, Spirodela,* and *Chara.* In larger muddy streams, other submersed and floating leaf aquatic plants, such as water lilies, can grow in protected sites such as backwaters and coves.

Many streams are intermittent. Depending on the season and duration of no flow periods, the stream bed may be colonized by herbaceous annuals, especially grasses (*Panicum*). Streamside vegetation reflects the situation through which the stream flows and is often a zone of wetland type plants. Streamside plants may range from herbaceous, particularly in upper reaches, to an extensive floodplain swamp in the lower reaches. In arid areas, tamarisk is common.

Soils. Stream substrates depend on the bedrock and the materials that are being carried and deposited in the floodplain. The streambed reflects the parent material through which the stream flows and the material being transported to that point, either on a daily or intermittent basis. As the flow slows and volume of materials being transported increases from erosion, unconsolidated sediments are deposited in the channel and along the banks.

Typically, higher streams have a rocky substrate because of the rapid flow and relatively small volumes. The bottom of smaller streams may be bedrock, rubble, gravel, sand, or mud. Larger streams and those flowing through arid environments generally have a mud bottom. Along the coast, streams cut through organic substrate.

Hydrology. The water of streams originates from rainfall, ground water seepage, springs, surface runoff, or wetland surface flow. Variation in stream flow depends on rainfall, soil conditions, slope of the landscape, vegetation in the catchment basin, temperature, humidity and evaporation.

Characteristic Fauna. Given the limited primary production within a stream, a characteristic of this ecosystem is that the food chain is based on materials being carried into and along the stream. Additionally, organisms are adapted to life in flowing waters.

The invertebrates present depend on the substrate and the type of food source. On bedrock and rubble, animals are generally adapted to attach firmly to the substrate. These include snails (*Physa*), clams (*Pisidium*), worms (*Limnodrilus*), caddisflies, and mayflies (*Ephemerella, Caenis*). On cobble, there are more interstices for hiding, and additionally this bottom can support midges (*Chironomus*), mosquitos (*Anopheles*), mussels (*Mytilus, Modiolus*), flatworms, leeches, and crustaceans. In sand, crustaceans increase. In mud, burrowing and mud specialists can occur, including the horse and deer flies (*Tabanus*), crayfish, snails (*Lymnaea*), and clams (*Sphaerium*).

Fishes in the upper reaches are the swift flow specialists, especially the salmonids (trout) and bottom dwelling fish (*Percina, Ammocrypta, Etheostoma*). In the lower reaches, additional species occur.

3.5 PONDS AND LAKES

Lakes and ponds are open, deep water systems occupying basins. There is no substantive difference between a pond and lake, except that lakes are larger, but even that relation differs among areas. Ponds are still waters, generally with submersed vegetation covering significant portions seasonally. Ponds and lakes may be formed artificially by damming rivers and flooding their valley. These reservoirs bear many similarities to natural lakes and ponds.

Geophysical Characteristics. Basins may develop in many ways, ranging from a meteor impact crater to an alligator den. Some of the largest and most interesting lakes in North America have diverse origins, such as glacial scouring (Great Lakes), remnants of a former sea floor depressions (Lake Okeechobee), volcanic action (Crater Lake), solution of limestone, and earthquakes (Reelfoot Lake). Salt lakes occupy fossil basins of former large lakes (Great Salt Lake). Smaller ponds and lakes result from glaciers, such as kettle lakes that formed in the depression left by melting blocks of glacial ice; oxbow lakes that form in the former channels of rivers; beaver lakes form when beavers dam streams; and the many depressions in various habitats that are deep enough to have open water.

Ecological Characteristics. *Vegetation.* Lakes and ponds generally have a zonation of vegetation following water depth gradients. Along the edges in shallow water if wind and wave exposure is limited, typical marsh or swamp vegetation usually develops. Submersed and floating leaf vegetation cover slightly deeper areas, and can seasonally cover an entire pond or small lake. In deeper basins, primary production depends on phytoplankton. The nutrient and energy relationships between the open water system and the fringing wetland vegetation are just now becoming better understood.

Soils. Substrates of lakes are variable, depending on the parent substrate and on the materials carried into the lake by streams or runoff. Substrates may be sand, clay, or organic sediment. In deep lakes, an anoxic organic, sediment forms. Often in the herbaceous zones of lakes and ponds, peat is deposited if erosional processes are not excessive.

Hydrology. The water supplies of lakes and ponds may come from rainfall, ground water, surface runoff, stream flow, or wetland surface flow. Most lakes have both input and output streams; many ponds lack surface outflow, although they may recharge groundwater. Reservoirs have outflow tightly controlled for flood control and water management. Outflow in these situations is often from depth rather than at the surface, which can effect downstream rivers and wetlands. Changes in the timing of flows from the reservoir can also substantially change the flooding regime of the floodplain and its wetlands.

Direct water input into a lake, such as by stream flow, can create density gradients within the lake. Density gradients are also maintained by temperature stratification, which effectively isolates aerobic layers from cool anaerobic bottom layers. Temperate lakes can overturn seasonally when water temperature equilibrates to depth. This turnover is a characteristic of the water quality of deep, temperate lakes. The water supply of ponds may be more local, and its water chemistry is very dependent on local conditions.

Characteristic Fauna. The fauna of lakes and ponds differs from that of wetlands because of its open water component. Phytoplankton and zooplankton (cladocerans, ostracods, copepods) generally form the basis of the food chain. Sediments are anaerobic and are suitable for only specialized animals such as oligochaetes (*Tubifex* spp.). Fish occupy most lakes and ponds. Since lakes are more stable than wetlands, fish can grow to larger size. Gizzard shad (*Dorosoma cepedianum*), bass (*Micropterus*), sunfish (*Lepomis* spp.), walleye (*Stizostedion lucioperca*) are some important lake fish in the U.S. Many lakes and reservoirs are stocked with native or non-native fish to enhance the recreational fishery.

4

Estuarine Wetlands

Robert J. Livingston

INTRODUCTION

This chapter will introduce the major types of estuarine wetlands, and their significant biological characteristics. Like all wetlands, estuarine wetlands are characterized by their hydrology, soils, and plants (see Chapters 1-3, and 5-6). More specifically, estuarine wetlands are those with relatively high, ocean-derived salinity (typically greater than 0.5 parts per thousand, ppt), although salinity is frequently subject to variation depending on freshwater input. Estuarine wetlands are typically found in coastal areas, usually within a few miles of the ocean.

There has been considerable scientific investigation of estuarine wetlands over the past 2-3 decades (for example, see Cutler, 1978; Livingston and Loucks, 1978; Tiner, 1984; Day et al., 1989; Montague and Wiegert, 1990). The broad diversity of classificatory approaches and interpretations in recent times attests to the extreme complexity of such systems (, see Martin et al., 1953; Shaw and Fredine, 1956; Bayly, 1967; Zoltai et al., 1975; Anderson et al., 1976; Stegman, 1976; Cowardin, 1978; Radford, 1978; Cowardin et al., 1979; Hedgepeth, 1987; Ecological Society of America, 1992). For purposes of this book, estuarine wetlands are tidally-influenced, saline coastal systems which can be divided into tidal marshes (both salt and brackish), mangrove swamps, and intertidal flats. Purely marine habitats, such as oceanic beaches, rocky shores, and intertidal coral reefs will be discussed only insofar as they impact ecological processes of estuarine systems.

Estuarine wetlands are subjected to periodic salinity changes due to differences in tidal fluctuations and freshwater inflow over relatively short periods of time. They are affected by factors such as freshwater runoff, groundwater input, precipitation, water currents, wind and tidal effects, and episodic events, such as storms. Based on average salinity, estuarine wetlands can be categorized as: (1) polyhaline, or strongly saline (18-30 ppt); (2) mesohaline, or areas of moderate salinity (5-18 ppt); and (3) oligohaline, or slightly brackish areas (0.5-5 ppt). In areas where estuarine wetlands are dissociated from regular tidal flushing (pannes), salinity may reach much higher values due to evaporation, and increased salinity may affect the distribution of salt-adapted plants (halophytes) in an estuarine wetland (see Wiegert and Freeman, 1990). Above salinities of 70 ppt, it has been shown that salt marsh plants, such as smooth cordgrass (*Spartina alterniflora*), experience cell wall permeability changes that reduce the influx of nutrients, and may induce osmotic stress with reduction of water and oxygen uptake (Adams, 1963; Linthurst, 1980; Linthurst and Seneca, 1981). Depth, substrate interactions and soil characteristics, input from oceanic sources, and various biological factors, such as competition and predation (see Pennings and Callaway, 1992; see also Section 2.2.2) are additional factors that are important in determining the distribution of estuarine wetland plants and animals.

As discussed in Section 2.3.1, the net productivity of estuarine wetlands is among the highest of all aquatic ecosystems (see Tiner, 1984). Estuarine food webs are inextricably associated with both freshwater, marine, and upland species in many ways. While there have been many studies of these relationships (see reviews in Nixon, 1980; Teal, 1986; Day et al., 1989; and Mitsch and Gosselink, 1993), most such studies have involved relatively generalized approaches to the association of estuarine wetlands to the broader estuarine and marine ecosystems, such that there remains considerable speculation and uncertainty as to the exact ecological significance of estuarine wetlands. Productivity of estuarine wetlands will be discussed in detail in Section 4.1.2.

Estuarine wetlands are particularly susceptible to influences by human activities. The most significant of these anthropogenic activities have included physical alterations caused by dredging, filling, enrichment by nutrients and organic carbon products, and the loading of toxic substances (see discussion in Section 4.4; see also Gerlach, 1981; Neilsen and Cronin, 1981; Day et al., 1989; Montague and Wiegert, 1990). It has been recognized for some time that estuarine and other coastal wetlands are ecologically and economically important. However, despite the popular conception of estuarine wetlands as highly productive and endangered areas, there has been relatively little scientific effort to place the ecological role of wetlands within the broader context of ecosystem process and function.

4.1 ESTUARINE WETLAND FUNCTIONS AND VALUES

4.1.1 Overview of Estuarine Wetland Functions and Values

Like some other wetland types, the significant values of estuarine wetlands include flood and erosion control, nutrient and pollutant processing, and high primary productivity associated with valuable habitat for both wetland and wetland-dependent plants and animals (see Section 2.3). Estuarine wetland-based food webs provide the foundation for many forms of harvestable crops, including many economically valuable fish and shellfish species. Despite the relatively harsh conditions of estuarine wetland ecosystems (fluctuating salinity, periodic disturbances by storms, etc.), habitat values of estuarine wetlands are generally quite high (Montague and Wiegert, 1990).

Estuarine wetlands perform important functions in shoreline anchoring and erosion control, although the exact nature of such control has not been well established for the different kinds of estuarine wetlands. Estuarine wetland vegetation (e.g., cordgrasses, *Spartina* spp., in salt marshes; or mangroves, *Avicennia* or *Rhizophora*, in mangrove swamps) serve to reduce shoreline erosion by: (1) binding sediments with root systems, and increasing the durability of soils; (2) dampening wave action through friction; and (3) reducing current velocity (see Dean, 1979; Knudson et al., 1982). In fact, the U.S. Army Corps of Engineers has successfully transplanted *Spartina alterniflora* to damaged shorelines as a means of reducing erosion (Woodhouse et al. 1976; Lewis, 1982), and mangroves have been transplanted with some success to reduce beach erosion along the Atlantic Coast (Lewis, 1982).

Estuarine wetlands probably serve to reduce flood damage as well (see discussion in Section 2.3.3). However, Sather and Smith (1984) presented an in-depth evaluation of the scientific basis for postulated wetland values, and found that there was a "significant" lack of information concerning the flood control effectiveness of a given wetland.

Nutrients and toxic substances of various types are processed by estuarine wetlands. Wiegert and Pomeroy (1981) found that salt marshes have the capacity to absorb nutrients and toxins, but in limited amounts and over the short term. Likewise, mangroves have the ability to trap and assimilate many nutrients and toxins, but only within limits (see Snedaker and Brown, 1981). It has been suggested recently that Florida mangrove stands in shallow water (< 1.0 m) can be used to intercept runoff from upland areas, and significantly reduce nutrient concentrations, but this idea remains largely untested and highly controversial (see Odum and McIvor, 1990).

Estuarine wetlands provide important habitats for various forms of commercially important finfish and shellfish, and are important feeding grounds for waterfowl (see Section 2.3). The specific values of estuarine wetlands for various forms of micro-organisms (plant and animal), macroinvertebrates, mammals, and non-game birds are less well understood, but have been reviewed recently by Pomeroy and Wiegert (1981), Day et al. (1989), Montague and Wiegert (1990), Odum and McIvor (1990), and Mitsch and Gosselink (1993).

4.1.2 Productivity

One of the most important characteristics of estuarine wetlands is their relatively high productivity. The role of wetland-derived nutrients has been the subject of considerable evaluation and debate in recent times. The diversity of the productive responses of coastal salt marshes was discussed by Bagur (1977). High productivity was evident regardless of the type of marsh under consideration. Marsh vegetation in the form of grasses, sedges, and rushes has been found to be responsible for organic production that exceeds most comparable terrestrial ecosystems.

In the early literature, it was thought that there was movement of nutrients and organic carbon from marshes into associated estuaries. Wetlands were viewed as an integral part of the year-round high production of the associated aquatic systems (Odum and De La Cruz, 1967). Vascular marsh vegetation, benthic algae, epiphytes of various types, and submerged aquatic vegetation (SAV) in the estuary proper were thought to be the basis of estuarine productivity. According to this conceptual framework, decomposition of the organic matter produced by the various plant species led to high generation of detritus which then entered the estuarine and coastal food webs via the microbes. The microflora enriched the nutritive value of the detritus along with major increases in the protein content. Although most recent evidence substantiates this model in various ways, other findings have modified scientific opinion regarding the wetlands input to associated systems (see Wiegert and Pomeroy, 1981; Livingston 1990b).

The export of detritus was considered to be an important source of energy for associated marine systems. Livingston and Loucks (1979) emphasized the interconnectedness of freshwater wetlands and estuarine systems with the need to carry out broad, ecosystem-level studies to determine the basis of the extremely complex connections among freshwater, estuarine, and marine food webs. The central theme of the early work was the ecological value of the estuarine and coastal wetlands as centers of production, and ultimate export of high value organic matter to associated estuarine and marine systems. Nixon (1980) gave a counter view to the idea of marshes as exporters of organic carbon with the suggestion that marshes are either sources or sinks in coastal marine nutrient cycles. Marshes appeared to be sinks for total phosphorus, and remobilization of phosphate occurred in the sediments, with salt marshes accounting for only a small part of the net export to nearshore systems.

Another line of research (Haines, 1979) indicated that the outwelling concepts of the salt marsh as a source of organic detritus should be revised. Stable carbon isotope ration analysis indicated that algal-derived organic matter (phytoplankton and benthic diatoms) had been underestimated as a source of carbon to estuarine/coastal systems. Salt marsh vegetation was viewed as being largely accumulated and consumed within the marsh and estuarine sediments. The complexity of estuarine food webs was considered to be greater than formerly believed, and salt marshes were seen as refuges and feeding habitats for estuarine organisms. Odum et al. (1979) put forward the idea that tidal wetlands may either export or import particulate organic matter on an annual basis depending on the geomorphology of the drainage basin and the relative magnitude of the tidal range and freshwater input from upland areas. More recent reviews (Dardeau et al., 1992) indicated that the net import of nutrients such as phosphate was quite complicated, with some salt marshes getting phosphate from coastal water (e.g., the Altamaha River). In other systems, such as North Inlet, North Carolina, there were net exports of phosphate from estuaries to coastal waters. Net exchanges of marsh-estuarine nitrogen were too small to influence the annual net budget of some coastal systems.

The nitrogen cycle in estuarine and other coastal waters is even more complex than that of phosphorus, with various systems having input by rivers with remineralization and exchange processes most active during summer periods (Livingston, 1984a). The impact of denitrification was thought to be less than sulfate reduction. However, because of the anaerobic soils in salt marshes, denitrification exceeded N-fixation with subsequent nitrogen depletion. In short, the nutrient and organic carbon exchanges within and between marsh sub-systems are complex and specific to local conditions.

Recent reviews (Livingston, 1991) confirm the complexity of the role of marshes in the nutrient dynamics of estuarine systems. Detrital input from forests adjacent to riverine marshes was not detectable by isotopic signatures in estuarine organisms of Sapelo Island, Georgia (Peterson and Howarth, 1987). Primary production, based on in-stream transport and utilization of nutrients and organic compounds, is now thought to be one of the most important functions in river-dominated estuaries (Howarth 1988; Baird and Ulanowicz, 1989). Riverine particulate carbon can be a major source of such matter to associated estuaries (Livingston, 1984a). This input, together with wind and tidal subsidies and associated phytoplankton productivity, are now thought to be an important part of the naturally high production of organic carbon in south temperate estuarine systems (Nixon, 1980, 1988). The relative importance of various sources of both organic carbon (dissolved and particulate) and inorganic nutrients can vary from estuary to estuary (Peterson and Howarth, 1987). Although freshwater and estuarine wetlands are buffers between upstream river systems and the receiving estuaries (Simpson et al., 1983), detailed knowledge of the linkage of nutrient fluxes between such areas is still lacking.

The significance of organic input to an estuary depends on the geomorphology of the drainage basin, the magnitudes of the tidal and freshwater input from upland areas, and biological components of the receiving area (Welch et al., 1982; Kemp and Boynton, 1984). The linkage between freshwater wetlands and estuaries is real (Livingston, 1984a), but the details of quantitative interactions through time remain largely unknown. There is insufficient information available to make broad generalizations concerning the role of estuarine and other coastal wetlands in the nutrient dynamics of coastal marine systems, but it is apparent that different marshes have very different functions in the overall ecological processes of coastal ecosystems. It is clear that such functions depend on the complex interactions of hydrological and biological processes.

The relationship of offshore marine systems to nutrient and organic carbon loading in barrier estuaries is also incompletely understood. Livingston et al. (unpublished data), working in the Perdido Bay system in the northeastern Gulf of Mexico, found that the input of nutrients and organic carbon from the lower bay and Gulf exceeded that of the upper bay, which was composed of alluvial river inputs and extensive marshes. Modelling efforts indicated that meteorological events in the Gulf region were responsible for transport conditions that were consistent with the movement of nutrients and organic matter up the bay from the populated sources. Modelling of known distributions of nutrients and organic carbon confirmed the lower bay as the major source for those compounds that were identified as responsible for the consumption of dissolved oxygen at depth. The landward lower layer flow from the lower bay was significant in comparison to the nutrients entering the bay from the Perdido River and upper bay sources. The average BOD5 (biological oxygen demand) entering Perdido Bay from the lower bay during the months of August and September was 21,000 lbs/day, versus 3,000 lbs/day from Elevenmile Creek, a stream that was heavily influenced by a pulp mill. For total nitrogen, the average mass loading from the lower bay was 9,300 lbs/day, 1,600 lbs/day for the Perdido River, and 400 lbs/day for Elevenmile Creek. Analogous loads for total phosphorous were 500 lbs/day from Inerarity Point, 125 lbs/day from the Perdido River, and 20 lbs/day from Elevenmile Creek. There is considerable evidence that if the influence of nutrient loading from marsh systems is to be understood, the loading to the system from all sources needs to be documented.

4.2 SALT MARSHES

Salt marshes are coastal, intertidal ecosystems of salt tolerant, rooted vegetation (halophytes), which are periodically inundated and drained by tides. Halophytic grasses (e.g., *Spartina*, *Juncus*, and *Distichlis*) are usually the dominant vegetation, although halophytic shrub vegetation (e.g., *Baccharis* and *Iva*) may occasionally replace the grasses. In addition, many types of algae, including several species of diatoms and blue-green algae, occur on the stems of emergent plants, on the surface of sediments, and suspended in the water column.

Coastal salt marshes typically exhibit complex patterns of zonation with respect to plants, animals, and microbes, which reflect patterns of fluctuation in such factors as salinity, water depth, and temperatures. These marshes are actually composed of interconnected, fluctuating passageways for the transport of fish, plankton, and nutrients into and out of the marsh. Because they form the boundary between open estuaries or the ocean on one side, and freshwater wetlands and uplands on the other, salt marshes form an important interface between marine, freshwater, and terrestrial ecosystems.

There have been numerous attempts to classify salt marshes. Chapman (1960) divided salt marshes worldwide into nine geographical categories based on physiography and composition of the flora, including one for "Eastern North America" (sufficient rainfall, muddy coasts, and broad marsh development), and another for "Western North America" (variable rainfall, rugged coasts, and limited marsh development). The U.S. Fish and Wildlife Service (FWS) uses a classification system based in part on relative salinity and the type of vegetation (Cowardin et al., 1979; see Figure 1.1). Several FWS publications have divided salt marshes (estuarine emergent wetlands) into (1) "salt marshes" (polyhaline), (2) "brackish marshes" (mesohaline), and (3) "oligohaline marshes" (for example, see Tiner, 1984, 1985a, 1985b; Metzler and Tiner, 1992). The FWS classification recognizes the dramatic effects of relative salinity on plant distribution, although the categories of marshes are somewhat arbitrary, such that overlap between marsh types is common. For purposes of this book, all estuarine marshes (as well as intertidal salt flats and some scrub-shrub wetlands) are discussed together as salt marshes, although inland salt marshes that may occur where evaporation exceeds freshwater input are excluded, since they are not tidally influenced.

4.2.1 Geophysical Characteristics

Coastal salt marshes develop in intertidal coastal areas where the slope is gentle, and some protection from wave action and storms is present. Sediments accumulate in these areas from upland runoff, upwelling from the ocean, and from biological production within the marsh itself.

Many salt marshes occur at or near the mouths of rivers and streams that flow into the ocean, such as in the delta marshes near the mouth of the Mississippi River (for examples of deltaic marsh development, see Mitsch and Gosselink, 1993). Other salt marshes develop along marine-dominated coasts where proper conditions for marsh formation occur (see Beeftink, 1977). The most extensive marine-dominated salt marshes occur in areas protected by sand bars or barrier islands, such as in the extensive salt marshes inland of the "outer banks" barrier island network along the Atlantic coasts of Virginia, Georgia, and the Carolinas. Other marine-dominated wetlands occur in bay areas, where there is sufficient protection from storms and tides for marsh development to occur, such as the extensive marshes in the Chesapeake Bay, or San Francisco Bay. Marsh formation in marine-dominated systems has been described by Chapman (1960). Coastal marshes also form along tidal rivers where salinities are highly variable with seasons. These marshes are called "brackish marshes."

The most extensive development of coastal salt marshes is in areas of muddy or sandy coasts, moderate temperatures, abundant rainfall, and gently sloping coastal relief. These conditions are best represented in the Southeast, where 78% of the total coastal wetlands in the United States are found (Gosselink and Bauman, 1980). The largest single area of coastal salt marshes is the delta of the Mississippi River in Louisiana, Mississippi, and Texas, with almost 40% of the coastal salt marshes in the coterminous United States (Day et al., 1989).

4.2.2 Wetland Criteria Characteristics

Vegetation. Salt marsh vegetation is generally dominated by grasses and flowering halophytes, which exhibit distinctive zonation based on relative salinity and frequency of flooding. The two most distinctive zones in most salt marshes as described by Tiner (1984) are: (1) regularly flooded marshes (low marsh), which are flooded at least once daily; and (2) irregularly flooded marshes (high marsh), which are flooded less often (see also Niering and Warren, 1980; Tiner, 1985a, 1985b, 1993c; Bertness, 1992; Metzler and Tiner, 1992; Pennings and Callaway, 1992; Wilen and Tiner, 1993). The degree of flooding of these zones varies depending on tides, season, and storm events. In addition to zonation, salt marsh vegetation is somewhat geographically variable between the northeastern U.S., Atlantic coast, coastal Gulf of Mexico, and the Pacific coast. Additional useful references on salt marsh vegetation are Zedler, 1982, Copeland et al. (1983), Josselyn (1983), Gosselink (1984), Livingston (1984), Zedler and Norby (1986), Conner and Day (1987), Lewis and Estevez (1988), Tiner and Finn (1986), Reed (1988a-1988f), Day et al. (1990), Montague and Wiegert (1990), and Mitsch and Gosselink (1993).

In the Northeast, salt marshes are generally restricted to the shores of coastal embayments and rivers, and to tidal creeks. Figure 4.1 shows relationships among the various zones within a northeastern salt marsh. Regularly flooded marshes are dominated by smooth cordgrass (*Spartina alterniflora*). Irregularly flooded marsh vegetation is more diverse, and includes salt marsh hay (*Spartina patens*), salt grass (*Distichlis spicata*), black grass (*Juncus gerardii*), alkali grasses (*Puccinellia* spp.), and baltic rush (*Juncus balticus*). Occasional salt

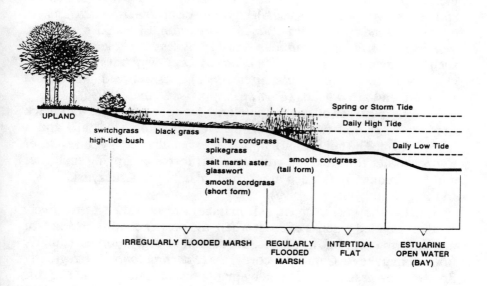

Figure 4.1: A cross-sectional diagram of a salt marsh in the northeastern U.S., showing zonation of marsh vegetation (from Tiner, 1984).

pannes and evaporation pools may contain stunted forms of *Spartina* and other grasses, blue-green algae, and sea lavender (*Limonium nashii*). Brackish marshes with lower salinity are encountered further upstream in larger rivers and streams, with vegetational transitions to saltmarsh bulrush (*Scirpus robustus*), Salt marsh hemp (*Amaranthus cannabinus*), rose-mallow (*Hibiscus palustris*), water parsnip (*Sium suave*), and soft-stemmed bulrush (*Scirpus validus*). Brackish marshes eventually give way to more typical freshwater marshes, which are often dominated by freshwater species, such as cattail (*Typha*).

From Delaware to northern Florida, salt marshes are the dominant coastal wetland type. The coast of peninsular Florida and much of the Gulf of Mexico coast is dominated by mangrove swamps (see Section 4.4), but occasional salt marshes occur in patches throughout the entire Atlantic and Gulf coast. In these areas, smooth cordgrass (*Spartina alterniflora*) is the dominant vegetation in regularly flooded marshes, often occurring in vast coastal marshes containing hundreds of acres. Irregularly flooded marshes are more diverse, and include black needlerush (*Juncus roemerianus*), glasswort (*Salicornia* sp.), salt grass (*Distichlis spicata*), high-tide bush (*Iva frutescens*), asters (*Aster* sp.), and switchgrass (*Panicum virgatum*). In brackish marshes, *Juncus roemerianus* is gradually replaced by less salt-tolerant species, such as big cordgrass (*Spartina cynosuroides*), wiregrass (*S. patens*), narrow-leaved cattail (*Typha angustifolia*), pickerelweed (*Pontederia cordata*), and arrowheads (*Sagittaria* spp.) (see Tiner, 1993c).

Two species of salt marsh plants, smooth cordgrass and black needlerush, exhibit two quite distinctive forms, both a "tall" and a "short" form. The tall forms of both species usually occupy zones closer to open water (low marsh), with the short forms occupying shallower and less frequently flooded areas (high marsh; see Kruczynski, 1982; Montague and Wiegert, 1990).

On the Pacific coast, salt marshes are restricted to relatively smaller, isolated patches of coast where the proper conditions exist for marsh formation. In these areas, the dominant vegetation in regularly flooded marshes is California cordgrass (*Spartina foliosa*). Irregularly flooded marshes are dominated by plants such as salt dodder (*Cuscuta salina*), perennial glasswort (*Salicornia virginica*), common glasswort (*S. subterminalis*), California sea blite (*Suaeda californica*), California sea lavender (*Limonium californicum*), spreading alkali weed (*Cressa truxillensis*), spiny rush (*Juncus acutus*), and tufted hairgrass

(*Deschampsia caespitosa*) (Tiner, 1984).

Among the important vegetational components of salt marshes are algae, which often form mats on mud surfaces. These mats form the bases of halophytic vegetation, particularly on extensive mudflats that often accompany salt marshes, and are otherwise largely unvegetated. These mats are dominated by bluegreen algae, diatoms, and green algae, but their ecology is poorly understood (see Montague and Wiegert, 1990; Wiegert and Freeman, 1990).

Soils. Salt marsh soils are peats, mucks, or sandy, clay, or silt-clay soils. Marsh substrate type has an important influence on vegetation patterns. The most productive marsh soils are organic soils, and clay or silt-clay mineral soils, which have greater density and higher nutrient concentrations than sandy soils. For example, DeLaune et al. (1979) found that the above-ground standing crop of *Spartina alterniflora* is correlated with soil density and soil mineral content. In addition, older marshes with greater accumulated below-ground biomass (organic materials and nutrients) tend to have higher productivity than younger marshes (see Bertness, 1987).

Salt marsh soils exhibit highly variable salinity and pH, both of which affect the distribution of marsh plants. Salinity tends to be highest in most irregularly flooded marshes, because evaporation allows salts to accumulate in the soils. In regularly flooded marshes, salts are usually flushed daily by tides. Salt marsh soils tend to be somewhat acidic, although an optimal pH of 6 has been shown for some marsh plants (see Day et al., 1989).

Hydrology. Unlike most freshwater wetlands, rainfall is only a minor source of water for salt marshes. The most important source of water for salt marshes is tidal input from adjacent estuaries or marine environments. This input can vary depending on the height of the tide, the time of year (i.e., Spring tides tend to be higher than Fall tides), and storm events. Brackish marshes, which depend on freshwater rivers and streams for water input, are more likely to be affected by periodic variation in rainfall. Coastal wetlands are hydrologically unique, with bi-directional water flow that changes with the tides.

4.2.3 Characteristic Fauna

Detailed lists of species that inhabit salt marshes are given by Weinstein (1979, 1980); Seliskar and Gallagher (1983), Stout (1984), Rozas and Odum (1987), Odum (1988), Day et al. (1990); Montague and Wiegert, (1990), Dardeau et al., (1992), and Mitsch and Gosselink (1993). Salt marshes are inhabited by large numbers of relatively few species of animals with high dominance and low species diversity. It is thought that such community organization is the result of high levels of stress caused by rapid changes of temperature, dissolved oxygen, and salinity. According to Montague and Wiegert (1990), certain more stable habitats exist within marsh systems among the stems and leaves of marsh grasses. Certain forms of marsh creeks can be physically stable, and salt marsh tidepools host large numbers of species of insects, spiders, and other fauna. The salt marshes of Sapelo Island, Georgia have been particularly well studied with respect to their microbes and meiofauna (microscopic animals; Christian et al., 1981), aquatic macroconsumers (Montague et al., 1981), and the grazing community on *Spartina* and their predators (Pfeiffer and Wiegert, 1981).

The dominant herbivorous marsh animals are insects of various types (herbaceous strata), and gastropod mollusks. Crustaceans such as isopods, amphipods, and decapods inhabit salt marshes, along with infaunal types such as polychaete worms. Tidal creeks support many juvenile organisms, including decapod crustaceans, sciaenid fishes, pinfish (*Lagodon*), and menhaden (*Brevoortia*). Among the fishes, topminnows (Cyprinodontidae and Poeciliidae), silversides (Atherinidae), pinfish, and mullet (Mugilidae) are usually dominant, especially in tidal creeks and associated estuarine areas. Wading birds, such as herons and egrets, also use the tidal creeks as primary feeding areas.

Relatively few species of reptiles, birds, and mammals are full-time residents of salt marshes, but there are many transients. Among the few permanent resident reptile species is the salt marsh snake (*Nerodia fasciata* ssp.), and the diamondback terrapin (*Malaclemys terrapin*), both of which occur in salt marshes along the southern Atlantic and Gulf coasts. Diamondback terrapins were once highly prized as a culinary delicacy, and were fished and cultivated along the mid-Atlantic coast during the 1930's (the mascot of the University of Maryland is still the "terrapin"). In extreme southern Florida, the endangered American crocodile (*Crocodylus acutus*) may occasionally be found in salt marshes, as are American alligators (*Alligator mississippiensis*).

Resident birds include the marsh wren (*Cistothornus palustris*), the clapper rail (*Rallus longirostris*), and the seaside sparrow (*Ammodramus maritimus*). Transient birds include cattle egrets (*Bubulcus ibis*), red-winged blackbirds (*Agelaius phoenicius*), insectivores such as swallows, and a variety of ducks, geese, gulls, and other water and shorebirds.

Among the mammals that are permanent residents of salt marshes are several species of rats (*Oryzomys, Microtus*), muskrats (*Ondontra zibethica, Neofiber alleni*), and the nutria (*Myocastor coypus*) of the Mississippi River delta. In Florida, the West Indian manatee (*Trichechus manatus*) is a transient in salt marshes, where it grazes on *Spartina alterniflora* along the banks of streams.

4.2.4 Wetland Values

The nursery value of salt marshes is indisputable, with various forms of commercially important populations utilizing such systems as larval and juvenile feeding grounds. Despite the controversy that still surrounds the issue of import and export of marsh nutrients and organic carbon, there is general agreement that marshes form the basis of organic carbon export in terms of the biomass of the various dominant populations that move out of the marshes after seasonal nursery foraging is ended.

Salt marsh vegetation is known to dampen the effects of wave action in coastal areas, particularly during storm events, and to reduce erosion (Wiegert et al., 1981; Tiner, 1984). Salt marshes also provide habitat for a variety of both transient and permanent resident animal species, and are useful for many forms of recreation (see Section 2.3).

4.2.5 Unusual Characteristics

Salt marshes are characterized by relatively high environmental stress, including extensive, short-term changes in temperature, salinity, and other physical/chemical factors. This relatively harsh environment is often characterized by relatively low biological diversity. Environmental stress is probably responsible for the zonation observed in marsh plants, which are thought to be dependent on salinity stress and sediment anoxia (Dardeau et al., 1992). With increased elevation or a lowering of interstitial salinity, the smooth cordgrass grades into black needlerush. On the coast along the northern Gulf of Mexico, tidal amplitudes are less than those on the Atlantic coast, and black needlerush is dominant in these areas.

4.3 MANGROVE SWAMPS

Among the more unique wetland ecosystems in the United states are mangrove swamps, a type of "estuarine scrub-shrub wetland" under the U.S. Fish and Wildlife Service classification scheme (see Cowardin et al., 1979; Wilen and Tiner, 1993). Mangrove systems are defined as "catch-all, botanically diverse" habitats represented taxonomically by over 50 species of about 12 families of plants (Odum et al., 1982). In a more narrow sense, the term "mangrove" includes assemblages of species of plants that are specifically adapted to the marsh environment.

Mangroves have various adaptations for living in anaerobic soils that are saturated with salts and other estuarine components, including shallow root systems, "prop roots" from the lower part of the stem, and "drop roots" from branches and the upper stem (Odum and McIvor, 1990). Other adaptations include diverse respiratory functions and various osmoregulatory devices. Reproduction is by viviparity (i.e., the embryo initiates germination and begins development while still on the tree). Dispersal of propagules (i.e., "seeds" without an intermediate resting stage) is by water flotation, with longevity of propagules lasting as much as 110 days. Mangroves stabilize sediments and provide a series of diverse habitats for other plants and animals. Primary production is high in mangrove systems which often serve as sinks (net accumulators) for substances such as nutrients, trace elements, and metals (Odum and McIvor, 1990). Litter fall is high so that mangroves provide organic carbon in various forms, which is then utilized by a variety of secondary consumers, including microbes and macroscopic organisms. Carbon fluxes associated with litter fall from leaves, twigs, and branches is continuous throughout the year, and is considered an important phenomenon in terms of providing particulate organic carbon to associated estuarine areas. Mangroves are thus an integral part of the trophic organization of associated ecosystems.

The interactions of mangrove recycling devices are complex and not well understood in the full range of mangrove habitats. The actual net fluxes of materials, such as dissolved organic carbon and various types of nutrients, depend on various factors including freshwater runoff, tidal flushing, localized response to upland nutrient inputs, and internal nutrient and recycling processes. Both dissolved and particulate carbon are transported from tidal mangroves into associated estuarine areas. However, the exact value of such export appears to be a function of local conditions within a given mangrove system.

4.3.1 Geophysical Characteristics

In the continental United States, mangroves are restricted to areas of the central and southern Florida coast, and intermittently along the Gulf of Mexico. Mangrove systems are limited by climate, substrate, salt water, and tidal fluctuations. According to Odum et al. (1982), mangroves grow in tropical areas where the annual average temperatures do not exceed 19°C. Temperature fluctuations exceeding 10°C, or temperatures below freezing are not tolerated. On a geographic scale, mangroves are located along tropical coastlines between 25° north latitude and 25° south latitude.

Mangroves do not need salt water, but they do not develop in exclusively freshwater environments. They may occur from hypersaline to nearly freshwater situations, although their maximum growth is in brackish waters. Tidal exchanges apparently enhance and direct development of mangrove systems for several reasons, such as nutrient exchanges, exclusion of potential competitors, and maintenance of appropriate salinity regimes.

4.3.2 Wetland Criteria Characteristics

Vegetation. In the United States, mangrove swamps are composed of three primary species: black mangrove (*Avicennia germinans*), red mangrove (*Rhizophora mangle*), and white mangrove (*Laguncularia racemosa*). Buttonwood (*Conocarpus erecta*) is often included in mangrove associations although this species does not have a root system that is adapted for saturated or saline soils (Odum and McIvor, 1990). As in salt marshes, there is distinct vegetational zonation in mangrove swamps. In general, the red mangrove is dominant in regularly flooded areas, the black mangrove dominates in most irregularly flooded areas, and the buttonwood is dominant in infrequently flooded areas (see Odum and McIvor, 1990; Wilen and Tiner, 1993; Tiner, 1993c).

Aside from the mangroves, few plants species grow within mangrove swamps, although mangrove ferns (*Acrostichum* spp.) are frequent understory plants. Salt marshes dominated by smooth cordgrass and black needlerush frequently occur adjacent to mangrove swamps. Additional useful references on mangrove vegetation include Lugo and Snedaker (1974), Odum et al. (1982), Day et al. (1989), Lugo et al. (1990), Odum and MvIvor (1990), and Mitsch and Gosselink (1993).

Soils. The most well developed mangrove systems occur along deltaic coasts or in estuaries that have fine silty sediments with high organic matter (Odum et al., 1982). However, mangroves may also grow in relatively sandy marine soils, and are known to stabilize sediments in a way that allows additional deposition by physical processes, such as water currents. In mature mangrove swamps, a layer of highly organic, nutrient-rich peat develops (see Twilley et al., 1986; Day et al., 1989; Odum and McIvor, 1990).

Hydrology. The hydrology of mangrove forests is dominated by marine currents, and tidal influences. As with salt marshes, tidal input can vary depending on the height of the tide, the time of year (i.e., Spring tides tend to be higher than Fall tides), and storm events. In brackish mangrove swamps, the influx of fresh water varies depending on stormwater runoff from adjacent uplands, as well as input from rivers and streams.

Regularly flooded and irregularly flooded zones occur in most mangrove swamps. The former is flooded daily by the tides, while the latter is flooded less often.

4.3.3 Characteristic Fauna

The detritus-based food webs associated with mangroves are fundamental to a variety of invertebrates and fishes, some of which have considerable value in regional sports and commercial fisheries. These coastal assemblages are located along salinity gradients, ranging from the basin mangrove forests through the fringing associations (riverine-estuarine-oceanic). The diverse mangrove habitat thus remains functionally contiguous for the developing stages of various species. Consumers such as copepods, amphipods, and decapod crustaceans, utilize particulate (and possibly dissolved) carbon sources (Odum and McIvor, 1990) that have been traced to mangroves. A broad range of fishes (up to 220 species) are also involved in these food webs. Amphibians and reptiles (24 species) utilize mangrove systems along with 181 species of (wading, probing shore, floating, and diving) birds and 18 species of mammals. Useful reviews of mangrove animals include Odum et al. (1982), Kaplan (1988), Odum and McIvor (1990), and Mitsch and Gosselink (1993).

Important mangrove mollusks include oysters (*Ostrea, Crassostrea*), and a variety of aquatic and terrestrial snails, and bivalves. Crustaceans include barnacles (*Balanus* spp.), the spiny lobster (*Panulirus argus*), the pink shrimp (*Panaeus duorarum*), and many species of crabs, including the fiddler crabs (*Uca* spp.). Other important invertebrate animals include sponges, flatworms, annelid worms, anemones, sea urchins, starfish, and tunicates, which grow on and among the roots and stems of mangroves within the intertidal zone.

Vertebrate animals that inhabit mangrove swamps include several protected species: the American crocodile (*Crocodylus acutus*), hawksbill sea turtle (*Eretmochelys imbricata*), Atlantic ridley sea turtle (*Lepidochelys kempi*), Atlantic salt marsh snake (*Nerodea fasciata taeniata*), indigo snake (*Drymarchon corais*), bald eagle (*Haliaeetus leucocephalus*), American peregrine falcon (*Falco peregrinus*), West Indian manatee (*Trichechus manatus*), and Florida panther (*Felis concolor coryi*). Other mammals which inhabit mangrove swamps and associated forests include the raccoon (*Procyon lotor*), mink (*Mustela vison*), river otter (*Lutra canadensis*), black bear (*Ursus americanus*), and striped skunk (*Mephitis mephitis*). Among the more visible bird species are the wood stork (*Mycteria americana*), white ibis (*Eudocimus albus*), roseate spoonbill (*Ajaia ajaja*), cormorants (*Phalacrocorax* spp.), pelicans (*Pelicanus* spp.), and egrets and herons. Important fish species in mangrove swamps include mullet (*Mugil cephalus*), tarpon (*Megalops atlanticus*), snook (*Centropomus undecimalis*), and mangrove snappers (*Lutjanus apodus*).

Figure 4.2: Highly diverse wetland on Dog Island, Florida, in the northeastern Gulf of Mexico.

Figure 4.3: Tidal creek system as part of the vast intertidal wetlands in the Apalachicola Bay system on the northeastern Gulf of Mexico.

Figure 4.4: Choctawhatchee River delta showing estuarine wetlands at the head of the Choctawhatchee Bay system along the north Florida Gulf coast.

4.4 WETLAND VALUES AND THE FUTURE OF ESTUARINE WETLANDS

Estuarine wetlands are represented by a broad diversity of habitats, which include mixed wetlands (Figure 4.2), tidal creeks in broad intertidal marshes (Figure 4.3), and alluvial river delta marshes and swamps (Figure 4.4). Such areas lie in the path of human activities, and can be adversely impacted by agricultural, municipal, and industrial activities. Combinations of chemical discharges, municipal sewage discharges, and stormwater runoff have led to some of the most extensive fish kills in the history of hypereutrophication (Figure 4.5). In such cases, the high natural productivity of wetland ecosystems is turned into a massive loss of estuarine habitat. This loss is usually caused by periodic hypoxia/anoxia associated with anthropogenic disruption of natural nutrient conditions, enhanced phytoplankton

Figure 4.5: Major fish kills in the early 1970s in Escambia Bay (northeastern Gulf coast) were associated with industrial and municipal releases of nutrients and organic matter that led to hypereutrophication and the loss of wetland habitats.

Figure 4.6: Sike's Cut, an artificial pass in the Apalachicola Bay system has been associated with habitat losses and salinity changes in the associated estuarine system.

Estuarine Wetlands 149

Figure 4.7: Municipal development on St. George Island in the Apalachicola Bay system showing direct construction impacts and dredging effects on local wetlands.

Figure 4.8: Marina development off Alligator Point, Florida. Note the oil slick that is moving out of the marina with the tide.

Figure 4.9: Construction of dikes and drainage ditches associated with agricultural activities in coastal wetlands have led to direct construction effects, altered freshwater drainages, and the release of nutrients and toxic agents.

productivity, and a resultant distrophic alteration of wetland foodwebs. Artificial cuts in barrier systems (Figure 4.6) add to such problems by altering salinity exchanges between the estuary and associated offshore systems, thus leading to enhanced salinity stratification and hypoxia at depth. Such passes are also associated with unstable land/water interactions and losses of local wetlands.

Municipal developments, which include associated dredging activities and the construction of bridges and roads (Figure 4.7), often lead to direct and indirect losses of natural estuarine wetlands through alterations of natural freshwater runoff. The massive development of coastal marinas along the Atlantic and Gulf coasts (Figure 4.8) has further increased the depletion of wetland productivity through the release and concentration of various oxygen-consuming and toxic substances. Agricultural activities such as ditching and draining (Figure 4.9), and clearcutting (Figure 4.10) add to the problems of coastal wetlands. Direct and indirect losses are caused by altered freshwater drainage, and pollution generated by agricultural activities and the draining of upland swamps and marshes. The cumulative effects of these changes often lead to gradual, long-term shifts in estuarine productivity and the loss of habitat, which can be translated into major reductions in the fisheries potential of estuarine wetlands.

According to Frayer et al. (1983) and Tiner (1984), by the mid-1970's only 5.22 million acres of estuarine wetlands existed in the lower 48 states. This represented only 0.3% of the land surface of this area. These estuarine areas included estuarine intertidal flats (0.75 million acres), estuarine emergent wetlands (3.9 million acres), and estuarine forested and scrub-shrub wetlands (0.57 million acres). By contrast, approximately 87 million acres or 94% of the total wetlands in the United States were represented by palustrine marshes and swamps. Thus, although the total acreage of estuarine wetlands is not great, the quality of the habitat is so high that the actual value transcends the relatively small expanse of such systems.

In recent times, coastal wetlands suffered the heaviest losses in such areas as California, Florida, New Jersey, Texas, and Louisiana. Urbanization was the predominant cause of such losses, along with increased groundwater withdrawals in most areas outside of Louisiana. Between the mid-1950's and mid-1970's, deepwater habitats (largely man-made reservoirs) made slight gains, while coastal wetlands decreased by a little less than 0.5%. In Louisiana, the change from marsh to open water was attributed to complex interactions of subsidence of the coastal plain, channelization, oil and gas extraction, levee construction, and natural rises in sea level (Tiner, 1984). Currently, in such areas as Florida, marsh regulation is relatively restrictive for most kinds of habitat alteration (Montague and Wiegert, 1990). Global concerns about sea level increases due to greenhouse effects may present a real problem to maintaining estuarine wetland values. According to Livingston (1990), sea level increases of 0.5 to 1.0 m over the next 100 years could be associated with the loss of major portions of Gulf estuarine and marine marshes. However, it should be emphasized that there is currently little scientific agreement concerning future sea level changes. According to Tanner (1960), relative sea level is currently rising in Florida at rates between 20 cm and 40 cm per century, which would probably be slow enough to allow marshes to advance inland with the sea (Montague and Wiegert, 1990). However, even without serious changes in sea level, rapid population growth in

Figure 4.10: Clearcutting of coastal wetlands may cause water quality alterations, hydroperiod disruptions, and changes in nutrient dynamics in associated estuarine wetlands.

coastal areas of Florida continues to present a very serious threat to the remaining estuarine marshes and mangroves. In addition, most agricultural and forestry activities in wetlands remain unregulated with respect to impacts from altered drainage and clearcutting.

Mangrove swamps in particular have been prone to human devastation. Although the overall reduction of mangroves through human activities is relatively small (9,522 ha out of a total of 190,000 ha in Florida; Odum and McIvor, 1990), high percentages of mangroves have been destroyed in highly populated areas along the lower east coast, the Florida keys, and portions of the Gulf coast. A future trend of further reduction remains difficult to predict although the high rate of coastal population growth does not bode well for the continued well-being of coastal mangroves. By the year 2010, coastal populations will have grown from 80 million to more than 127 million, an increase of almost 60 percent (NOAA, 1990; Dennison, 1993).

5

Field Recognition and Delineation of Wetlands

Ralph W. Tiner

INTRODUCTION

Since passage of environmental laws beginning in the 1960s, certain land use controls on private property and public lands have required that the presence of wetlands and their boundaries be determined (see Chapter 7). These laws recognize that wetlands possess significant functions and values, including flood storage, water quality improvement, fish and shellfish nursery grounds, and important wildlife habitats (see Section 2.3). Wetland identification and delineation is a key component in the regulatory framework aimed at preserving these functions and values. Failure to identify and delineate the wetlands properly can lead to wetland destruction and loss of valued public services because many wetland areas would then escape regulatory oversight.

Until relatively recently, no manuals or guidebooks existed for wetland recognition and delineation. Without well-defined standards and trained personnel to implement them, wetland delineation is fraught with inconsistency, chaos, and uncertainty for the landowner, regulator, and general public alike. The 1987 Corps of Engineers Wetlands Delineation Manual (U.S. Army Corps of Engineers, 1987) was perhaps the first federal manual officially released for use in identifying wetlands subject to regulation under § 404 of the Clean Water Act (see Chapter 7 for discussion of this regulatory program). Today, this manual represents the federal regulatory standard for wetland determinations. Other manuals, including the technically based and politically controversial 1989 Federal Manual for Identifying and

Delineating Jurisdictional Wetlands (Federal Interagency Committee, 1989), are used primarily by northeastern states for administration of state wetland protection statutes.

The purpose of this chapter is to introduce readers to the approach and techniques presented in the 1987 Corps Manual (which have been further clarified by recent Corps headquarters guidance) for identifying federally regulated wetlands. Also, primary sources of information that are helpful in making wetland determinations will be reviewed. Other wetland delineation techniques will not be discussed; for information on them consult additional references (Tiner, 1989, 1993a, 1993b).

5.1 PURPOSE AND USE OF THE CORPS MANUAL

The 1987 Corps Manual was developed to provide technical guidelines and methods for using a multi-parameter or three-parameter approach to identifying and delineating wetlands subject to regulation under § 404 of the Clean Water Act (CWA). It was designed primarily as a working document for Corps field personnel, and it provides supportive technical information useful for applying the technical guidelines.

Wetlands regulated under the CWA are defined as "areas that are inundated or saturated by surface or ground water at a frequency and duration sufficient to support, and that under normal circumstances do support, a prevalence of vegetation typically adapted for life in saturated soil conditions" (33 CFR 328.3(b)). Thus, wetlands from the CWA perspective must be vegetated wet habitats (Figure 5.1), including marshes, bogs, swamps, fens, certain bottomland hardwood forests, and prairie potholes (see Chapters 3 and 4 for detailed discussions of these wetland types).

Nonvegetated wetlands, such as tidal mudflats, are not viewed as wetlands, but are considered "other waters of the United States" for CWA regulatory purposes. These and other areas (sanctuaries, refuges, wetlands, vegetated shallows, coral reefs, and riffle-pool complexes) are collectively recognized as "special aquatic sites" which are included with deepwater habitats as "waters of the U.S." subject to regulation through Section 404 of the CWA. Artificial wetlands intentionally or accidentally created by human activities and exempted under Corps regulations or policies are not intended to be included as wetlands using

Figure 5.1: The term "wetland" suggests "land" that is "wet" for significant periods. Most wetlands are periodically inundated by surface water and/or saturated at or near the surface by seasonally high groundwater levels.

the 1987 Corps Manual.[1]

The 1987 Corps Manual emphasizes its purpose of identifying Section 404 wetlands and points out the difference between it and the U.S. Fish and Wildlife Service's (FWS) scientifically based definition of wetland: the latter "requires that a positive indicator of wetlands be present for any one of the three parameters, while the technical guidance for wetlands [in the Corps manual] requires that a positive wetland indicator be present for each parameter (vegetation, soils, and hydrology), except in limited instances identified in the manual" (U.S. Army Corps of Engineers, 1987). The Corps' interpretation of the CWA

---

[1] The Corps and EPA, jointly responsible for administering § 404 of the CWA, had different interpretations of the degree of evidence required to identify CWA wetlands, which led to differences in the extent of wetlands and the location of wetland-nonwetland boundaries in many areas. See U.S. Army Corps of Engineers, 1987; and Sipple, 1988.

definition of wetland was also significantly different from EPA's interpretation, which typically required evidence of less than three parameters (Sipple, 1988). Moreover, the Corps Manual states that wetlands mapped by the FWS's National Wetlands Inventory as temporarily flooded or intermittently flooded should be "viewed with particular caution since this designation is indicative of plant communities that are transitional between wetland and nonwetland." In practice, these communities have typically been considered to be nonwetlands according to the Corps Manual. They also represent the predominant wetland type in many areas, especially along the Coastal Plain south from New Jersey.

To define the limits of federal jurisdiction in wetlands for the CWA, the 1987 Corps Manual necessarily embodies both technical and policy considerations. It clearly does not attempt to identify all areas that may qualify technically as wetlands, but seeks to define only those wetlands that should be subject to the CWA. The preface of the manual states that "the manual is not intended to change appreciably the jurisdiction of the Clean Water Act (CWA) as it is currently implemented [in 1987]. Should any District find that use of this method appreciably contracts or expands jurisdiction in their District as the District currently interprets CWA authority, the District should immediately discontinue use of this method and furnish a full report of the circumstances to the Office of the Chief of Engineers" (U.S. Army Corps of Engineers, 1987). These statements suggest that the manual was simply intended to provide some technical guidance to Corps districts on how to identify and delineate wetlands, but not to set a consistent national standard for doing so, nor to ensure technically supportable wetland determinations by the districts. Moreover, the document provides substantial latitude for interpretation by the districts and others using the manual.

From 1987 to 1991, use of the manual by Corps personnel was not mandatory, so each district still could continue to use other procedures for making jurisdictional wetland determinations. This practice meant that there was intentionally no national standard and perhaps even no standardization within Corps districts. Consequently, determining the extent of jurisdictional wetlands remained at the discretion of individual districts.

In 1989, the federal government published the Federal Manual for Identifying and Delineating Jurisdictional Wetlands (Federal

Interagency Committee, 1989). This manual was developed by an interagency committee from the Corps, the Environmental Protection Agency (EPA), the FWS, and the Soil Conservation Service (SCS). This manual was intended to be strictly technically based for multiple applications: regulatory (including but not limited to the CWA), as well as non-regulatory (e.g., identifying the limits of vegetated wetlands for inventories or other studies). Such a document could serve the needs of many agencies including federal, state, and local governments. It could conceivably become the national standard, so that a single wetland-dryland boundary could be produced, rather than having a different boundary for different jurisdictions (federal-state-local).

On January 10, 1989, the four federal agencies (Corps, EPA, FWS, and SCS) officially agreed that the interagency manual would serve as "the technical basis for identifying and delineating jurisdictional wetlands in the United States" (Federal Interagency Committee, 1989). A "jurisdictional" wetland was simply a vegetated wetland potentially subject to some form of government regulation or policy. This new manual did not change the existing definitions of wetland used by each agency, but recognized the similarities in these definitions for vegetated wetlands and focused on providing a scientifically sound and consistent procedure for identifying and delineating such wetlands. The interagency manual, unlike the 1987 Corps Manual, was not intended as a policy instrument for the CWA and did not establish policy regarding agency actions in wetlands, although it was clearly recognized by all members of the Federal Interagency Committee for Wetland Delineation that the CWA regulatory program would be the principal user of the document, at least at the outset. Each of the signatory agencies had to decide how it would use this manual for their own programs. For using the manual in the CWA wetland regulatory program, the Corps and EPA had to develop a memorandum of agreement (MOA).

On January 19, 1989, the Corps and EPA signed a MOA that specified that the interagency manual would be used to identify the geographic extent of wetlands subject to the provisions of the CWA, among other things. This agreement established for the first time, a national standard for determining the limits of federally regulated wetlands under the CWA. All Corps districts and EPA regions were required to use this manual. As one might expect, such mandatory use of this manual (or any similar standard including the 1987 Corps

Manual) had the potential to expand federal jurisdiction to wet areas that individual Corps districts had not previously regulated for various reasons, provided these areas met the new criteria. The effect varied regionally. For example, in the Southeast where wetlands are extremely abundant, this effect was particularly pronounced, especially in pine flatwoods and other seasonally saturated wetlands not previously regulated. Yet, in the Northeast, there was little noticeable effect. The extent of areas identified as wetlands by the interagency manual was quite similar (except, perhaps, for farmed wetlands) to the vegetated wetlands mapped by the FWS's National Wetlands Inventory and to wetlands identified following EPA's wetland delineation manual (Sipple, 1988) which had been developed strictly for CWA use.

The apparent increased scope of jurisdiction through the mandated use of the interagency manual and concurrent policy changes, particularly how farmed wetlands were regulated, created a tremendous outcry from the regulated community, especially farmers and developers. They effectively lobbied Congress to ban the Corps' use of the interagency manual for determining CWA authority. In 1992, Congress passed the Johnston amendment to the 1992 Energy and Water Development Appropriations Act which specified that the Corps must not use the interagency manual for identifying any land as a water of the U.S. or any other manual that has not fulfilled the requirements for notice and public comment in accordance with the Administrative Procedures Act, or else the Corps would be denied millions of dollars in operating funds.

As of August 17, 1991, the Corps began to use its 1987 manual for wetland determinations.[2] In a change from previous Corps practices, all Corps districts had to use the Corps manual, so the Corps manual became the new regulatory standard for wetland delineation. Additional guidance has since been provided for using the Corps manual to clarify some significant points of interpretation.[3] On

---

[2] U.S. Army Corps of Engineers, Memorandum from John Elmore on wetlands delineation and the 1992 Energy and Water Development Appropriations Act (August 23, 1991).

[3] U.S. Army Corps of Engineers, Memorandum from John F. Studt on questions and answers on 1987 manual (October 7, 1991); and U.S. Army Corps of Engineers, Memorandum from Major General Arthur E. Williams on clarification and interpretation of the 1987 manual (March 6, 1992).

January 4, 1993, EPA and the Corps signed an amendment to the January 19, 1989 MOA, specifying that the 1987 Corps manual would be used by both agencies to determine the geographic extent of wetlands subject to CWA regulations (38 Fed. Reg. 4993 (Jan. 19, 1993)). Further joint agency guidance for clarifying and interpreting the manual is expected to be forthcoming, since EPA has oversight responsibilities for administrating the CWA program.

5.2 WETLAND INDICATORS

The CWA regulatory definition of wetland emphasizes two conditions: 1) permanent or periodic inundation or saturation by surface or ground water and 2) under normal circumstances, a prevalence of vegetation typically adapted for life in saturated soil conditions. The hydrology in the former must be of sufficient duration and frequency to support the latter: hydrophytic vegetation. Hydrophytic vegetation is defined as all of the "macrophytic plant life growing in water or on a substrate that is at least periodically deficient in oxygen as a result of excessive water content" (U.S. Army Corps of Engineers, 1987). The phrase "under normal circumstances" was included to deter people from removing wetland vegetation with the intent to get the Corps to change the designation of a wet area from wetland to nonwetland, which would remove the area from federal jurisdiction. If the current land use of a wetland is agricultural, and as a result of cultivation and farming practices, hydrophytic vegetation is not normally present, then that area would not be classified as wetland for CWA purposes. Farmed wetlands of this type are typically considered nonwetlands by the 1987 Corps Manual.

The Corps' interpretation of the CWA wetland definition led to the development of the multi-parameter approach by which positive wetland indicators of vegetation, soils, and hydrology typically must be found to make a positive wetland determination, with a few noted exceptions. Deepwater aquatic habitats differ from wetlands in that they are only permanently inundated areas (mostly deeper than 6.6 feet mean annual water depth), lack rooted emergent or woody plants, and possess nonsoil substrates. Nonwetlands can be recognized by plants that are typically adapted for life in aerobic soils, nonhydric soils, and/or the duration of inundation or soil saturation is not sufficient to preclude establishment of nonhydrophytic plant species.

EPA's interpretation of the CWA wetland definition was broader. It allowed identification of wetlands based solely on the dominance of plant species unique to wetlands (obligate hydrophytes) as long as the area was not significantly drained or similarly hydrologically modified. When such vegetation was not present, EPA's manual required positive indicators of hydric soils and wetland hydrology for wetland recognition (Sipple, 1988). This approach was consistent with the scientifically based definition of wetlands developed by the FWS for inventorying U.S. wetlands.

Differences between the Corps and EPA interpretations of the CWA wetland definition meant that the boundaries of federally regulated wetlands could vary depending on the manual used. This was a significant issue, especially for enforcement cases involving violation of CWA regulations. The 1987 Corps Manual was clearly more conservative, in that only the wetter wetlands would be regulated if desired by a Corps district, while the drier wetlands (e.g., those whose vegetation was more transitional or lacking certain indicators of wetland hydrology) would not. Because the 1987 Corps Manual required evidence of three parameters rather than less than three, federal jurisdiction in wetlands following this manual was less extensive than if based on the EPA manual. This eventually created great confusion among the regulated community, since EPA and the Corps jointly administer the CWA regulatory program. The differences in these two manuals also led to the appearance of a significant expansion of federal regulation following the Corps' and EPA's adoption of the interagency manual for use in the CWA program, despite the latter manual's similarity to the EPA manual. The Corps' lack of consistency and standardization in identifying CWA wetlands throughout the country prior to January 1989, however, was probably more responsible for the expansion of regulated areas following the federal government's adoption of the interagency manual for the CWA program.

The Corps multi-parameter approach usually requires finding evidence or positive indicators of three parameters (vegetation, soils, and hydrology) to substantiate a wetland determination. Except in limited cases, such signs must be present at the time of observation. The following subsections review indicators for each of the parameters.

5.2.1 Hydrophytic Vegetation

Hydrophytic vegetation is macrophytic plant life growing in water (permanently inundated areas) or in areas subject to periodic flooding or permanent or periodic saturation near the soil surface and at least periodically deficient in oxygen due to excessive water. Simply stated, hydrophytic vegetation is an assemblage of plants growing in water or at least periodically anaerobic (oxygen-deficient) wet soils.

The FWS with cooperation from the Corps, EPA, and SCS has compiled a list of plant species that occur in U.S. wetlands. Several versions of the list have been published with the most recent being the 1988 list: "National List of Plant Species that Occur in Wetlands: National Summary" (Reed, 1988a). Out of the more than 20,000 plant species that grow in the U.S. and its possessions, about one-third (or 6,728 species) have been reported in wetlands. Not all plants found in wetlands are exclusive to wetlands. In fact, only about 27% of the listed plants are species that are restricted to wetlands (Tiner, 1991a). The majority of the nation's wetland plant species also grow in both wetlands and nonwetlands (uplands) to varying degrees. Consequently, in compiling the wetland plant list, various indicator status categories reflecting a species' affinity for wetlands were assigned to each species.

Five indicator status categories are recognized: 1) obligate wetland (OBL; nearly always occurs in wetlands - frequency of occurrence in wetlands exceeds 99%), 2) facultative wetland (FACW; occurs mostly in wetlands - frequency of occurrence in wetlands ranges from 67-99%), 3) facultative (FAC; frequency of occurrence in wetlands ranges from 34-66%), 4) facultative upland (FACU; occurs mostly in nonwetlands - frequency of occurrence in wetlands ranges from 1-33%), and 5) obligate upland (UPL; nearly always occurs in nonwetlands - frequency of occurrence in wetlands is less than 1%).[4] The OBL species are exclusive to wetlands, and are therefore the best vegetative indicators of wetlands. The facultative-types represent species that occur in wetlands and nonwetlands to varying degrees. Of these, the FACW species are the most reliable indicators of wetland, since they are mostly found in these habitats. The FAC plants are neither indicative of wetlands nor uplands, being nearly equally abundant in both

---

[4] A "+" sign or a "-" sign is assigned to the facultative categories when a species exhibits an affinity for the "wetter" or "drier" end of the frequency range, respectively.

habitats. FACU species are more characteristic of nonwetlands, but also grow in wetlands where they may even be a dominant species. This reality in plant ecology makes it difficult to establish just what types of plants should be considered hydrophytes or wetland plants and raises the question as to whether the species is the appropriate level in plant classification to use for determining hydrophytic vegetation. The ecotype, subspecies, or variety of a species may be a more precise indicator (see Tiner, 1991a for a review of the concept of a hydrophyte for wetland identification). The 1987 Corps Manual considers only the species level and those species with an indicator status of OBL, FACW, and FAC (excluding FAC-) represent hydrophytic vegetation - plants typically adapted for life in anaerobic or saturated soils. This concept fails to acknowledge the existence of wetland ecotypes of "dry-site" species that dominate or are common associates in certain wetlands (see Table 5.1). Consequently, FACU-dominated wetlands (e.g., hemlock swamps) are not accounted for in the Corps manual, except if one uses sound professional judgment based on knowledge of the wetland literature. Certain Corps districts routinely classify such areas as wetlands, while others do not.

Basic rule for determining hydrophytic vegetation. To assess the presence of such vegetation, the plant community in question must be evaluated and the dominant species determined. There are at least two acceptable ways to establish dominants: 1) select the three most dominant species from each stratum (trees, saplings/shrubs, herbs, and woody vines) or five from each stratum if only one or two strata are present, or 2) select dominants from each of five strata (trees, saplings, shrubs, herbs, and woody vines) using ecologically based methods including the method described in the 1989 interagency manual.[5] Table 5.2 shows examples for determining dominant species following both approaches. Measures of dominance include percent areal cover, stem density, and basal area.

Once dominants are established, the basic rule is that hydrophytic vegetation is present when more than 50% of the dominant species from all strata in the plant community have an indicator status of OBL, FACW, and FAC (excluding FAC-) on the appropriate

[5] U.S. Army Corps of Engineers, Memorandum from Major General Arthur E. Williams on clarification and interpretation of the 1987 manual (March 6, 1992).

Table 5.1: Examples of FAC- and FACU plant species that are common or dominant in wetlands in the northeastern United States.

Plant Species	Indicator Status
White pine (*Pinus strobus*)	FACU
Pitch pine (*P. rigida*)	FACU
Loblolly pine (*P. taeda*)	FAC-
Red spruce (*Picea rubens*)	FACU
White ash (*Fraxinus americana*)	FACU
American holly (*Ilex opaca*)	FACU+
Bunchberry (*Cornus canadensis*)	FAC-
Virginia creeper (*Parthenocissus quinquefolia*)	FACU
American beech (*Fagus grandifolia*)	FACU
White oak (*Quercus alba*)	FACU-
Canada mayflower (*Maianthemum canadense*)	FAC-
Spring beauty (*Claytonia virginica*)	FACU
White avens (*Geum canadense*)	FACU
Multiflora rose (*Rosa multiflora*)	FACU
Wintergreen (*Gaultheria procumbens*)	FACU
Black haw (*Viburnum prunifolium*)	FACU
Pawpaw (*Asimina triloba*)	FACU+
Staggerbush (*Lyonia mariana*)	FAC-
Eastern hemlock (*Tsuga canadensis*)	FACU
Tulip poplar (*Liriodendron tulipifera*)	FACU
Persimmon (*Diospyros virginiana*)	FAC-
Bittersweet Nightshade (*Solanum dulcamara*)	FAC-
Japanese honeysuckle (*Lonicera japonica*)	FAC-
Partridgeberry (*Mitchella repens*)	FACU

regional wetland plant list (Figure 5.2).

Figure 5.2: Geographic scope covered by regional wetland plant lists (from Reed, 1988a).

Table 5.2: Calculating dominant species for a hypothetical plant community following two approaches: (1) 1987 Corps Manual (4 strata: overstory, understory, herbs, and woody vines; select 3 most dominant species in each stratum); and (2) 1989 Interagency manual (5 strata: tree, sapling, shrub, herb, and woody vine; dominant species in each stratum represent more than 50% of the total measure for dominance for the stratum, plus any species that represents 20% or more of the total dominance measure for the stratum; this may be called the "50/20 dominance rule".)

Hypothetical Plant Community:

Trees: Red maple (70% areal cover), Green ash (30%), Loblolly pine (10%), Sweet gum (5%) (Total cover = 115%)
Saplings: Red maple (35% areal cover), American holly (5%) (Total cover = 40%)
Shrubs: Sweet pepperbush (25% areal cover), Southern arrowwood (20%), Highbush blueberry (5%) (Total cover = 50%)
Herbs: White grass (25% areal cover), Canada mayflower (10%), Cinnamon fern (5%) (Total cover = 40%)
Woody Vines: None

---

1. Dominants following the 1987 Corps Manual:

Overstory (Trees - greatest relative basal area*): Red maple, Green ash, Loblolly pine
Understory (Saplings/Shrubs - greatest height**): Red maple, American holly, Southern arrowwood (tallest shrub)
Herbs (greatest percent areal cover): White grass, Canada mayflower, Cinnamon fern

Number of dominants: 9

Comments: The top 3 dominants were selected for each stratum.

2. Dominants following the 1989 interagency manual:

Table 5.2: (Continued)

Trees: Red maple, Green ash
Saplings: Red maple
Shrubs: Sweet pepperbush, Southern arrowwood
Herbs: White grass, Canada mayflower

Number of dominants: 7

Comments: Two tree species are dominant: Red maple with 70% cover represents more than 50% of the total areal cover (50% of 115% = 57.5%); Green ash is also a dominant because its cover of 30% represents 26% of the total cover (remember, any species comprising 20% or more of the total dominance measure, in this case, areal cover, is also a dominant species). Two shrubs are dominant: Sweet pepperbush with 25% areal cover equals but does not exceed 50% of the total shrub layer, so the next most abundant species is also a dominant which brings in Southern arrowwood. The remaining shrub, highbush blueberry, had only 5% cover which represents only 2.5% of the total shrub cover, so it is not a dominant species. White grass is dominant because it represents 62.5% of the total cover of the herb stratum (25/40 = 62.5), and Canada mayflower is also a dominant because its cover (10%) represents 25% of the total herb cover.

CONCLUSIONS: Different dominants may be selected; this potentially could produce a difference in verifying the presence of hydrophytic vegetation which could affect the wetland determination. The Corps approach is simple, easier to employ, but the outcome is not based on weighted values of chosen dominance measures. The Interagency method takes more time, is more quantitative and appears more scientifically sound, and dominants are determined by their weighted rankings.

[*] Greatest relative basal area happened to produce the same dominants as percent areal cover, but this is not always the case.
[**] If areal cover were used, dominants would be different. Red maple, Sweet pepperbush, and Southern arrowwood have the most areal cover of understory species.

FAC neutral option. This option simply considers FAC species as neutral in establishing the presence of hydrophytic vegetation. These species have no interpretive value and are therefore given no weight in making a wetland determination. The decision whether an area has hydrophytic vegetation is based on comparing the number of dominants with an indicator status wetter than FAC versus the number of dominants drier than FAC. Hydrophytic vegetation is present when there are more dominants wetter than FAC. If all dominants are FAC or the number of dominants wetter than FAC equals the number drier than FAC, then nondominants must be considered. The FAC neutral option may be used to make a wetland determination where available evidence of wetland hydrology or hydric soil is weak.[6] The use of this option is discretionary, so each Corps district must be consulted regarding application of this option.

Other hydrophytic vegetation indicators. If the basic rule is not met, other signs of hydrophytic vegetation may be useful for identifying hydrophytic FAC and FACU plants including: 1) actual observations of these species growing in areas of prolonged inundation and/or soil saturation, 2) the presence of morphological adaptations to wetland hydrology, 3) knowledge of physiological adaptations for life in saturated soils, 4) knowledge of reproductive adaptations for life in wetlands, and 5) literature citations of species common to wetlands. All of these indicators require use of sound professional judgment. The first indicator requires periodic observations over several years and documentation of the findings. Table 5.3 lists common plant adaptations used for wetland determinations. Hydrophytic vegetation is present if two or more dominant species possess such adaptations and most of the individuals must exhibit the morphological adaptation. Reference to scientific literature may also be a valid means of verifying hydrophytic vegetation, but one must be certain that the conditions observed in the field are comparable to those reported in the literature.

Vegetation inconclusive. If all the considered species are FAC or the number wetter than FAC equals the number drier than FAC, the wetland determination must be made on the basis of soil and hydrologic

---

[6] *Id.*

indicators. This is a case where positive indicators of less than two parameters is permitted for a wetland determination.

5.2.2 Hydric Soil

Hydric soil is soil that is saturated, flooded, or ponded long enough during the growing season to develop anaerobic conditions in the upper part of the soil (U.S. Soil Conservation Service, 1991). Soils must be inundated for usually one week or more or saturated for at least 15 consecutive days during the growing season. The term "hydric soil" was coined by the FWS in its wetland definition (Cowardin et al., 1979). Subsequently, the FWS asked the SCS to develop technical criteria and a national list of these soils for applying its wetland definition and classification system.

Technical criteria for hydric soil. These criteria were adopted from the criteria established by the National Technical Committee for Hydric Soils. In general, hydric soils include: 1) all organic soils (except Folists), 2) poorly drained and very poorly drained nonsandy soils with a seasonal water table within 1.0 - 1.5 feet from the soil surface depending on soil permeability, 3) somewhat poorly drained, poorly drained, and very poorly drained sandy soils with a seasonal water table within six inches of the surface, and 4) other soils inundated for one week or more during the growing season. Table 5.4 lists the specific technical criteria for hydric soils. While hydric soils may be drained, when they are effectively drained, they no longer meet the water table requirement in the technical criteria and are therefore not hydric. However, the 1987 Corps Manual allows both drained and undrained hydric soils to be a positive indicator of hydric soils, while calling the undrained ones "wetland soil."

Soil wetland indicators. The Corps manual lists the following indicators which are presented in order of decreasing reliability. These indicators should be found within 10 inches of the soil surface or immediately below the A-horizon, whichever is shallower. Plates VII-X (see endpiece) illustrate some of the more common diagnostic hydric soil properties (see Tiner, 1988, 1991a; and Tiner and Veneman, 1987 for additional color photographs of soils demonstrating these properties).

Table 5.3. Plant adaptations or responses to flooding and waterlogging (Tiner, 1993a). The morphological features may be most useful for recognizing hydrophytic plants, although some are not unique to wetland species.

Morphological Adaptations

Stem hypertrophy (exaggerated swelling of trunks and stems at base)
Fluted bases of tree trunks
Aerenchyma tissue in roots and stems
Hollow stems
Shallow root systems (roots at or near soil surface; sometimes form conspicuous hummock at base of woody plants)
Adventitious roots
Pneumatophores (e.g., cypress knees)
Swollen, loosely packed root nodules
Soil water roots
Succulent roots
Aerial root tips
Hypertrophied (enlarged) lenticels
Relatively pervious cambium in woody plants
Heterophylly (submerged leaves differ from emergent leaves)
Succulent leaves
Floating leaves

Physiological Adaptations

Transport oxygen to roots (evidence = oxidized rhizospheres)
Anaerobic respiration
Increased ethylene production
Reduction of nitrate to nitrous oxide and nitrogen gas
Malate production and accumulation
Reoxidation of NADH
Metabolic adaptations

Table 5.3: (Continued)

Reproductive Adaptations

Pollination and seed germination under water
Viviparous seeds

Other Adaptations

Growth dormancy during inundation
Elongation of roots, stem or petioles
Root mycorrhizae in upper soil layer
Change in root or shoot direction (horizontal or upward)
Long-lived seeds
Breaking of flood induced dormancy by multiple stem buds
(produces multiple stems or trunks)

* Although a common attribute of many wetland plants, especially conspicuous in trees, upland plants in rocky soils and severely eroded sites may also possess shallow roots for reasons other than wetness.

For nonsandy soils:
1) organic soils (Plate VII), 2) histic epipedons (organic surface layers usually 8- to 16-inches thick; Plate VIII), 3) sulfidic material (rotten egg odor of hydrogen sulfide), 4) aquic or peraquic moisture regime (periodic or permanent reducing environment), 5) observed reducing conditions (as detected by a colorimetric test), 6) soil colors (gleyed layer immediately below the surface horizon; soils with a subsoil matrix of chroma 2 or less with bright mottles found within 10 inches of the soil surface, as shown in Plate IX; soils with a subsoil matrix of chroma 1 or less lacking bright mottles within 10 inches), 7) soil listed as a hydric soil, and 8) iron and manganese concretions within 3 inches of the soil surface.

For sandy soils:
1) high organic matter content in the surface layer, 2) organic streaking in the subsoil (Plate X), and 3) organic pans.

Table 5.4: National technical criteria for hydric soils (September 1991).

1. All Histosols except Folists, or

2. Soils in Aquic suborders, Aquic subgroups, Albolls suborders, Salorthids great group, Pell great group of Vertisols, Pachic subgroups, or Cumulic subgroups that are:

   A. Somewhat poorly drained and have a frequently occurring water table at less than 0.5 ft from the soil surface for a significant period (usually more than two weeks) during the growing season, or

   B. Poorly drained or very poorly drained and have either:

   (1) frequently occurring water table at less than 0.5 ft from the surface for a significant period (usually more than two weeks) during the growing season if textures are coarse sand, sand, or fine sand in all layers within 20 inches, or for other soils:

   (2) frequently occurring water table at less than 1.0 ft from the surface for a significant period (usually more than two weeks) during the growing season if permeability is equal to or greater than 6.0 inches per hour in all layers within 20 inches, or;

   (3) frequently occurring water table at less than 1.5 ft from the soil surface for a significant period during the growing season if permeability is less than 6.0 inches per hour in any layer within 20 inches, or;

Table 5.4: (Continued)

3. Soils that are frequently ponded for long duration (one week or more), or long duration (one month or more) during the growing season, or;
4. Soils that are frequently flooded for long duration or very long duration during the growing season.

Note: "Frequently" means that the condition occurs more than 50 years out of 100, or has a probability of more than 50% of occurring in any given year. The difference in water table height requirements relate to differences in capillary action of sands, silts, and clayey soils, and its effect on the capillary fringe (the saturated tension zone above the water table).

When examining the soil for hydric properties, the 1987 Corps Manual recommends looking at the soil within 10 inches of the soil surface, or immediately below the A-horizon (surface layer), whichever is shallower. Such guidance can create misidentification problems in certain soils, especially plowed soils, soils at the toe of eroding slopes, and soils with strongly leached subsoils due to podzolization. Consequently, the Corps recently clarified that this rule was intended for general guidance and recognizes the certain problem soils may require assessment of properties lower in the soil profile.[7] Typically, soil properties below the A-horizon and within 20 inches of the surface in nonsandy soils and usually within 12 inches of the surface in sandy soils should be examined.

The 1987 Corps Manual suggests using the national list of hydric soils and soil surveys and confirmation that the soil in question is the one mapped. SCS has recently published local lists of hydric soil mapping units on a county-basis. These lists take precedence over the national list, but the soil in question still must be verified in the field. Both lists, however, include some soils that are hydric in depressional areas and nonhydric in other places (e.g. more sloping terrain). Simple reliance on the lists and hydric series without regard to the properties reflecting wetland conditions versus nonwetland conditions can lead to

---

[7] *Id.*

erroneous determinations in the field. Recent Corps guidance has attempted to rectify this problem by emphasizing that mapped hydric soils should exhibit indicators listed in the manual to be considered a positive indicator of hydric soil.[8]

5.2.3 Wetland Hydrology

All habitats have hydrology, since the hydrology of an area is simply the properties, distribution, and movement of water in the area. So what then is wetland hydrology? The Corps manual defines it as "the sum total of wetness characteristics in areas that are inundated or have saturated soils for a sufficient duration to support hydrophytic vegetation." The duration must, therefore, be sufficiently long to produce oxygen-deficient soil conditions that favor hydrophytes. Unfortunately, wetland science is currently unable to determine with precision what this length of time might be because there has been inadequate scientific study to date.

Flooding or saturating a soil for a period of time will create anaerobic conditions. Studies have shown that oxygen is removed rapidly by flooding and within a day, most soils become anaerobic (Ponnamperuma, 1972). If one week of inundation or two weeks of soil saturation near the surface (e.g., within one foot in nonsandy soils and within six inches in sandy soils) at a frequency of more than 50 years out of 100 years is enough to classify as soil as hydric, this hydrology should also be long enough to promote the establishment of hydrophytic vegetation. The hydrologic conditions for hydric soils emphasize conditions during the "growing season" (defined as when the soil temperature at 20 inches is above 41°F - "biologic zero"), yet certain soils in the Arctic never reach this temperature although they display hydric properties (Ping, et al., 1990; Tiner, 1991b). However, root growth and soil microbial activity have been reported occurring below this temperature (Lyr and Hoffman, 1967; Lopushinsky and Max, 1990; Ping et al., 1990; and Tiner, 1991b). The use of 41°F for biological "zero" appears to be unsupported by the literature. Microbial activity and root growth have been observed below this temperature (Van Cleve and Sprague, 1971; Boyd and Boyd, 1972; Flanagan and Veum, 1974; Billings et al., 1977; Schlentner and Van Cleve, 1985; and

---

[8] *Id.*

Ping et al., 1990), which, combined with the fact that many wetland functions continue year-round throughout the country, lead to the conclusion that perhaps year-round hydrology should be considered in assessing wetlands rather than using conditions during the growing season (Tiner 1991b).

The 1987 Corps Manual originally recommended using the frost-free period to approximate the growing season. This invariably led to the exclusion of thousands and perhaps millions of acres of wetlands. In an attempt to correct this oversight, recent guidance from Corps headquarters states that the growing season dates used for wetland hydrology are when the air temperatures are above the 28°F or lower threshold at a frequency or probability of occurrence of more than five years in ten. This information can be derived from local soil surveys (usually tables 2 or 3 of the modern reports). An important exception to this threshold is made for the South, where temperatures above the 32°F threshold may be used if the Corps district chooses to use it. This policy-driven discretion makes it possible to minimize the extent of regulated wetlands in the South where wetlands are very abundant. Many wetlands in these and other areas, especially in the eastern half of the country, are wettest in winter and winter hydrology is not considered in making a wetland determination. In Berkeley County, South Carolina, defining the growing season by the average period of the last freeze date in Spring to the first freeze date in Fall using the 28°F threshold versus the 32°F threshold accounts for a difference of approximately one month, with the former's growing season extending from March 16 to November 13, and the latter's going from April 3 to November 2 (Long, 1980). Plants are actively growing before and after either period.

Today, areas inundated and/or saturated to the surface for a consecutive period covering more than 12.5% of the growing season are considered wetlands according to the 1987 Corps Manual. Areas wet between 5.0% and 12.5% of the growing season in most years may or may not be wetlands. Although these percentages were derived from studies of southeastern bottomland hardwood forests (Clark and Benforado, 1981), they are applied to all types of nontidal wetlands. Limits for tidal wetland hydrology are not defined in the manual. Presumably, professional judgment must be exercised in tidal areas.

Wetland hydrology indicators. A list of examples of these indicators is presented in the 1987 Corps Manual: 1) recorded data; 2) visual observations of inundation and/or soil saturation (within the major portion of the root zone which is usually within 12 inches of the soil surface); 3) water marks; 4) drift lines; 5) sediment deposits; and 6) wetland drainage patterns. Although it is clearly stated that indicators of wetland hydrology are not necessarily limited to these "primary" examples, in practice, users of the manual typically have not considered any other indicators. Recent Corps headquarters guidance acknowledges four secondary indicators of wetland hydrology on the new field data form for routine determinations: 1) oxidized root channels in the upper 12 inches; 2) water-stained leaves (in foreground of Plate V); 3) local soil survey data (hydrology data); and 4) FAC neutral option.[9] Two or more secondary indicators are required to verify wetland hydrology. The use of oxidized rhizospheres around living roots is permitted on a case-by-case basis provided they are "reasonably abundant and within the upper 12 inches of the soil profile."[10] They also must be supported by other indicators of hydrology such as the FAC neutral option (see Section 5.2.1) if hydrology evidence is weak. For systems driven by ground water hydrology, use of hydrology data for soil series (found in SCS county soil survey reports) can be used in conjunction with other hydrologic indicators such as the FAC neutral option to verify wetland hydrology. Soils must be confirmed to the series level to accomplish this.

5.3 FIELD PROCEDURES FOR WETLAND DELINEATION

Wetland delineation is the process of identifying and marking a specific wetland-nonwetland boundary line in the field. It always requires an onsite inspection and usually close examination of the site for positive indicators of hydrophytic vegetation, hydric soils, and wetland hydrology. Field observations must be sufficient to allow an investigator to accurately locate and mark the wetland-nonwetland boundary. Depending on site characteristics, wetland recognition may

---

[9] *Id.*

[10] *Id.*

be easy or difficult. The wetter the wetland, the easier it is to identify since there is often strong evidence of wetness (e.g., surface water or soils wet underfoot) plus conspicuous and highly wetland-specific vegetation (OBL species either dominant or common in association with FACW species). The average citizen can readily identify these areas as wetlands. Yet, the majority of wetlands have fluctuating water levels and may or may not be inundated for brief periods. The absence of standing water and saturated soils often makes wetland recognition more difficult, since the corresponding vegetation is usually not unique to wetlands. Periodically anaerobic soils typically develop specific and readily observable properties that reflect site wetness (Tiner, 1991c). Consequently, in areas not subject to significant hydrologic modification, hydric soil properties are invaluable for identifying the presence of wetland.

The 1987 Corps Manual requires that, with a few noted exceptions, areas identified as wetland must have positive indicators of hydrophytic vegetation, hydric soils, and wetland hydrology, rather than relying on the presence of one or more highly reliable wetland indicators (e.g., plant community dominated by OBL species or undrained organic soils). This basic rule virtually ensures that a wetland determination made by Corps personnel or others using the manual will be unquestionably a wetland and eliminates the possibility of erroneously classifying nonwetland areas that may have less than all three positive indicators. Unfortunately, this approach also eliminates wetland areas having positive indicators of hydrophytic vegetation and hydric soils but lacking a positive indicator of wetland hydrology and yet no sign of significant drainage or similar alteration. Thus, the strength of the 1987 Corps Manual, the verification of each of three parameters, also creates its greatest weakness: namely, the omission of many vegetated wetlands that are not listed as specific exceptions to the rule. It is recognized that one can use professional judgment to identify such areas, but this appears an inadequate way of addressing the matter. It is advisable either to provide criteria or protocols for recognizing these areas as wetlands, or simply expand the list of exceptions to include other regional wetland types that fail to meet the strict three-parameter test.

5.3.1 General Guidance for Field Work

Before conducting field work, the investigator should review existing information about the site such as National Wetlands Inventory maps, county soil survey reports, and aerial photographs (see Section 5.4 for discussion of these and other useful sources of information). These data sources provide a good overview of the project site, so the investigator has some idea of what to expect in the field.

If possible, more than one visit to the project site is recommended: once at the beginning or during the wettest part of the growing season (usually to assess actual site wetness for positive wetland hydrology indicators) and later during the peak of the growing season (to assess vegetation, especially the herbaceous species which are frequently better indicators of site wetness than the woody species). In practice, however, site inspection is usually restricted to one season - the time the landowner needs a wetland determination.

When conducting field work, one should avoid undertaking investigations after periods of extreme flooding, or during periods immediately following heavy rains because low-lying uplands may be inundated at these times. Experienced wetland specialists can usually make an accurate wetland determination by a single visit provided the soil is not frozen. Winter determinations in northern snow-covered regions are usually preliminary and should be refined by a second inspection during the growing season. Early Spring and mid- to late Fall inspections have an advantage over other seasons, because in most temperate regions, the foliage is absent from the deciduous woody plants. This makes it much easier to observe subtle changes in topography that typically reflect wetland-nonwetland boundary areas. This also allows one to mark the boundary with less flags, thereby making follow-up surveying by site engineers more efficient and less costly. Moreover, annoying insects are often less abundant at these times.

A tract of land often has more than one plant community and may have wetlands scattered throughout a largely upland area. Each plant community, regardless of size, should be treated separately for wetland determination purposes. Homogeneous stands in similar landscape positions (e.g., depressions, flats, slopes, and ridges) should be identified for evaluation. An initial walk through the project site should reveal these different communities. Each plant community will be inspected for positive indicators of hydrophytic vegetation, hydric soils, and wetland hydrology to determine if any community is a wetland. If an extensive network of drainage ditches or other significant hydrologic alteration is observed or known to occur that greatly affects the site's hydrology, the hydrology parameter will require a detailed examination following specific procedures presented in the manual. In many such cases, it may be necessary to consult an expert to determine the current hydrology.

In areas where the topography changes abruptly, a marked change in the plant community usually takes place. A low-lying wetland dominated by OBL, FACW, and/or FAC species may give rise to a community dominated by FAC, FACU, and UPL species. The latter community is represented by nonhydrophytic vegetation, so the community is not wetland unless it happens to be one of the problematic wetlands (see Chapter 6 for discussion).

In many locales, however, topographic relief is more gradual and obvious differences in vegetation patterns are lacking. FAC species may dominate both wetlands and contiguous uplands. This happens in low-lying coastal plains, floodplains of major river valleys, and on gentle slopes in areas of groundwater discharge (seepage slopes). In these and similar situations, including the upper edges of many wetlands, a mixed plant community of FACW, FAC, and FACU species often forms what some people call the "transition zone." Signs of wetland hydrology may be difficult to find in these places, since flooding may be brief or soils may be saturated near the surface only during the early part of the growing season. Many of these wetlands, especially in the eastern U.S., are wetter longer during the non-growing season. In those cases where hydrology is not apparent or weakly expressed and the vegetation is inconclusive, the soil properties reveal important information about site wetness. It is widely acknowledged that soils reflect the long-term hydrology and therefore, hydric soil properties are highly reliable indicators of wetland in the absence of significant hydrologic modification (Tiner and Veneman, 1987; Sipple, 1988; Federal Interagency Committee, 1989).

5.3.2 Corps Wetland Delineation Procedures

The 1987 Corps Manual describes two basic approaches for making wetland determinations: 1) routine determinations and 2) comprehensive determinations. The routine determination methods are intended to be used in all, but the most controversial or difficult situations. The comprehensive method should only be used when rigorous documentation of site characteristics is required or when the project area is very complex. In practice, the comprehensive method is used when the agency anticipates possible litigation. Procedures for atypical situations and problem areas are also included in the Corps manual.

Before proceeding with any method in the field, it is imperative that existing information about the project site be assembled and reviewed. These data sources include topographic maps, National Wetlands Inventory maps, soil surveys, stream and tidal gage data, environmental impact statements and similar documents, aerial photographs, and any site survey plan or engineering design. This information gives the investigator a good sense of likely site characteristics and helps in the selection of a method for making wetland determinations. This data review is best performed in the office prior to conducting field work.

Each method is outlined or discussed below. The actual steps for each method have been simplified for this presentation. Some procedures are only briefly discussed.

Routine determination methods. Depending on the size of the project area or vegetative complexity, two routine methods are available. If the site is greater than five acres in size or diverse in landscape patterns, a transect method is recommended. If the site is smaller or the landscape is simple and homogeneous, wetland determinations can be made without transects, simply by identifying plant communities. The procedures for each method are generally outlined below. Remember that the steps are condensed for simplicity. The investigator should follow the actual steps in the manual when conducting a field inspection.

For areas generally five acres or less, or areas that are larger but homogeneous in vegetation, do the following:

1. Locate the project area in the field and walk the site. In doing so: (a) determine the number and location of plant communities on the site; (b) determine if there is any evidence of natural or human alteration to vegetation, soil, and/or hydrology that would make

wetland determinations difficult (for these areas, special procedures for "Atypical Situations" should be followed); and (c) determine whether seasonal changes in hydrology or temperatures may pose a serious problem for making a wetland determination according to the three-parameter method (follow guidance in "Problem Areas" section).

2. Evaluate each plant community for positive indicators of hydrophytic vegetation, wetland hydrology, and hydric soils, respectively. Pick one community to begin your assessment. Select one or more observation points (as necessary) that typify the community and mark the locations on a map.

3. Characterize the plant community. Determine dominant species in each stratum (two options: 4-strata or 5-strata maximum) by visual observation. Record the indicator status of each dominant and determine whether hydrophytic vegetation is present (more than 50% of the dominants are OBL, FACW, and FAC, excluding FAC-). If it is not, the area is usually nonwetland, unless it is a problematic wetland for identification following the 1987 Corps Manual (contact local Corps district for guidance).

4. Evaluate the site's hydrology by looking for positive indicators of wetland hydrology. Any community having a positive wetland indicator has wetland hydrology. If no indicators are present (including those mentioned in recent Corps guidance memoranda), the area is not a regulated wetland.

5. If all dominants are OBL, or OBL and FACW, and the wetland boundary is abrupt, soils do not need to be examined - hydric soils are assumed to be present and the community is wetland. Other areas require examination of the soils.

6. Determine whether hydric soils are present. Dig a hole about one foot wide and two feet deep and look for positive indicators of hydric soil immediately below the A-horizon (usually within one foot of the soil surface). If such indicators are present, the community is a wetland. If not, it is typically a nonwetland, unless it is a problematic wetland. Remember any area presently or normally having wetland indicators of all three parameters is a wetland, and areas lacking one are typically nonwetlands (see Figure 5.3 for an example of a completed data form).

7. Delineate the boundary between wetland and nonwetland plant communities. The 1987 Corps Manual advises drawing it on a map, however, in practice, the boundary is marked on the ground,

usually following a contour that marks the limit of where positive indicators of all three parameters are found. A survey crew will later plot all wetland boundary flag locations on a detailed map.

For areas larger than five acres, or having diverse assemblages of plant communities, do the following:
 1. Same as 1 above.
 2. Establish a baseline for setting up transects. The baseline should be parallel to the major river course or perpendicular to the hydrologic gradient.
 3. Determine the number and position of transects. If the baseline is one mile long or less, locate at least three transects equal distance from one another. If the baseline is one to two miles long, three to five transects are needed. If longer, more transects are required; transects should be no more than 0.5 mile apart. Since all communities must be included, it may be necessary to reorient a transect to insure that all communities are on at least one transect (see Figure 5.4).
 4. Walk each transect, identify different plant communities along the transect as you go, and when you reach the end of the transect, turn around to return to the baseline, and sample each plant community on your way back.
 5. In each community, establish an observation point that best represents the plant community and evaluate each parameter in the following sequence: hydrophytic vegetation, hydric soils, and wetland hydrology.
 6. Sample vegetation within a 30-foot radius plot for trees and woody vines, and plot for herbs and saplings/shrubs within a 5-foot radius. Plot size and shape may be adjusted to match the site conditions (e.g., ridges and swales or pits and mounds). Determine dominants in all strata (4- or 5-strata maximum) by visual observation. Record indicator status of all dominants and determine whether hydrophytic vegetation is present. If it is not, the area is typically a nonwetland or at least not a regulated wetland.
 7. If all dominants are OBL or all are OBL and/or FACW (with at least one OBL dominant), hydric soils are assumed to be present. If not, then soils must be examined for hydric soil indicators usually within one foot from the soil surface. If such indicators are present, the community has hydric soils. If not, the area is typically nonwetland unless it is a problematic wetland for identification purposes.

DATA FORM
ROUTINE WETLAND DETERMINATION
(1987 COE Wetlands Delineation Manual)

Project/Site: **Andrew's Horse Farm** Date: **6/26/93**
Applicant/Owner: **A.J. Gallop** County: **Middlesex**
Investigator: **I.C. Waterman** State: **MA**

Do Normal Circumstances exist on the site? **Yes** / No   Community ID: **A**
Is the site significantly disturbed (Atypical Situation)? Yes / **No**   Transect ID: ___
Is the area a potential Problem Area? Yes / **No**   Plot ID: ___
 (If needed, explain on reverse.)

VEGETATION

Dominant Plant Species	Stratum	Indicator	Dominant Plant Species	Stratum	Indicator
1. Red Maple	tree	FAC	9.		
2. White Ash	tree	FACU	10.		
3. Highbush Blueberry	shrub	FACW-	11.		
4. Winterberry	shrub	FACW+	12.		
5. Tussock Sedge	herb	OBL	13.		
6. Cinnamon Fern	herb	FACW	14.		
7. Poison Ivy	woody vine	FAC	15.		
8.			16.		

Percent of Dominant Species that are OBL, FACW or FAC (excluding FAC-): **(6 of 7)   86%**

Remarks: **Used 50/20 rule to establish dominants.**

HYDROLOGY

Recorded Data (Describe in Remarks):
___ Stream, Lake, or Tide Gauge
___ Aerial Photographs
___ Other
✓ No Recorded Data Available

Field Observations:

Depth of Surface Water: **0.5** (in.)
 (Local - wet spots)
Depth to Free Water in Pit: **10** (in.)
Depth to Saturated Soil: **1** (in.)

Wetland Hydrology Indicators:
Primary Indicators:
 ✓ Inundated **(in scattered depressions)**
 ✓ Saturated in Upper 12 Inches
 ___ Water Marks
 ___ Drift Lines
 ___ Sediment Deposits
 ✓ Drainage Patterns in Wetlands **(depressions)**
Secondary Indicators (2 or more required):
 ___ Oxidized Root Channels in Upper 12 Inches
 ✓ Water-Stained Leaves
 ___ Local Soil Survey Data
 ___ FAC-Neutral Test
 ✓ Other (Explain in Remarks) **(Peat mosses)**

Remarks: **Localized depressions still have standing water, but most of plant community lacks surface water at this time. Soils saturated near the surface, however. Clumps of Sphagnum present in the depressions.**

Figure 5.3: Example of a completed data form for a wetland plant community.

SOILS

Map Unit Name (Series and Phase):	Unknown (not available)		Drainage Class:	NA
Taxonomy (Subgroup):	NA		Field Observations Confirm Mapped Type?	Yes No

Profile Description:

Depth (inches)	Horizon	Matrix Color (Munsell Moist)	Mottle Colors (Munsell Moist)	Mottle Abundance/Contrast	Texture, Concretions, Structure, etc.
0-4					Organic (muck)
4-10		10 YR 2/1			Silt Loam
10-18		10 YR 5/2	7.5 YR 5/8	Common/Prominent	Silt Loam
18-20		2.5 Y 4/2	5 YR 4/6	Common/Prominent	Fine Sandy Loam

Hydric Soil Indicators:

___ Histosol
___ Histic Epipedon
___ Sulfidic Odor
___ Aquic Moisture Regime
___ Reducing Conditions
✓ Gleyed or Low-Chroma Colors
___ Concretions
___ High Organic Content in Surface Layer in Sandy Soils
___ Organic Streaking in Sandy Soils
___ Listed on Local Hydric Soils List
___ Listed on National Hydric Soils List
___ Other (Explain in Remarks)

Remarks:

WETLAND DETERMINATION

Hydrophytic Vegetation Present?	(Yes) No (Circle)			(Circle)
Wetland Hydrology Present?	(Yes) No			
Hydric Soils Present?	(Yes) No		Is this Sampling Point Within a Wetland?	(Yes) No

Remarks: Red maple - white ash swamp; mapped as PFO1E on National Wetlands Inventory map.

Approved by HQUSACE 3/92

Figure 5.3: (Continued)

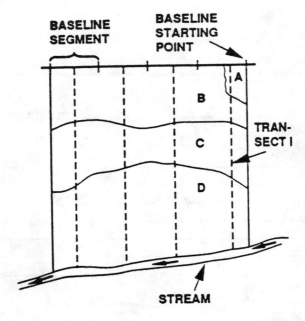

Figure 5.4: Positioning transects for evaluation of each plot (A, B, C, and D) at a project site. All transects start at the midpoint of each baseline segment, except transect #1, which was repositioned to include community A. Transects should be perpendicular to the stream, since plant communities usually change with increasing slope due to differing environmental conditions.

    8. Examine the community for positive indicators of wetland hydrology. If none are present, the community is a nonwetland. If one or more indicators are found or known to normally occur, then the area is a wetland.

    9. Proceed to the next community along the transect and repeat steps 6-8, making a wetland or nonwetland determination for each community.

    10. Determine the boundary between any wetland and nonwetland along the transect.

    11. Complete evaluation for all transects.

    12. Connect wetland boundary lines between transects by following the contour that best reflects the wetland-nonwetland boundary, periodically evaluating all three parameters to ensure that the boundary is accurately represented by this contour (Figure 5.5). Use

Field Recognition and Delineation   185

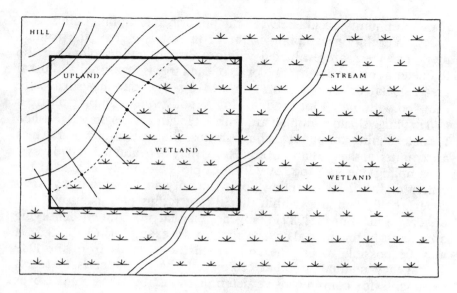

— Project Area
— Transect Line
• • • Wetland–Upland Boundary Point
- - - - Designated Wetland Boundary

Figure 5.5: The wetland-nonwetland boundary is established between transects by following the contour that represents the boundary points determined along each transect.

flagging tape to mark the boundary for survey crews and others to find later.

Comprehensive determination method. The major difference between this method and the routine methods is in the emphasis on vegetational analysis. In general, the steps follow the transect approach of the routine method, except that: 1) sampling is supposed to be done at fixed intervals based on the length of the transect rather than within recognizable plant communities along the transect; and 2) vegetation sampling is more rigorous. If the transect is less than 1000 feet, the sampling interval is 100 feet; if length is 1000-5000 feet, interval is 100 to 500 feet; if length is 5000-10,000 feet, interval is 500 to 1000 feet; if transect length is greater, then sampling interval is 1000 feet. In practice, however, investigators may sample at shorter intervals when a distinct plant community occurs between the prescribed sampling points.

Vegetation sampling is more quantitative than the routine methods. Trees are evaluated within a 30-foot radius plot, but the diameter at breast height (dbh or d) of each tree is measured and basal area calculated: $BA = \pi \times d^2/4$, or $3.1416 \times d^2/4$. Saplings/shrubs are sampled within a 10-foot radius plot. Woody vines climbing all trees and saplings/shrubs sampled are tallied. Height classes are established for determining dominant saplings/shrubs, while stems of woody vines are counted at the ground surface. Herbs may be evaluated in two ways: 1) sampling one 3.28 foot by 3.28 foot plot commonly located at the center of the fixed sampling point (identifying dominants by areal cover); or 2) sampling within the 30-foot radius plot using multiple quadrats following the 1989 interagency manual procedures (identifying dominants by mean areal cover estimates; see examples in Table 5.3). It may be possible to use the comprehensive methods from the 1989 interagency manual or other ecologically based methods to assess dominants for comprehensive determinations, but the recent Corps headquarters guidance is not totally clear on this (consult the local Corps district).

Atypical situations. These situations involve conditions where positive indicators of one of the parameters are not present due to recent human activities or natural events. Human actions include unauthorized activities (dredging, filling, removal of vegetation, or hydrologic alteration) and artificial wetlands purposefully or unintentionally created (e.g., impoundments, irrigation projects, and stream channel realignments). It must be emphasized that this section of the manual is not intended to extend federal jurisdiction to manmade wetlands that are exempted from CWA requirements by Corps regulations or policy. Natural disturbances may create new wetlands or alter existing wetlands. These actions include beaver, fire, avalanches, volcanic eruptions, and changing river courses. In considering the above, the 1987 Corps Manual stresses that the approximate date of the alteration must be established to determine whether the event occurred prior to implementing Section 404 of the CWA.

To evaluate atypical situations, the previous condition of each altered parameter (vegetation, soils, or hydrology) needs to be established. This requires identifying the nature of the alteration, reviewing existing site information (aerial photos, maps, soil surveys, previous inspection reports, etc.) and, when necessary, conducting an onsite inspection or assessing conditions on a similar undisturbed site, usually adjacent to the site in question. Documentation of positive indicators is still required.

For man-induced wetlands, indicators of hydric soils are usually absent and therefore not expected to occur, so the wetland determination may be based on the presence of hydrophytic vegetation and wetland hydrology (provided there is documented evidence that the hydrology changed so recently that soils could not develop hydric properties). The manual requires that current Corps regulations and policy be considered. If the wetland is exempt, no further action is required. If the hydrology that maintains hydrophytic vegetation is strictly man-induced and could be terminated (e.g., irrigation), the area is not a regulated wetland.

Problem areas. The manual included several types of wetlands or conditions where positive indicators of one or more parameters may not exist at certain times of the year. These are considered "problem areas." A list of representative examples is provided in the manual and includes: 1) wetlands on drumlins, 2) seasonal wetlands (in arid and semiarid regions), 3) prairie potholes, and 4) vegetated flats. These and other examples are discussed in Chapter 6 of this book.

Although presented as a list of examples, many users of the manual tend to view those listed as the only bona fide problem areas. Recent Corps headquarters guidance has affirmed that the list is a list of examples and that the situation may be applied to other wetlands, but the guidance unfortunately does not present other examples. The procedures for wetland determinations in these situations, however, still emphasize that wetland indicators of all three parameters must normally be present during part of the growing season to be a wetland. Otherwise, the area is a nonwetland.

5.4 MAJOR WETLAND DATA SOURCES

Many types of information are available that may be used to assist in making wetland determinations. Existing maps, including U.S. Geological Survey topographic maps, U.S. Fish and Wildlife Service National Wetlands Inventory maps, U.S. Soil Conservation Service soil survey maps, and state or local wetland maps, can provide useful background information on site characteristics. Hydrologic data from stream or tidal gages or groundwater wells may be used to determine whether an area is sufficiently wet to be considered a wetland. Aerial photographs and, to a lesser extent, satellite imagery show features that aid in wetland recognition for site-specific projects. These sources and other materials should be a part of any office's reference collection and should be reviewed by wetland delineators prior to conducting onsite investigations.

Available information varies across the country, so contact federal, state, and local government agencies to learn what they have. The following discussion will focus on primary data sources that should be most useful in locating wetlands.

5.4.1 Topographic Maps

The U.S. Geological Survey has produced a series of large-scale (typically 1:24,000 for the coterminous U.S.) maps showing topographic, hydrologic, and planimetric (e.g., roads, houses, and cities) features. These maps show landform contours (topography), various water bodies (estuaries, lakes, rivers, streams, ponds, and intermittent ponds and streams), and many wetlands (identified by a marsh or swamp symbol). Besides the wetlands shown, likely places for the occurrence of other wetlands (e.g., depressions, flats along slopes, and lowlands along rivers) can be located by considering the local topography on these maps. The wetlands actually depicted are in most cases conservatively mapped, since many wetlands are not shown. In general, those mapped tend to be the wetter, more conspicuous wetlands. It must be emphasized that the purpose of these maps was not to identify wetlands, but for topographical, hydrological, and cultural features.

5.4.2 National Wetlands Inventory Maps

The U.S. Fish and Wildlife Service initiated a wetlands inventory in the mid-1970s. This effort is producing a series of large-scale maps showing U.S. wetlands, and is preparing acreage summaries and various reports on the status of wetlands and recent trends. The National Wetlands Inventory (NWI) maps show the location, size, and type of wetlands within defined geographical areas for the entire country. These maps are scheduled to be completed for the lower 48 states by the end of 1998. Available maps can be ordered by calling 1-800-USA-MAPS.

NWI maps are useful tools for first identifying the likely presence of wetland in a given area. Wetlands are categorized according to the Service's official wetland classification system: "Classification of Wetlands and Deepwater Habitats of the United States" (Cowardin et al., 1979). Figure 5.6 shows the general hierarchy for classification: system, subsystem, class, and subclass. Five systems are recognized: marine (ocean), estuarine (brackish tidal waters and associated wetlands), riverine (rivers and streams), lacustrine (lakes and reservoirs), and palustrine (inland marshes, bogs, and swamps). Major wetland classes include emergent wetland (marshes, fens, and meadows), scrub-shrub wetland (woody plants less than 20 feet tall), forested wetland, aquatic bed, and unconsolidated shore (mudflats and beaches). These and other descriptors permit classification of wetlands by hydrology, water chemistry, vegetation, soil or substrate, alterations by humans or beaver, and other parameters.

The NWI maps use alpha-numeric codes to represent the classification system. A sample of a portion of an NWI map is presented in Figure 5.7. In this example, various codes can be seen. The code PFO1E represents a nontidal wooded swamp: P - Palustrine (System), FO - Forested Wetland (Class), 1 - Broad-leaved Deciduous (Subclass), and E - Seasonally Flooded/Saturated (Water Regime modifier). This wetland could be a red maple swamp. Some other types of forested wetland are designated as PFO4E or PFO2E. The former may include evergreen swamps dominated by northern white cedar, black spruce, white pine, hemlock, and balsam fir, while the latter code represents a tamarack or larch swamp. Mixed forested wetlands of evergreen and deciduous trees are shown as PFO4/1E and PFO1/4E. Shrub swamps include PSS1E (deciduous shrubs like alders and willows), PSS3Ba (leatherleaf bog: P - Palustrine, SS - Scrub-shrub Wetland, 3 - Broad-leaved Evergreen, B - Saturated, and a - Acid), and PSS1/3E (mixed deciduous and evergreen shrubs). Ponds are designated by the codes PUBH, PUBHx, and PUBFx: Palustrine, Unconsolidated Bottom, Permanently or Intermittently Exposed, and x - excavated (Special modifier). Freshwater marshes (palustrine emergent wetlands) are represented by PEM1F and PEM1E. Lakes are indicated by the code L1UBH (Lacustrine Limnetic Unconsolidated Bottom

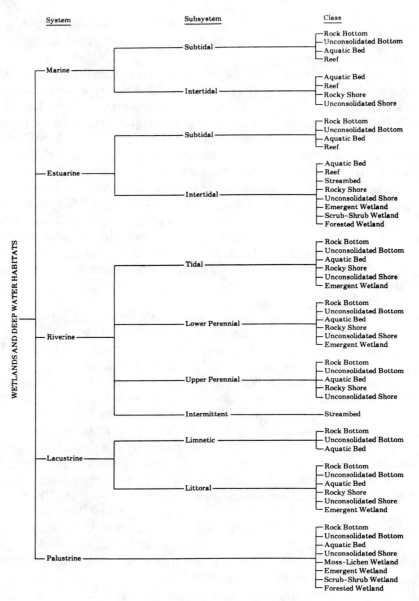

Figure 5.6: Classification hierarchy for wetlands and deepwater habitats, according to Cowardin et al. (1979). Most wetlands fall within the estuarine and palustrine systems.

Permanently Flooded). Although not shown in this example, estuarine wetlands begin with the letter E (e.g., E2EM1P for an intertidal salt marsh or E2SS4N or E2FO4N for a mangrove shrub swamp or forest), marine wetlands start with the letter M (e.g., M2USN for an intertidal beach or flat), and riverine wetlands begin with an R (e.g., R2ABH for a riverine aquatic bed). A legend at the bottom of each map explains the code symbology.

The maps are compiled mainly from interpretation of medium to high-altitude aerial photographs and supplemented by field inspections for verification of photo-signature patterns. Consequently, they have certain limitations, including the following. Depending on the scale of aerial photos used and the type of wetland, the minimum mapping unit (mmu) varies from about 0.25 acres to about 5.0 acres in size. The smaller mmu is associated with conspicuous wetlands (e.g., prairie potholes and ponds) and 1:40,000 color infrared photography, while the larger mmu is associated with many forested wetlands and 1:80,000 black and white photography. Where 1:58,000 color infrared photography was used, the mmu is about an acre for most forested wetlands. Due to the mmu, small forested wetlands and linear wetlands occurring as a thin fringing band along watercourses are usually not mapped. Most farmed wetlands are also not designated; exceptions include cranberry bogs, farmed potholes, and some diked former tidelands now in cultivation. Since the maps were based on one set of aerial photographs, they reflect wetness conditions at that time. This means that if the conditions were drier than average, some wetlands will not be identified. In general, the maps are conservative in mapping the drier wetlands for they are among the most difficult to photointerpret as well as to identify on the ground (Tiner, 1990). Wetland boundaries shown on the maps are in most cases general and need further refinement on the ground, except perhaps where topographic changes are abrupt or soil wetness conditions change dramatically for other reasons. The date of the aerial photos used establishes the effective date of the inventory. In all cases, the maps cannot and do not reflect changes that have occurred since the photo date. It is important for the map user to be aware of these limitations and many of these are simply limitations inherent to any map. A site inspection should always be conducted in order to accurately identify and delineate wetlands for specific projects.

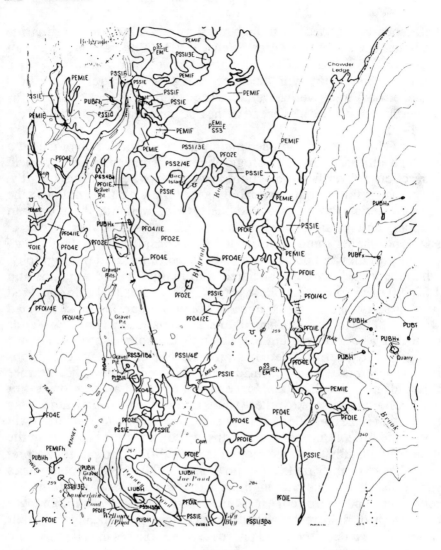

Figure 5.7: Example of a portion of a National Wetlands Inventory map (from Tiner, 1991c). Alpha-numeric codes signify different wetlands and deepwater habitats: forested wetlands (PFO4E, PFO2E, PFO1/4E, etc.), marshes (PEM1F, PEM1E, etc.), shrub swamps (PSS1E, PSS2/4E, PSS4Ba, etc.), ponds (PUBHx, PUBH), and lakes (L1UBH).

5.4.3 Soil Surveys and Hydric Soil Lists

The U.S. Soil Conservation Service (SCS) conducts surveys of the soils in various geographic areas. These soil surveys are often prepared on a county basis, especially in the eastern United States. Soil scientists examine soils associated with different landforms within a specific geographic area, describe the various soil types, and correlate changes in soil types with changing landforms and vegetation patterns. To prepare the soil maps, soil scientists combine field observations with aerial photo interpretation of landforms and plant communities to delineate different soil map units. The results of a soil survey are published in a report which contains invaluable information about the county and its soils and includes a set of large-scale (often 1:20,000) photo-based maps showing the location and configuration of individual soil map units. These units typically represent the predominant soil type (series, complexes, or associations) or land type (e.g., alluvial land, marsh, swamp, or made-land). Soil map units are usually designated by a letter code (e.g., Ac or HaB) or by a number referenced on a legend preceding the maps. An example of a soil survey map is shown in Figure 5.8.

Lists of soil map units containing hydric soils are available from SCS county offices. These lists reference both map units dominated by hydric soil series and those dominated by nonhydric soils but having possible hydric inclusions. The latter are usually by name and landscape position (e.g., depressions or sloughs). These lists are most helpful in identifying which soil map units are most likely to contain hydric soils and wetlands. Field work, however, is required to verify the presence and limits of hydric soils based largely on soil morphology and landscape position.

When the hydric soil list is used in conjunction with the county soil survey, the soil survey maps can help identify the likely presence of a wetland in a specific area. Yet as with any map, there are limitations. In general, the minimum map unit ranges from 1.5 to 10 acres depending on landscape diversity and survey objectives (Tiner, 1991a). Soil map units have inclusions of other soils, so that a large map unit may contain substantial acreage of other soils. Some wet spots within drier soils may be indicated by a wet symbol. Many soil surveys do not differentiate between drained and undrained hydric soils, making it virtually impossible to separate current wetlands from historic wetlands. Some of the more modern surveys are attempting to separate flooded phases and drained phases.

There is also a limitation of the mapping based on how soils are classified taxonomically. Soil series were not defined with the intention of separating hydric or wetland soils from nonhydric soils. The concept of hydric soil evolved well after the designation of most soil series. As a result, some members of hydric soil series do not possess hydric properties and are not coincident with wetlands, while other members of the same series do have hydric properties. These "facultative" hydric soil series may or may not be hydric depending largely on landscape position. Those in depressions are typically hydric, while those slightly upslope (better drained) are not hydric. The soil maps characteristically do not distinguish between them.

Although these limitations are widely understood by soil scientists, most nonspecialists have failed to recognize these points and have simply used the acreage of hydric soil map units to signify the extent of wetlands potentially subject to government regulation. This has led to grossly exaggerated reports on acreage subject to federal and state wetland regulations. Soil survey maps like NWI maps are a good source of background information to help identify potential wetland areas, but site inspections are still essential to accurate identification of wetlands and their boundaries.

5.4.4 Aerial Photographs

Interpretation of aerial photographs can provide a good representation of the different plant communities in a given project area and reveal signs of wetness (e.g., flooding, surface saturation, natural drainage patterns, and water-stressed crops) that can help detect wetlands (Figure 5.9). To use this data source most effectively, one should have some training in photointerpretation. Leaf-off photography (i.e., when the leaves are not on deciduous trees) is optimal for detecting forested wetlands because soil wetness is not obscured by the tree canopy. Unfortunately, this does not improve the detection of evergreen forested wetlands which retain their leaves. The drier wetlands are also difficult to photointerpret (Tiner, 1990). Aerial photos are particularly valuable for detecting changes in land use over time, some of which may have affected local hydrology and caused a

Figure 5.8: Example of a portion of a soil survey map (Tiner, 1991c). Alpha codes represent soil mapping units. Hydric units include TO (Togus peat), VA (Vassalboro peat), Bo (Biddeford mucky peat), and ScA (Scantic silt loam).

Figure 5.9: Aerial photograph showing coastal marshes behind a barrier beach (lower left), and palustrine wetlands adjacent to large lake, along rivers and streams, and in depressional areas.

loss or gain in wetlands. Sources of aerial photography include the U.S. Geological Survey, U.S. Agricultural Stabilization and Conservation Service, state transportation departments, local governments, and private aerial survey firms.

5.4.5 Recorded Hydrologic Data

Recorded data on the distribution and flow of water is available for many rivers and tidal waters, but is limited for wetlands. From existing stream and tidal gage stations, data on flood stage levels and seasonal flows can be obtained which might aid in estimating the flood duration and frequency of contiguous wetlands. The U.S. Geological Survey publishes annual reports on water resources data for each state. The reports present gage data for surface water discharge, surface water quality, and groundwater levels. Contact state offices of the U.S. Geological Survey for this information.

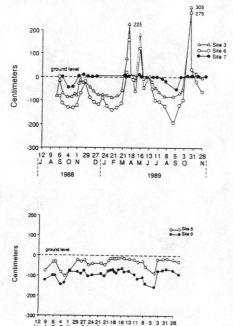

Figure 5.10: Observed fluctuations in the watertable and surface water in floodplain soils (from Veneman and Tiner, 1990). Sites 3, 6, 7, and 8 are wetlands, and site 9 is nonwetland.

Installation and periodic observation of shallow ground water or water table wells in wetlands can provide specific data on the fluctuations of the water table over time (Figure 5.10). The USGS and various state and local agencies may be collecting water table data, but usually their sites are not located in wetlands. Investigations by wetland research scientists, soil scientists, and consulting scientists for wetland permit applicants may provide hydrologic data from wells in wetlands. Such data can be interpreted in various ways. Long-term monitoring studies covering the full range of local weather conditions (dry years, wet years, and "normal" rainfall years) will provide the most useful results. Short-term monitoring (e.g., 1- or 2-year studies) can also be useful especially if they do not encompass or succeed an extended abnormally wet or dry period and sufficient baseline information is available from other ground water wells to establish that water table conditions are normal in the locale.

6

Problem Wetlands for Delineation

Ralph W. Tiner

INTRODUCTION

Wetlands subject to regulation under § 404 of the Clean Water Act (CWA) typically must meet a multi-parameter test under the 1987 Corps Manual (U.S. Army Corps of Engineers, 1987); they must have positive indicators of three parameters (vegetation, soils, and hydrology) that are associated with wetlands, with a few noted exceptions. For an area to be designated as a potentially regulated wetland, it must possess at the time of inspection or be expected to normally have hydrophytic vegetation, hydric soils, and certain signs of wetland hydrology (see discussion in Chapter 5). Since wetlands characteristically are subjected to variable hydrologic regimes, and since most wetlands are not permanently wet, the wetland hydrology parameter is usually the most difficult to verify during a single site inspection. Since water is not always present in wetlands, surrogates must be used to confirm wetland hydrology in the three-parameter test. The list of positive indicators of wetland hydrology, therefore, becomes critically important for ensuring accurate wetland identification and delineation. While finding acceptable wetland hydrology indicators is the most common problem in wetland delineation following the three-parameter approach, there are also wetlands that fail to have positive indicators of the other parameters. For example, some wetlands are represented by plant communities that are dominated by FAC- and/or FACU plant species or have soils that fail to display typical hydric soil properties, despite being wet frequently and long enough to meet the Corps wetland hydrology parameter. In addition, certain situations may be encountered in the field that complicate wetland delineations.

Collectively, the wetlands and situations referred to above may be called "problem wetlands." These wetlands are not necessarily less valuable or less a wetland than other types, but simply present problems for identification when applying the Corps' three-parameter test.

The purpose of this chapter is to make readers aware of these types of wetlands and to provide some insight into how such areas may be recognized and delineated. The 1987 Corps Manual lists only five types and two situations as examples: wetlands on drumlins, seasonal wetlands, prairie potholes, vegetated flats, man-induced wetlands, unauthorized activities (illegal deposition of fill material), and natural events (e.g., beaver dams, fires, avalanches, volcanic activity, mudslides, and changing river courses). These and other problem wetlands are discussed in this chapter. Since these areas may be contentious from the regulatory perspective, it is emphasized that the chapter is intended solely for discussion purposes and that the ultimate decision on whether such areas should be regulated rests with individual Corps districts and EPA regions. Contact the appropriate Corps district and EPA region for specific direction on how to address these problem areas.

6.1 PROBLEMATIC WETLAND PLANT COMMUNITIES

Certain wetlands fail to meet the 1987 Corps Manual's basic rule for hydrophytic vegetation which states that hydrophytic vegetation is present when more than 50 percent of the dominant species have a wetland indicator status of OBL, FACW, and/or FAC, excluding FAC-. These wetlands may be dominated by FAC- and/or FACU species either year-round (throughout the growing season), seasonally (usually during late Summer or the driest season), or during extremely dry periods, such as prolonged droughts. Some vegetated wetlands may lack plant cover at certain times which also makes verification of hydrophytic vegetation difficult. All of these plant communities can be recognized by considering landscape position (e.g., depressions, sloughs, floodplains, drainageways, broad flat plains, and slopes below springs or groundwater seepage sites), hydric soils, and other signs of wetland hydrology when necessary.

While communities dominated by FACU-dominated communities characterize many nonwetlands, they are sometimes found in wetlands and may actually typify certain types, such as hemlock swamps (Plate V; Federal Interagency Committee, 1989). Hydric soils and signs of wetland hydrology may be used to distinguish the latter from the former. Wetlands dominated by FAC- and/or FACU plants may exist in drier situations along floodplains or along wetland boundaries. Another type of FACU-dominated wetland is not restricted to drier conditions. Certain evergreen tree species that dominate wetlands experiencing prolonged seasonal saturation and/or inundation are more characteristic of nonwetlands and have been assigned an indicator status of FACU. The individual species growing in wetlands

Figure 6.1: Pitch pine (*Pinus rigida*) can grow under a variety of soil moisture conditions, ranging from seasonally inundated wetlands (shown) to xeric sandy uplands. In the Pine Barrens of New Jersey, the species dominates certain forested wetlands locally called pitch pine lowlands.

are clearly adapted in some way for life in periodically anaerobic soils and should be considered hydrophytic plants (Figure 6.1). The species level in plant classification or taxonomy does not appear the best for identifying these hydrophytes (Tiner, 1991a). Examples of these plants include the following: red spruce (*Picea rubens*), white spruce (*P. glauca*), eastern white pine (*Pinus strobus*), pitch pine (*P. rigida*), jack pine (*P. banksiana*), and eastern hemlock (*Tsuga canadensis*) in the Northeast and Midwest, longleaf pine (*Pinus palustris*) in the Southeast, and Engelmann spruce (*Picea engelmannii*), Sitka spruce (*P. sitchensis*), ponderosa pine (*Pinus ponderosa*), lodgepole pine (*P. contorta*), western hemlock (*Tsuga heterophylla*), Pacific silver fir (*Abies amabilis*), white fir (*A. concolor*), and subalpine fir (*A. lasiocarpa*) in the West and Alaska (Federal Interagency Committee, 1989). Other examples of FAC- and FACU species common or dominant in wetlands in the northeastern U.S. are shown in Table 5.1 At some locales these plants may actually occur more frequently in wetlands than in nonwetlands. For example, in southeastern Massachusetts, eastern hemlock typically grows in wetlands, so its predominance usually indicates wetlands.

The 1987 Corps Manual acknowledges the existence of two types of wetlands whose vegetation varies seasonally and refers to them as seasonal wetlands and prairie potholes. In the former type, perennial OBL and FACW species are normally dominant during the wetter part of the growing season, while UPL annuals may dominate during the drier season. Prairie potholes are marshes and wet meadows that have formed in glacially-formed depressions in the upper Midwest (e.g., the Dakotas and western Minnesota; Figure 6.2). Prolonged droughts typify the Pothole region as they do other semiarid and arid regions where evapotranspiration demands exceed annual precipitation, thus creating a water budget deficit. During these droughts, perennial upland plants (UPL) can colonize and eventually dominate certain wetlands. In agricultural areas such as the Prairies of the Dakotas, for example, many wetlands are tilled and planted with crops during dry periods (Kantrud et al., 1989). Wetland basins that are not cultivated may be colonized by weedy FACU and UPL annuals. Consequently, the vegetation is significantly disturbed and not useful in verifying the hydrophytic vegetation parameter. This situation applies equally to playa wetlands in the Southwest and to groundwater wetlands on the Cimmaron Terrace of Oklahoma and Kansas (Taylor et al., 1984).

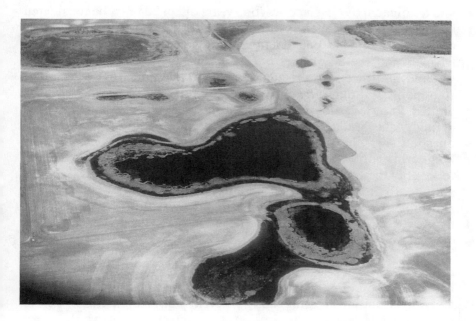

Figure 6.2: Prairie pothole wetlands are glacially-formed basins in the upper Midwest. The semiarid climate with prolonged droughts leads to enormous cyclical changes in vegetation patterns.

Hydric soil properties reflect the long-term hydrology and are the most useful indicators of wetlands under these circumstances, provided the area has not been effectively drained.

Wetlands with vegetation that varies greatly over time due to wide fluctuations in annual water availability typical of the regional climate may be referred to as "cyclical wetlands" (56 Fed. Reg. 40,446 (Aug. 14, 1991)). Prairie potholes, playas, and California vernal pools are examples. During wet periods of the natural hydrologic cycle, these wetlands possess positive indicators of hydrophytic vegetation, hydric soils, and wetland hydrology. Yet, during dry periods, the only indicator of their periodic wetness may be the hydric soils. These conditions pose an interesting question for regulatory agencies and the scientific community alike. What frequency and duration of inundation and/or soil saturation is sufficient to separate wetlands from nonwetlands in these situations? Should an area that is extremely wet and dominated by OBL species for 5 years out of 20 years be considered a wetland? If so, what about an similar area that is wet for 5 years out of 30 years and so on? Eventually, a point is reached where the frequency of the event may not be sufficient to consider the area a wetland, both from the regulatory perspective and from a technical standpoint, although the endpoints may not be the same. Presently, limited knowledge of wetland hydrology, especially in regards to defining the wetland-nonwetland boundary, cannot answer this question. Yet, the surrogate of wetland hydrology that best reflects long-term hydrology, that is, hydric soils, may be most useful in helping to identify these cyclical wetlands and their limits (Tiner, 1993b; Tiner and Veneman, 1987). In the absence of human alteration of local hydrology by drainage ditches or similar measures, the presence of hydric soil indicators should be sufficient to verify the existence of most cyclical wetlands.

California's vernal pools are cyclical wetlands whose vegetation changes markedly during each year in response to fluctuating hydrologic conditions. The hydrology varies widely seasonally and annually, resulting in significantly different plant communities. Vernal pools achieve maximum size during unusually wet years when individual pools merge to form large pool complexes (Zedler, 1987). Pools range in size from 10 acres to 10 feet wide. Impermeable layers of clay or iron-silica cement underlie the pools, forming perched water tables. Winter rains inundate the pools and increased evapotranspiration in Spring removes the surface water. The typical cycle of vernal pool development may be represented by four phases: 1) wetting phase, 2) aquatic phase, 3) drying phase, and 4) drought phase. Hydrophytic vegetation is evident during the first two phases, but upland species may invade during the last two phases. Vernal pools may be identified by one or more of the following features: a cemented hardpan, a confining clay layer in the subsoil, algal encrustations on the soil surface, other evidence of algal remains (periphyton), pit and mound relief, and ridges and swales (Figure 6.3).

Alaska's permafrost wetlands may be considered cyclical wetlands dependent on the permafrost barrier near the surface. In the subarctic, the permafrost is greatly affected by recurring wildfires with a cycle ranging from 50 to 200 years (Ping et al. 1990). Severe fires are known to destroy the insulating and confining permafrost layer that is principally responsible for creating wetland hydrology. Loss of the

Problem Wetlands for Wetland Delineation 205

Figure 6.3: Vernal pool complex in California (Jepson Prairie) during the drying phase.

shading from the spruce forest and loss of the insulating moss layer allow the soils to warm, thereby destroying the permafrost. This can change poorly drained soils to well drained soils because the confining permafrost layer is no longer present to perch and retain surface water. Over time, however, with vegetation succession, the permafrost layer again develops, restoring wetland hydrologic conditions and, presumably, wetland functions. Mechanical alteration of the vegetation and soils can have similar effects. The thawed permafrost soil retains its low chroma mottles and still appears to have hydric properties, but the area no longer supports hydrophytic vegetation. Exactly how such areas should be treated from the CWA regulatory perspective is an interesting question for the Corps and EPA to answer.

Some of the wettest wetlands lack vegetation during certain times of the year, typically in Winter. These wetlands, which include regularly flooded fresh tidal marshes and exposed river bars, typically are colonized by OBL species in Spring and Summer, yet appear as nonvegetated wetlands at other times. The 1987 Corps Manual refers to these areas as "vegetated flats." Inspections during the growing season would detect and confirm the presence of hydrophytic vegetation, while observations at other times would not (unless one looks for overwintering perennial plant parts such as underground tubers). Fortunately, because of their obvious wetness at the latter times, there is little problem recognizing these areas as wetlands, or at least as waters of the U.S. for regulatory purposes.

6.2 PROBLEMATIC HYDRIC SOILS

While soils of most wetlands tend to exhibit diagnostic properties reflecting hydric soil conditions (e.g., accumulation of organic matter, gleyed matrices, and low chroma matrices with high chroma mottles), a fair number of hydric soils are not easily recognized and even pose problems to soil scientists. The 1987 Corps Manual mentions hydric soils derived from red parent materials such as found in the Red River valley, and sandy soils (e.g., accreting sandbars) as soils in which typical hydric soil properties are not to be expected. The 1989 Interagency Wetland Delineation Manual and recent Corps headquarters guidance list others: Entisols (floodplain and sandy soils), Spodosols (evergreen forest soils), and Mollisols (prairie and steppe soils).[1] In addition, hydric Vertisols with high clay content and shrink and swell potential may also be difficult to separate from nonhydric forms. Soils derived from gray parent material may create similar problems. Moreover, since the concept of hydric soils is still evolving, there is some debate among soil scientists over the appropriate indicators to verify the hydric soil criteria. Although several versions of the national list of hydric soils have been published, the USDA Soil Conservation Service (SCS) has not published an official list of hydric soil indicators (i.e., soil properties reflecting seasonal high water tables associated with wetlands). This list is in development and until it is published, there will continue to be much debate over whether certain properties meet the hydric soil criterion and indicate wetland hydrology.

---

[1] See Federal Interagency Committee, 1989; and U.S. Army Corps of Engineers, Memorandum from Major General Arthur E. Williams on clarification and interpretation of the 1987 Corps Manual (Mar. 6, 1992).

Be sure to contact the SCS state soil scientist for the latest information on hydric soils.

In general, landscape position or landform reveals much information on the likelihood of finding hydric soils and wetlands. The existence of excellent vegetation indicators of wetlands, especially OBL species, in suitable landscapes further support a wetland determination, despite the presence of soils that do not possess typical hydric soil properties. Observing signs of wetland hydrology (e.g., watermarks, silt deposits, and water-stained leaves) yield even more support. Unless one is trained in soil morphology and classification, he or she will usually be unable to detect the subtle properties separating hydric from nonhydric soils in these confounding situations. If questions arise in the field, it may be advisable to examine the site during the wettest time of the year. The following discussion is offered to provide an overview of the more frequently encountered problematic soils. Remember, however, that the majority of hydric soils are readily distinguishable from nonhydric soils.

Hydric entisols develop mostly along floodplains, glacial outwash plains, and tidal embayments. Both sandy and finer textured floodplain soils are included in this grouping, while the hydric entisols of outwash plains and tidal flats are sandy soils. Many of the finer textured hydric entisols possess typical hydric soil properties. Sandy soils do not possess such properties since significant accumulations of organic matter usually do not occur and silt and clay materials which best reflect gleization are virtually absent. These hydric entisols (textures of loamy fine sand and coarser sand within 20 inches of the surface) may be recognized by a matrix chroma of 3 or less provided the hue is between 10YR and 10Y and distinct or prominent mottles are present (Soil Survey Staff, 1975, 1990). Newly formed soils of river bars may not have these colors. Data on local hydrology must be consulted for verification of the hydric nature of these soils. Reference to local soil surveys may also be helpful. Hydric sandy soils associated with tidal marshes can be easily recognized by the obvious nature of this wetland type with OBL hydrophytes and the smell of rotten eggs (hydrogen sulfide).

Spodosols are common in northern temperate and boreal regions of the U.S. and along the Coastal Plain from New Jersey south (soil surface). Evergreen forests of hemlocks, spruces, and pines dominate these regions, but larch and oaks are also associated with spodosol formation (Soil Survey Staff, 1975; Buol et al., 1980). The breakdown of the organic leaf litter from these forests creates organic acids that leach downward through the soil stripping clean the sand grains in the subsoil layer just below the soil surface. This leached layer called the E-horizon is often gray-colored, since it is usually free of organic matter, aluminum, and iron oxides under well drained conditions. The latter materials are deposited lower in the subsoil in a horizon called the spodic horizon. This soil-forming process is termed podzolization and the soils were formerly called podsols. The dominant gray colors immediately below the surface layer may suggest hydric soil properties, but these colors in nonhydric spodosols are not the product of reduction from wetness. Hydric spodosols may have one or more of the following properties: a thin surface layer of peat or muck (not leaf litter); a thick, black-colored sandy surface layer; streaks of organic matter in the E-horizon or masses of sand grains coated with organic material giving the horizon a blotchy appearance; high chroma mottles and/or oxidized rhizospheres within the E-horizon, and usually within 12 inches of the soil surface; iron concretions within the E-horizon or spodic horizon; a partly or wholly cemented spodic horizon (orstein), usually within 18 inches of the soil surface; and mottling within the spodic horizon.

In arid and semiarid interior regions of the country, precipitation is not sufficient to support vast forests, so grasslands (native prairies) and grasslands mixed with shrubby vegetation (steppes) predominate. Here forests tend to be limited to higher elevations in the mountains and to floodplains (riparian corridors). Grasslands characterize the middle of the country from eastern Illinois to Montana, south to Texas. Native grassland soils called mollisols have formed in these areas. Mollisols typically have thick, dark-colored surface layers due to the accumulation of organic matter from the breakdown of extensive root systems associated with prairie and steppe grasses (Soil Survey Staff, 1975; Buol et al., 1980). These dark colors may suggest hydric soils, but are attributed to root decomposition and mixing of this organic matter with the soil by burrowing animals such as earthworms, ants, moles, and prairie dogs. In typical wetland landforms (e.g., depressions and sloughs), hydric mollisols develop under conditions of prolonged anaerobiosis and reduction. Hydric mollisols may be differentiated from nonhydric forms by one or more of the following properties: a thin layer (at least 0.25 inch thick) of peat or muck; high chroma mottles (including oxidized rhizospheres) within 12 inches of the soil surface; iron and/or manganese concretions within the surface layer; low chroma mottles or matrix immediately below the surface layer, and the crushed color is chroma 2 or less; and the remains of aquatic invertebrates within 12 inches of the surface in pothole-like depressions (Tiner, 1993b).

In the Southwest, from Texas to southern California, soils with high shrink and swell potential due to high content of montmorillonite clays have developed. These soils called vertisols create a gilgai microtopography of low mounds and shallow depressions. Ponding may occur in the depressions during rainy periods. Vertisols are well mixed, due to the high clay content which swells upon wetting and shrinks upon drying. Large cracks form when the soil shrinks during the dry season, soil falls into the cracks, and when the soil is rewetted during the rainy season, the soil expands causing a churning process. Consequently, the surface layer is often very thick and dark-colored due this churning of organic-enriched soil near the surface. These colors may suggest hydric soil properties, but they are not produced by wetland hydrologic conditions; the high clay content and regional climate create these properties in both well drained and poorly drained soils. Hydric vertisols may be distinguished from their nonhydric counterparts by one or more of the following properties: iron and/or manganese nodules or concretions (sometimes resembling buckshot pellets) in a thick, blackish colored surface layer (A-horizon); gray-colored (low chroma) mottles within the A-horizon; and dominant low chroma colors immediately below the A-horizon and within 20 inches of the soil surface. In some cases it might be necessary to make direct observations of hydrology during the wet season to assess site wetness.

Soils derived from red parent materials (e.g., strongly weathered clays and Triassic sandstones and shales) occur in many places in the country. These soils are common in the Southeast (e.g., red clays - Ultisols), but also are frequent in the Midwest, Southwest, and West (Alfisols), in the tropics (Oxisols), and in glaciated landscapes where red sandstone and shale formations are exposed and eroding. The red colors are attributed to the dominance of the iron mineral - hematite. These colors are redder than 10YR (on the Munsell charts) and obscure low chroma colors that normally develop under anaerobic, reducing conditions. Some hydric soils with red colors may have low chroma mottles present within 1.5 feet of the soil surface, but many do not. Vegetation, other signs of wetland hydrology, and landform position may provide the best clues for recognizing these areas as wetlands. Wet season field inspections may be required to verify the hydrology parameter.

6.3 HYDROLOGICALLY DIFFICULT WETLANDS

The wetness threshold of the wetland hydrology parameter of the 1987 Corps Manual is derived from studies of southeastern bottomland hardwood swamps (see Section 5.2.4). As a result, there are many types of wetlands that are not inundated or saturated to the surface for 12.5% of the growing season. While the manual acknowledges that some wetlands are only wet for 5.0-12.5% of the growing season, specific guidance is lacking, except for providing a list of positive indicators of wetland hydrology.

The following wetland types may be difficult to verify using guidance in the 1987 Corps Manual: seasonally saturated pine flatwoods along the Coastal Plain in the Southeast; seasonally saturated wet meadows; temporarily flooded floodplain wetlands, including riparian wetlands along western rivers; and other temporarily flooded and/or seasonally saturated wooded swamps. These areas generally fail to be wet enough to satisfy the wetland hydrology requirement in the manual. Most of these wetlands are either wet during the "non-growing season" or not inundated or saturated to the surface long enough. Most riparian habitats in the arid region of the country fail to meet the hydrology parameter because they are not flooded frequently enough for sufficient duration, given regional rainfall patterns. Yet when rainfall is significant, these areas provide many functions that wetlands normally do. While many of these areas are not wet enough to qualify as wetlands, they may be significant for water quality maintenance and may, therefore, be worthy of regulation under the Clean Water Act (CWA). They are, without question, critical habitat for many wildlife species. All the types listed above may be potentially regulated wetlands, so consult with the appropriate Corps district and EPA region for specifics, including how to identify them using the 1987 Corps Manual.

Hydrologically altered wetland systems are equally difficult to interpret. In these cases, the hydrology has been significantly changed by drainage ditches, tile drains, groundwater withdrawals, regulated riverflows, surface water diversions, and similar actions. This essentially negates the interpretive value of vegetation and soils for determining the presence of wetlands, forcing one to use other features to verify wetlands. The 1987 Corps Manual provides limited guidance for these difficult and spatially significant problem areas. The 1989 Federal Interagency Committee manual provides more guidance, but more is needed from the regulatory agencies. Installation and monitoring of shallow groundwater wells may be required to determine the current hydrology. A series of such wells and piezometers installed at different elevations within the area in question should provide the best results. Alternatively, assessment of the extent of water table drawdown and lateral influence of drainage structures may be achieved by applying the ellipse equation or similar hydrologic modelling equation (Federal Interagency Committee, 1989).

6.4 PROBLEMATIC FIELD CONDITIONS

In the field, an investigator may encounter one of several situations that complicate making a wetland determination. Complex landscapes with high interspersion of wetlands and nonwetlands, rocky areas, and significantly altered wetland systems may be particularly difficult to interpret. The 1987 Corps Manual offers guidance on how to handle some of these circumstances, but interpretations may vary among Corps districts and EPA regions, so consult with the appropriate offices for specific protocols to employ in the field.

Certain landforms such as ridge and swale topography, pit and mound relief, and linear drainageways, force the question: How small or narrow an area should be delineated for regulatory purposes? Also, in complex landscapes pockmarked with small wetlands and small uplands, making it practically impossible to separate wetlands from drylands, specifically pit and mound terrain, what ratio of wetlands to nonwetlands (drylands) should be used to regulate the entire mosaic parcel - 51:49, 66:34, 25:75, etc.? What procedures should be used to determine these ratios in the field - line intercept transects, point intercept transects, belt transects, plot samples, or other methods? These significant questions can only be answered by the regulatory agencies seeking to protect certain wetland functions and values.

Rocky areas (e.g., wetlands on till in glaciated landscapes and caprock limestone wetlands in South Florida) make it impossible to find positive indicators of hydric soils. This also applies to newly established wetlands where formerly nonhydric soils are now flooded sufficiently to create wetland conditions, yet the soils have not been anaerobic and reduced for a long enough period to form and display typical hydric soil properties. The time required to do so may be tens or hundreds of years, although low chroma mottles may start to form in mineral soils within a few years under more optimal conditions (e.g., much available organic matter). Prominent high and low chroma mottles were evident within three years in newly deposited floodplain soils along the Connecticut River in western Massachusetts (Veneman and Tiner, 1990). Although the 1987 Corps Manual recognizes man-induced wetlands, it is important to note that current Corps policies may exempt many such wetlands (especially those dependent on irrigated waters) from CWA regulations.

Disturbed areas with vegetation and soils removed and/or hydrology significantly altered are among the most difficult situations under which to perform a wetland delineation. Positive indicators of two parameters are generally required to verify wetlands in these cases (see Section 5.3.2 and the 1987 Corps Manual for specifics).

7

The Regulatory Framework

Mark S. Dennison
and
James F. Berry

INTRODUCTION

As earlier chapters have clearly demonstrated, scientific delineation of wetlands is a complex process dependent on several important variables. Even more difficult, however, has been development of a legal definition of wetlands that gives due consideration to the scientific understanding of wetlands. As is the case with most environmental laws, science and technology form the underlying basis for the regulatory framework. No good regulatory scheme can emerge for wetlands unless scientific study and technological advance are taken fully into account.

What follows in this chapter is a description of the current regulatory framework for protecting the Nation's wetlands. The process is complicated, cumbersome, costly, and fraught with interagency conflict. However, until a better system emerges, this is the one that you need to understand when addressing legal issues concerning wetlands.

7.1 CLEAN WATER ACT § 404 PROGRAM

7.1.1 Background

The principal federal statute regulating activities in wetland areas is the Federal Water Pollution Control Act, commonly referred to as the Clean Water Act (33 USC § 1251 *et seq.*) Section 404 of the Act is the primary federal authority for protecting the Nation's wetlands (33 USC § 1344). The Army Corps of Engineers is primarily charged with oversight of the § 404 program with guidance from the U.S. EPA.

The Corps was first given authority to regulate construction activities involving dredging, filling, or obstructing "navigable waters" under the Rivers and Harbors Act of 1899 (33 USC § 401 *et seq.*), however, this authority did not expressly extend to wetlands.[1] The Rivers and Harbors Act had very limited impact on the protection of wetlands because such areas are usually outside the mean high water mark (see *Borax Consolidated, Ltd. v. City of Los Angeles*, 296 U.S. 10 (1935), discussion of mean high water boundaries). For this reason, the Corps operated its permit program for almost seventy years while paying little attention to wetlands protection.

It was with passage of the National Environmental Policy Act of 1969 (NEPA; 42 USC § 4321 *et seq.*) that the Corps' power to consider environmental factors in its permitting process was strengthened. Under NEPA, all federal agencies are required to consider the possible environmental impact of their proposed actions and projects (42 USC § 4332). The first test of Corps environmental protection powers came in *Zabel v. Tabb*,[2] in which two developers attempted to build a mobile home park on eleven acres of wetlands in Boca Ciega Bay, Florida. The developers applied to the Corps for a permit to fill the proposed site. Even though the Corps concluded that the development would not impede navigation, it denied the permit because the proposed construction would have had a detrimental impact on marine life in the bay. The U.S. Court of Appeals for the Fifth Circuit upheld the Corps decision and concluded that the Corps could refuse dredge and fill permits on the basis of environmental considerations.

Congress passed the Federal Water Pollution Control Act Amendments of 1972, which created the present day § 404 program. The express language of § 404 is very limited on its face to requiring permits for the "discharge of dredged or fill material" into "navigable waters" (33 USC § 1344). Under Corps regulations, dredged material is defined

---

[1] The original purpose of the Rivers and Harbors Act regulation of dredge-and-fill activities was to protect and promote navigation. 33 USC § 403.

[2] 430 F.2d 199 (5th Cir.1970), cert. denied, 401 U.S. 910 (1971).

as "material that is excavated or dredged from waters of the United States" (33 CFR 323.2(c)). The regulations define fill material as "any material used for the primary purpose of replacing an aquatic area with dry land or of changing the bottom elevation of a waterbody" (33 CFR 323.2(e)). The draining of wetlands, which is a major source of wetland losses, is not expressly regulated or prohibited by § 404.

The "navigable waters" language was, however, later construed to mean all "waters of the United States" after the EPA and other public interest groups had sought an expanded definition of the term. This expanded definition was consistent with the definition of navigable waters found in the Clean Water Act (33 USC § 1362(7); see *Natural Resources Defense Council, Inc. v. Callaway*, 392 F. Supp. 685 (D.D.C. 1975)). For a more detailed discussion of the meaning of "waters of the United States" see Section 7.1.3.

The Corps initially refused to expand its § 404 wetlands jurisdiction, relying upon prior judicial decisions under the Rivers and Harbors Act, that construed "navigable waters" as limited to the mean high water mark. The Corps reluctance to expand its jurisdiction can only be understood as an effort to limit the scope of the law so that it would not have to oversee activities in up to 150 million acres of wetlands not previously regulated by the Corps. However, in response to judicial interpretation, the Corps prepared new regulations to expand its authority under the § 404 program. The Corps now defines "waters of the United States" to mean "[a]ll waters which are currently used, or were used in the past, or may be susceptible to use in interstate or foreign commerce, including all waters which are subject to the ebb and flow of the tide...." (33 CFR 328.3(a)(1)). This definition also includes "[a]ll other waters such as intrastate lakes, rivers, streams (including intermittent streams), mudflats, sandflats, wetlands, sloughs, prairie potholes, wet meadows, playa lakes, or natural ponds, the use, degradation or destruction of which could affect interstate or foreign commerce...." (33 CFR 328.3(a)(3)).

The Corps' regulations also encompass wetlands "adjacent" to waters associated with interstate commerce, and have been interpreted to include jurisdiction over certain "isolated wetlands" (33 CFR 328.3(a)(5),(7); see cases discussed in Section 7.1.5.). Several courts have upheld various aspects of the Corps' expansive interpretation of its wetland jurisdiction (see *United States v. Riverside Bayview Homes, Inc.*, 474 U.S. 121, 123 (1985), discussion of meaning of "waters of the

United States"); and cases discussed in Section 7.1.4).

7.1.2 Wetland Jurisdiction

Under the § 404 program, the Corps and EPA have concurrent jurisdictional authority over the dredging and filling of waters of the United States, including wetlands.[3] The Secretary of the Army, acting through the Chief of Engineers, is authorized to issue individual permits for the discharge of dredged or fill material into the waters of the United States, which includes wetlands (33 USC § 1344(a); Corps regulations governing individual permits are found at 33 CFR 323; see Section 7.2).

In some circumstances, the Corps may issue "nationwide permits" for certain activities in jurisdictional wetlands that are deemed to have minimal environmental impacts (33 USC § 1344(e); Corps regulations governing the Nationwide permit program are found at 33 CFR 330; see Section 7.3.3). Although the Corps' field personnel are responsible for making the initial decision to grant or deny permits, the EPA is responsible for formulating the § 404(b)(1) Guidelines used by the Corps to make the permit decisions (Clean Water Act § 404(b)(1) Guidelines; Correction, 55 Fed. Reg., 9210, 9211 (Feb. 7, 1990). The EPA is also empowered to veto or overrule the granting of permits by the Corps (33 USC § 1344(b)-(c)). Despite this veto authority, however, EPA has rarely overruled a Corps decision to issue a permit (see U.S. General Accounting Office, 1988).

The Corps regulations set forth extensive procedures for the permit process (33 CFR Parts 325, 323, 320; the permit process is fully explained in Section 7.2). The application form must describe the purpose, scope and need for the proposed activity, its location and the names and addresses of adjoining property owners. Following submission of a permit application for activity in a wetland area, the

---

[3] Although the Clean Water Act is essentially silent on which agency has authority to make jurisdictional determinations under the Section 404 Program, the EPA and Corps have formulated agreements detailing their respective jurisdictional responsibilities. EPA/Department of Defense, Memorandum of Understanding on "Geographical Jurisdiction of the Section 404 Program (MOU)," 45 Fed. Reg. 45,018 (July 2, 1980); Department of the Army/EPA Memorandum of Agreement Concerning the Geographic Jurisdiction of the Section 404 Program and the Application of the Exemptions under Section 404(f) of the Clean Water Act (MOA)" (January 19, 1989).

Corps must decide whether to grant the permit and, if granted, whether any conditions should be placed on the permit (33 CFR 320.4 lists the criteria for evaluating a permit application.). In evaluating a permit application, the Corps is required to consider the recommendations of the U.S. Fish and Wildlife Service and the National Marine Fishery Service (33 CFR 320.4(c)). Under authority of the Fish and Wildlife Coordination Act, the FWS and the National Marine Fisheries Service review applications for these federal permits and provide comments to the Corps on the environmental impacts of proposed work. In addition, the FWS has been conducting an inventory of the Nation's wetlands and is producing a series of National Wetlands Inventory maps for the entire country.

Comments and objections from certain state agencies must also be considered. For instance, the Coastal Zone Management Act permits a state to object to a proposed permit if the state has an approved Coastal Zone Management Program (CMP) and the state determines that issuance of the permit will be inconsistent with the goals of the state CMP (16 USC § 1456(c)(3)(A)). Although various state and federal agencies may object to a permit application, the Corps may decide to issue a permit over the objections of other agencies (33 CFR 325). However, if a state objects to issuance of a permit application because of inconsistency with the state's approved CMP, the Corps cannot issue the permit unless the state's consistency objection is overruled by the Secretary of the Interior, or until the applicant modifies the proposal so that it is consistent with the state CMP (16 USC § 1456(d)). Still, even if the Corps issues a permit, the applicant may need to secure additional state and local wetland permit approvals (see discussion of state and local wetland regulation in Section 7.9.).

7.1.3 Jurisdictional Determinations

The Clean Water Act does not specify which agency (the Corps or EPA) has authority to make jurisdictional determinations concerning wetlands, nor does it define how to proceed when the two agencies disagree on whether a particular site is subject to regulation under § 404. From 1972 to 1979, the Corps routinely made jurisdictional determinations as part of its permitting program.

On September 5, 1979, responding to a dispute between EPA and the Corps, then U.S. Attorney General Benjamin Civiletti issued an opinion stating that EPA has "the final administrative authority" for determining the reach of waters subject to regulation under § 404. Thereafter, EPA and the Corps entered into a Memorandum of Understanding (MOU) under which they agreed that the Corps would continue to make most jurisdictional determinations. EPA makes jurisdictional determinations in "special cases" involving significant technical and/or policy issues, and the two agencies work together in enforcement situations (EPA/Department of Defense, Memorandum of Understanding on "Geographical Jurisdiction of the § 404 Program (MOU)," 45 Fed. Reg. 45,018 (July 2, 1980)). On January 19, 1989, the agencies entered into a new Memorandum of Agreement (MOA) that essentially retains the division of responsibility set out in the 1980 MOU and also makes it applicable to determinations on whether a particular activity is exempt from § 404 regulation by virtue of § 404(f) (Department of the Army/EPA Memorandum of Agreement Concerning the Geographic Jurisdiction of the § 404 Program and the Application of the Exemptions under § 404(f) of the Clean Water Act (MOA)" (January 19, 1989)). Under the 1977 Clean Water Act Amendments to § 404, Congress exempted normal agricultural, forestry and ranching operations from the permit requirements (33 USC § 1344(f)(1)(A)).

While Congress intended the Clean Water Act to reach a broad range of waters and wetlands, Congress never intended to regulate all wetlands as "waters of the United States." Congress' primary concern was waters used in interstate commerce and waters that together form a "hydrologic chain" (see 42 Fed. Reg. 37,122 (1977)). Recognizing that Clean Water Act jurisdiction is not limitless, both EPA and Corps regulations assert jurisdiction over wetlands only if destruction or degradation of the wetland could affect interstate commerce (see 40 CFR 230.3(s); 33 CFR 328.3(a)). These regulations call for a case-by-case determination concerning whether a given wetland has the required effect on commerce. EPA has recognized that "small, isolated wet areas may not be waters of the United States . . . because, even if wetlands, their destruction or degradation would not have any effect on interstate commerce" (see 45 Fed. Reg. 33290, 33398 (1980)). However, given the expansive interpretation of activities that affect commerce, most areas that qualify as wetlands will also meet the effect-on-commerce test (see discussion of "adjacent" and "isolated" wetlands in Sections 7.1.4 and 7.1.5.).

The major issue is how to distinguish jurisdictional wetlands from nonjurisdictional wetlands. The precise boundaries of federal wetland jurisdiction under the Clean Water Act are unclear. The Corps initially construed the term "navigable waters" in its traditional sense to include waters that are subject to the ebb and flow of the tide, or waters that are used, have been used, or are susceptible for use in interstate commerce (see 39 Fed. Reg. 12115 (1974)). Environmental groups immediately challenged this interpretation, and a federal district court agreed that it was too restrictive. The court held that Congress did not intend the term "navigable waters" to be "limited to the traditional tests of navigability" (*Natural Resources Defense Council, Inc. v. Callaway*, 392 F. Supp. 685, 686 (D.D.C. 1975).).

In response to this ruling, the Corps promulgated new "interim final" regulations providing for a phased expansion of its jurisdiction. The new regulations first covered traditional navigable waters, then primary tributaries of navigable waters, and finally "any navigable water," including wetlands (see 40 Fed. Reg. 31319 (1975)). During congressional debate over the 1977 amendments to the Clean Water Act, various attempts were made to confine federal authority to traditional navigable waters, but these efforts failed.

7.1.4 Jurisdiction Over "Adjacent" Wetlands

In 1985, the U.S. Supreme Court issued a significant decision in the case of *U.S. v. Riverside Bayview Homes* (474 U.S. 121 (1985)), finding that the Corps' jurisdiction over wetlands extended to areas that are adjacent to navigable waters. In its decision, the Court observed: "Congress evidently intended to repudiate limits that had been placed on federal regulation by earlier water pollution control statutes and to exercise its powers under the Commerce Clause to regulate at least some waters that would not be deemed 'navigable' under the classical understanding of that term" (see also *U.S. v. Hobbs*, 21 Envt'l L. Rep. (ELI) 20,830 (E.D. Va. Aug. 24, 1990); and see 40 CFR 230(s) and 33 CFR 328.2). The Court upheld the Corps' regulation of wetlands adjacent to navigable waters because these are "waters that together form the entire aquatic system" (474 U.S. at 133, quoting from 42 Fed. Reg. 37,122, 37,128 (1977)). The Court reasoned that adjacent wetlands would "affect the water quality of the other waters within that aquatic system."[4]

Since the 1985 decision in *Riverside Bayview Homes*, courts have ruled on the adjacency issue in several other cases. For example, in *United States v. Lee Wood Contracting, Inc.* (529 F. Supp. 119, 120 (E.D. Mich. 1981)), it was held that a wetland area met the statutory definition of an adjacent wetland because it was connected to a river by a slough and, therefore, was contiguous.

Wetlands that are not contiguous but that affect the water quality and aquatic ecosystems of navigable waters may also be considered "adjacent wetlands." For example, in *Conant v. United States*, (786 F.2d 1008 (11th Cir. 1986)), the wetland trapped undesirable pollutants and sediments before they reached a navigable river, and was, therefore, held to be "adjacent."

7.1.5 Jurisdiction Over "Isolated" Wetlands

In addition to jurisdiction over adjacent wetlands, Corps regulation covers Clean Water Act jurisdiction over certain "isolated" waters. The Clean Water Act's jurisdiction also is extended by regulation over certain intrastate waters not part of a surface tributary system, that is "isolated" waters, if their "use, degradation or destruction ... could affect interstate or foreign commerce" (33 CFR 328.3(a)(3); see also 33 CFR 330.2(e)).

Although the "isolated" wetlands issue seldom arose during the first fifteen years of the § 404 Program, when it did, the Corps or EPA generally determined jurisdiction according to the effect the proposed wetland activity might have on interstate commerce. To complete this analysis, they generally considered whether the site in question served an interstate market or was visited by out-of-state residents (either for recreation or study) and whether the proposed work in the wetlands area would be likely to affect such visits (see *United States v. Byrd*, 609 F.2d 1204 (7th Cir. 1979); Memorandum of the Acting General Counsel, "Clean Water Act Jurisdiction Over Springs in Ash Meadows, Nevada" (July 5, 1983)).

Other examples of the kinds of wetland activities that could

---

[4] Id. at 134. The regulations define "adjacent" as meaning "bordering, contiguous, or neighboring..." but do not establish a distance for "adjacency." 33 CFR 323.2(d).

affect interstate or foreign commerce include wetlands from which fish or shellfish are taken and sold in interstate or foreign commerce and waters that are used or could be used for industrial purposes by industries in interstate commerce.

There has been little judicial analysis of what types of isolated waters are included under the Corps' jurisdiction (see Geltman, 1989). In *Riverside Bayview Homes*, the Supreme Court specifically left open the question whether "isolated wetlands" (i.e., wetlands that do not have a hydrological connection to "waters of the United States") are within the scope of jurisdiction under the § 404 Program (474 U.S. at 131).

In *Leslie Salt Co. v. United States* (896 F.2d 354 (9th Cir. 1990), *cert. denied*, 111 S. Ct. 1089 (1991)) the court found that artificially-created wetlands, which formed in crystallization basins and calcium chloride pits that had previously been used for salt production, could be subject to Corps jurisdiction even though isolated from the tidal arm of San Francisco Bay by a quarter of a mile, as long as those waters had sufficient ties to interstate commerce. The Ninth Circuit Court of Appeals rejected the district court's assertion that Corps jurisdiction extends only to natural formations, holding that "the Corps intends to exempt from its own jurisdiction only those artificially created waters, which are currently being used for commercial purposes, and that even those waters are subject to such jurisdiction on a 'case-by-case' basis of review."

The salt pits in question in *Leslie Salt Co. v. United States* had been abandoned for decades. The Ninth Circuit Court of Appeals rejected the lower court's conclusion that the pits and crystallizers were not waters because water only collected in them during the rainy season. The court noted that "the seasonal nature of the ponding presented no obstacle to Corps jurisdiction," because intermittent streams and playa lakes are specifically enumerated as types of isolated waters over which the Corps may have jurisdiction. The court remanded for a determination of the existence of sufficient connections to interstate commerce, noting, however, that the use of the waters by migratory birds and endangered species would probably satisfy this requirement.

7.1.6 Migratory Bird Link to Interstate Commerce

The Corps and EPA have used migratory birds to establish an

effect on commerce for purposes of jurisdiction over isolated wetlands. The essence of the migratory bird rule is that if a bird might use an isolated wetland as a stop-over or nesting grounds while on its migratory flight, any activity disturbing the bird's wetland habitat could affect interstate commerce since bird watchers would be unable to observe the bird during interstate travels. The agencies have based this interpretation of the § 404 Program's jurisdictional reach on an internal legal memorandum prepared by EPA's general counsel in 1985, however, they have never incorporated the "migratory bird memo" into the published regulations (see 51 Fed. Reg. 41217 (Nov. 13, 1986)).

The migratory bird rule has generally been met with disfavor in the courts. In *Tabb Lakes v. United States* (No. 87-635-N (E.D. Va., Nov. 7, 1988)), the court rejected the Corps' use of migratory birds to assert jurisdiction because the Corps had not published a formal rule establishing migratory birds as a jurisdictional base. On appeal, the Fourth Circuit upheld the lower court, finding that the "migratory bird rule" was a substantive rather than an interpretive rule, and that the provision could not be effective without a prior opportunity for notice and comment (*Tabb Lakes, Ltd. v. United States*, No. 89-2905 (4th Cir. Sept. 19, 1989) (unpublished)).

The most recent case to analyze the migratory bird reasoning is *Hoffman Homes, Inc. v. Administrator, United States Environmental Protection Agency* (__ F.2d __, 1993 WL 264673 (7th Cir. 1993)). Initially, EPA's Chief Administrative Law Judge held that the Agency did not have jurisdiction over an isolated wetland pocket (.8 acre in size) because there was no evidence that birds actually used the site. EPA then appealed the decision to its Chief Judicial Officer (CJO) who reversed, holding that the effect on commerce could be established by potential migratory bird usage of the site (*In re The Hoffman Group, Inc.*, CWA Appeal No. 89-2 (Nov. 19, 1990)). Hoffman appealed this administrative decision to the Seventh Circuit Court of Appeals, which held that § 404's prohibition against discharges of dredged or fill materials into "navigable waters" without a permit did not give EPA jurisdiction over intrastate, nonadjacent or "isolated" wetlands. The court agreed with the CJO that the language of the EPA regulations grants jurisdiction to those wetlands whose degradation or destruction "could" affect interstate commerce (40 CFR 230.3(s)(3)), including the "potential" use of wetlands by migratory birds. However, the court reversed the CJO due to a lack of substantial evidence in the record of actual or even potential use of the particular wetland in question by migratory birds. Although some property owners may see the Hoffman decision as a victory (see Albrecht and Isaacs, 1992), the court specifically recognized the validity of the migratory bird rule.

7.1.7 Discharges Into Waters of the United States

Once there has been a determination of jurisdiction over a wetland, the next inquiry involves determining whether the activity requires a permit. Until recently, EPA, the Corps, and courts have held that § 404 regulates only physical discharges of dredged or fill material into navigable waters (see *United States v. Lambert*, 589 F. Supp. 366 (M.D. Fla. 1984)). On its face, this language indicates that many activities that destroy wetlands are not regulated under the Clean Water Act. This is consistent with the plain language of § 404, which regulates only "discharges" of pollutants (33 USC § 1311(a)). This narrow "discharge rule" has, however, been eroded by recent regulatory and judicial decisions.

For instance, in *Avoyelles Sportsmen's League, Inc. v. Marsh* (715 F.2d 897 (5th Cir. 1983)), the court held that certain land-clearing activities, which resulted in redeposits of material taken from the wetland, constituted a discharge of a pollutant and, therefore, required a dredge-and-fill permit. Significantly, the government argued that "mere removal" of vegetation from a wetland would not constitute a discharge because "the term discharge is defined as the 'addition' of pollutants, not the removal of materials" (*Id.* at 922). Noting that the case involved more than "mere removal," the court declined to rule on the government's argument but did hold that the bulldozer used for clearing constituted a "point source" and that material discharged as a result of the landclearing activities constituted "fill material." Corps regulations define "fill material" as "any material used for the primary purpose of replacing an aquatic area with dry land or of changing the bottom elevation of a waterbody" (see 33 CFR 323.2(e)).

In a Regulatory Guidance Letter (RGL; No. 90-5, July 18, 1990), the Corps abandoned its position in *Avoyelles* and its previous RGLs on landclearing, stating that all landclearing activities using mechanized equipment are subject to 404 permit requirements. However, landclearing activities that are carried out with chain saws are not subject to regulation.

In 1990, a Texas federal district court held for the first time that an activity need not involve a discharge to be regulated under Section 404. In *Save Our Community v. EPA* (741 F. Supp. 605 (N.D. Texas 1990)) an environmental group challenged an EPA/Corps determination that they did not have legal authority to require a permit when the only activity conducted on a jurisdictional wetland was drainage through mechanical means. No discharges were made during the draining process. Significantly, the court found that draining a "legally designated" wetland property is a regulated activity under Section 404(b), requiring a permit when the draining would alter or destroy the wetland.

Finally, in a Regulatory Guidance Letter (No. 90-8, Dec. 14, 1990), the Corps addressed the applicability of Section 404 to projects constructed with pilings. In that RGL, the Corps stated its intention to regulate projects placed on pilings when the placement of such pilings "is used in a manner essentially equivalent to a discharge of fill material in physical effect or functional use and effect."

7.1.8 Federal Manual For Delineating Wetlands

In order to exercise jurisdiction over a wetland, the area must first be delineated as a jurisdictional wetland. At the federal level, four agencies are principally involved with wetland identification and delineation: Army Corps of Engineers (Corps), Environmental Protection Agency (EPA), Fish and Wildlife Service (FWS), and Soil Conservation Service (SCS). While the SCS has been involved in wetland identification since 1956, it has recently become more deeply involved in wetland determinations through the "Swampbuster" provision of the Food Security Act of 1985, and its 1990 amendments (16 USC § 3801 *et seq.*).

In 1989, the Corps, EPA, FWS and SCS agreed to use one approach for delineating areas as wetlands under the jurisdiction of the Section 404 program. Previous varied agency approaches and lack of standardized methods during the first seventeen years of the 404 Program had resulted in inconsistent determinations of wetland boundaries for the same type of area (see Section 5.1). This created confusion and identified the need for a single, consistent approach for wetland determinations and boundary delineations. The four agencies reached agreement on technical criteria for identifying and delineating wetlands and merged their methods into a single wetland delineation manual, which was published on January 10, 1989 as the "Federal Manual for Identifying and Delineating Jurisdictional Wetlands." This established a national standard for wetland identification and delineation, and terminated previous locally implemented approaches that were not, in some cases, scientifically-based or consistent.

During the following two years, the 1989 Manual was used by the agencies for wetland delineation, chiefly for identifying and delineating wetlands subject to federal regulations under the Clean Water Act. Use of the manual coupled with policy changes (e.g., farmed wetlands) led to a significant increase in federal jurisdiction in many areas of the country. Unfortunately, during this time misconceptions concerning use of the 1989 Manual (e.g., classifying any area mapped with hydric soil as wetland without considering other criteria), also led to exaggerated claims of expanded federal regulation. This atmosphere of misinformation and public concern over regulation created an obvious need to review the 1989 Manual and revise it as necessary. The four agencies recognized that additional clarification and/or changes might be required, and had planned to review the situation after the first year of implementation..

Thus, in May 1990, the agencies initiated an evaluation of the 1989 Manual, concluding that while the manual represented a substantial improvement over pre-existing approaches, several key issues needed to be re-examined and clarified. Some of the key technical issues needing re-examination were: (1) wetland hydrology criterion; (2) the use of hydric soil for delineating the wetland boundary; (3) the assumption that facultative vegetation indicated wetland hydrology; and (4) the open-ended nature of the determination process, which created opportunities for misuse.

In August 1991, EPA issued proposed revisions to address the inadequacies of the 1989 Manual (56 Fed. Reg. 40,446 (Aug. 14, 1991)). As mentioned earlier, considerable controversy surrounds the proposed revisions. EPA received a virtually unprecedented 80,000+ public comments, reflecting conflicting public and private interests on the proposed revisions. While a new delineation manual is being developed, Congress has forbidden use of the 1989 Manual (Pub. L. No. 102-377, 106 Stat. 1315, 1324 (1992)). For now, EPA and the Corps have agreed to use the previous 1987 Corps Manual to delineate wetlands (58 Fed. Reg. 4995 (Jan. 19, 1993)). Discussion of wetland delineation criteria in

Chapters 5 and 6, as well as throughout this book is therefore necessarily based on use of the 1987 Corps manual (see Chapters 5 and 6 for a full discussion of the wetland delineation criteria).

7.1.9 State Assumption of the § 404 Program

Section 404(g) of the Clean Water Act authorizes individual states to assume the responsibilities for administering the § 404 dredge and fill permit program by applying to EPA for approval of its program (33 USC § 1344(g)-(i)). Once its program has been approved, a state, rather than the Corps, may issue § 404 permits directly. Copies of all § 404 permit applications must be sent to EPA, and EPA retains authority to veto a permit, or to withdraw its approval of the state's § 404 program, if statutory and regulatory requirements are not followed (33 USC § 1344(i)-(j)). However, the agency is authorized to waive the notification requirement for categories of permit applications specified by EPA (33 USC § 1344(k)).

To date, the only state that has assumed responsibility for the § 404 program is Michigan (see Harrington, 1986; Brown, 1989). There are several reasons why more states have not assumed § 404 responsibilities, including inadequate supportive state legislation, a perception that assumption would not streamline the permit process, and funding shortfalls by the states (Davis, 1991). In order to make assumption more palatable to individual states, EPA issued new regulations in 1988 that established a more lenient system for waiver of notification for many § 404 permit applications, such that virtually every category of application can now be waived (53 Fed. Reg. 20,764 (June 6, 1988)). Several states are currently considering assumption under the more lenient EPA regulations, including New Jersey and Florida (Dix and Denson, 1993).

The § 404 assumption program has not been without its detractors (see Wood, 1989, 1990), and a recent controversy involving Michigan's § 404 assumption program suggests that many problems still exist. In the Michigan controversy, a developer of a condominium resort called "The Homestead" in the northwestern corner of the state's lower peninsula, applied for a § 404 permit to build a golf course on 3.5 acres of wetlands adjacent to the Crystal River. The virtually pristine river is adjacent to Sleeping Bear Dunes National Lakeshore, and is a popular canoeing and fishing stream.[5] The Michigan Department of Natural Resources (MDNR) attempted to issue a § 404 permit despite the objections of the Corps, EPA, FWS, and the National Park Service, as well as MDNR staff, and an impartial panel of wetlands experts. Region V of EPA (based in Chicago) attempted to block the permit, but then-EPA Administrator William Reilly transferred the agency's authority over the permit from Region V to LaJuana Wilcher, then-Assistant Administrator for Water in Washington, D.C. Wilcher and her Washington staff reviewed the documentation prepared for the case, and overruled the Chicago office by determining that EPA should withdraw its objections to the permit, and transfer permitting authority to MDNR (for detailed reviews of these events, see Jones, 1992; and Dreher, 1992).

Wilcher's decision to transfer authority back to MDNR prompted a coalition of environmental groups to file suit against EPA, alleging that EPA could not withdraw its original objection to the permit. In *Friends of the Crystal River v. U.S. Environmental Protection Agency* (794 F.Supp. 674 (W.D. Mich. 1992)), the district court held that, under § 404(j) of the CWA, EPA's objection to the permit immediately shifted permitting authority to the Corps, such that EPA could not later withdraw its objection. The case is currently on appeal.

The Homestead golf course project remains on hold at this writing as the various legal maneuvers are played out, but it serves as a warning of the kinds of problems state assumption might hold.

7.2 DREDGE AND FILL PERMITS

Under § 404(a) of the Clean Water Act, the Secretary of the Army (acting through the Chief of Engineers, 33 USC § 1344(d)) may issue permits for the discharge of dredged or fill material into the navigable waters (33 USC § 1344 (a)). As discussed in Section 7.1, "navigable waters" has been construed as including wetlands.

The process by which a person, called the "applicant" in Corps regulations, applies for a permit is described in Corps regulations at 33 CFR 320-344 (see Want, 1993). The process itself is complex, and

---

[5] Figure 2.6 shows wetlands adjacent to the Crystal River within a few yards of the proposed golf course site.

228   Wetlands

replete with snares for the unprepared. The cautious applicant may avoid unnecessary difficulties and delays by preparing carefully the necessary materials for the application. Many applications, particularly those that are large or controversial, require the assistance of scientific and legal consultants.

7.2.1 The Application Process

Who Must Apply? The requirement for dredge and fill permits under § 404 of the Clean Water Act and § 10 of the Rivers and Harbors Act applies to any private or governmental individual, entity, organization, or agency (33 CFR 323.2(a) and (b)), although certain federal projects which are specifically approved by Congress are exempted under § 404(r), if pertinent information is supplied in an environmental impact statement (33 USC § 1344(r); see Section 7.6.2).

The Corps need not apply for a § 404 permit from itself, but it is required nevertheless to comply with the same laws and follow the same procedures as any other applicant (33 CFR 209.145, 322.3(c)(1), 323.3(b), 336(1)(a), and 40 CFR 230.2(a)(2); see *Minnesota v. Hoffman*, 543 F.2d 1198 (8th Cir. 1976)). Certain parties involved in work on Corps projects may qualify for an abbreviated application process (Regulatory Guidance Letter No. 88-9, "Corps Civil Works Projects" (July 21, 1988)).

The Application Form. The next step in the process involves obtaining the permit application itself. An applicant may obtain the proper Department of the Army permit form (ENG Form 4345, OMB Approval No. OMB 49-R0420; see 33 CFR 325.1(c)) from the office of the local district engineer to whom the application will be submitted. A permit form is also found as Appendix A to 33 CFR 325 (see Want, 1993 for an example of a completed application form). Some District offices require the use of forms with some slight variations in order to facilitate local coordination with other federal, state, or local agencies (e.g., applicants in Illinois complete a joint form from the U.S. Army Corps of Engineers and the Illinois Department of Transportation; applicants in Florida fill out a joint form from the Corps, the Florida Department of Environmental Regulation, and the Florida Department of Natural Resources).

Regulatory guidance on the contents of the application is found at 33 CFR 325.1(d). The general rule is that an application will be considered complete "when sufficient information is received to issue a public notice" (33 CFR 325.1(d)(9); see Section 7.3.1). These requirements are met when the following information from 33 CFR 325.1(d)(1) is provided:

1) *A complete description of the proposed activity, including necessary drawings, sketches, or plans sufficient for a public notice. Detailed engineering plans and specifications are not required.* Although there is some variation among Corps offices, it has been our experience that excessive information (e.g., multiple copies of similar engineering and site plans) may actually serve to delay an application. Three maps must be submitted: (1) a vicinity map (a road map, or U.S. Geological Survey map, scale 1:24,000); (2) a plan view that shows the development as it would appear if one were looking straight down on it from above, and that indicates water bodies (names, water levels, depths, and dimensions), dimensions of the activity, and various other details; and (3) an elevation or cross-section view that is a scale drawing of the project that shows side, front or rear views (see Want, 1993).

2) *The location, purpose and need for the proposed activity.* The statement of location must be complete and accurate, but need not include a formal legal definition unless it is specifically requested, or if there is a possibility of confusion with other property. The statement of purpose and need for the activity should be complete but concise, and should include information that is likely to be important at a subsequent public hearing (i.e., public health and welfare issues, safety and environmental issues, and other potentially controversial issues are often included).

3) *Scheduling of the activity.* Dates during which the various phases of a project are scheduled for completion, along with a project completion date, must be included.

4) *The names and addresses of adjacent property owners.* This information may not be easily or immediately available, but the applicant must make every effort to obtain it. Local governmental bodies (such as taxing bodies or districts) are often useful in determining ownership.

5) *The location and dimensions of adjacent structures.*

6) *A list of authorizations required by other federal, interstate, state or local agencies for the work.* This list must include all approvals already received and denials already made. Since most projects are located in areas of overlapping governmental authority, it is important to include *all* other agency approvals in this list. This is typically the responsibility of an applicant's consultant or attorney.

7) *Preliminary jurisdictional determination.* Although it is not specifically required, most applicants also include a preliminary determination of Corps jurisdiction (or lack of jurisdiction) over the subject wetland. Preliminary determinations are typically performed by scientific consultants, and are not binding on the Corps. Wetland determinations and delineation are discussed more fully in Section 7.1, and Chapter 5.

8) *Signature.* The application must be signed by the person who desires to undertake the activity. If the applicant is represented by an agent (e.g., a contractor, consultant, or attorney), that information must be included with the application (33 CFR 325.1(c)(7)).

Other Application Requirements. Corps regulations also list additional requirements that may apply to a particular application. If an activity will involve dredging in navigable waters, the application must include a description of the type, composition, and quantity of the material to be dredged, as well as the method of dredging, and the site and plans for its disposal (33 CFR 325.1(c)(3)). If there will be a discharge of dredged or fill materials (or transportation of these materials for ocean discharge), then the application must describe the source of material; the purpose of the discharge; the type, composition, and quantity of material; the method of transportation and disposal; and the site of disposal (33 CFR 325.1(d)(4)). Certification is also required from the EPA under § 401 of the Clean water Act (33 USC § 1341), the National Pollutant Discharge Elimination System (NPDES).

If the activity includes construction of a filled area, pile, or float-supported platform, then the application must describe the use of, and specific structures to be erected (33 CFR 325.1(d)(5)). If the project includes the construction of an impoundment structure, the application may be required to demonstrate compliance with state dam safety criteria (33 CFR 325.1(d)(6)).

At the discretion of the district engineer, an applicant may be required to submit additional information as may be necessary to make a public interest determination (see Section 7.3). Additional information may include environmental data, or information on alternate methods and sites that may be necessary to comply with required environmental documentation (33 CFR 325.1(e)).

The most controversial application requirement is contained in Corps regulations at 33 CFR 325.1(d)(2):

> "All activities which the applicant plans to undertake which are *reasonably related* to the same project and for which a DA [Corps] permit would be required should be included in the same permit application. District Engineers may reject, as incomplete, any permit application that fails to comply with this requirement." [33 CFR 325.1(d)(2), emphasis added.]

There has been considerable disagreement as to what constitutes "reasonably related" projects. For pragmatic reasons, applicants may wish to submit separate applications for two related projects and escape close Corps scrutiny of the cumulative impacts of the two projects taken together. In *Russo Development Corp v. Thomas* (735 F.Supp. 631 (D.N.J. 1989)) a federal district court rejected the Corps' attempt to require a single application for two separate properties on the basis that they were contiguous (on one property the developer sought a permit to construct buildings, and on the other he sought an after-the-fact permit for a building already constructed).

A related question occasionally arises when an applicant wishes to change a project after permit approval, but without filing a new permit application. Such changes are not uncommon, and are generally approved by the Corps if the scope of wetland fill and environmental impacts is the same. In the famous (if somewhat bizarre) case *Missouri Coalition for the Environment v. Corps of Engineers* (866 F.2d 1025 (8th Cir. 1989)), a federal appellate court held that a developer's change from an industrial park to a football stadium did *not* require a new permit, because the scope of wetland fill and potential impacts were similar.

7.2.2 Notice, Comment, and Conflict Resolution

Upon receipt of a § 404 permit application, the district engineer must immediately assign it an application number for identification purposes, and advise the applicant of receipt of the application and assignment of the identification number (33 CFR 325.2(a)(1)). The application number is of particular significance to the applicant, since it is through this number that an application's progress through the application process can be determined.

Within 15 days of receipt of an application, the district engineer must review the application and respond in one of two ways: (1) either issue a public notice, or (2) advise the applicant that additional information is necessary for a complete application (33 CFR 325.2(a)(2)). In practice, Corps offices are nearly always late despite the clear mandate of the regulations. Occasional contact with the Regulatory Division of the local Corps office may serve to satisfy an applicant that his application is receiving timely treatment.

The Corps is also required to evaluate each permit application to determine if it meets the requirements for a nationwide permit, or would meet those requirements after reasonable modification or conditions (33 CFR 330.2(f); nationwide permits are discussed more fully at Section 7.3.3). If the district engineer determines that an application meets the nationwide permit criteria, then the applicant must be so informed.

Public Notice. Corps regulations contain specific requirements for the contents of the public notice, as detailed in 33 CFR 325.3. In general, the contents of the public notice must "include sufficient information to give a clear understanding of the nature and magnitude of the activity to generate meaningful comment" (33 CFR 325.3(a)). Despite the plaintiff's contention that insufficient information was contained in the public notice for proper evaluation, the federal appellate court in *Environmental Coalition of Broward County, Inc. v. Myers* (831 F.2d 984 (11th Cir. 1987)) upheld the Corps' issuance of a permit because the Corps had followed the regulatory procedures for public notice.

In general, the Corps attempts to make knowledge of the application available as generally as possible. Under Corps regulations, public notices are to be distributed to the public in general by posting in post offices "or other appropriate public places in the vicinity of the site of the proposed work ..." (33 CFR 325.3(3)(d)(1)). Notices are sent specifically to the applicant, adjacent property owners, appropriate local officials and state agencies, Indian tribes, and to concerned federal agencies. Notices are also sent to concerned business interests, environmental organizations, state and regional clearing houses for such information, and local news media. In addition, the Corps will send a notice to "all parties who have specifically requested copies of public notices" (33 CFR 325.3(3)(d)(1)). Our experience has been that any person (or organization) with an interest in dredge and fill issues in a particular area is well advised to request from the local Corps office that they be sent any public notices for § 404 permit applications received by the office.

Each public notice contains a time limit for receipt of comments by the local Corps office (e.g., 30 days from the release date of the notice). Comments received after that date need not be considered by the Corps in subsequent deliberations. The district engineer may issue a supplemental, revised, or corrected public notice at his discretion if he feels that there has been a change in the application or application data that would affect the public's view of the proposal (33 CFR 325.2(2)), and time limits may be extended for this and other reasons.

The public notice *may* specify that a public hearing will be held on the application (33 CFR 327.4(b)), but it is far more common for the Corps to decide if a public hearing will be held after comments are received. Public hearings will be discussed in Section 7.3.1.

Comments. Any interested person or organization may make comments on a § 404 permit application. The district engineer acknowledges receipt of comments "if appropriate," and will make them a part of the administrative record of the application (33 CFR 325.2(a)(3)). In cases where the issuance of a § 404 permit (or failure to issue one) is challenged in court, the administrative record may be of critical importance to the litigants.

Comments to the Corps on § 404 permit applications take many forms. Very general comments often receive more cursory treatment than thoughtful comments directed to specific aspects of the project. Likewise, overly lengthy, rambling, or repetitive comments may receive less careful scrutiny than carefully drafted, concise comments. It is important for the person who wishes either to oppose or support an application to give careful thought in drafting comments.

The district engineer must "consider all comments received in response to the public notice in his subsequent actions on the permit application" (33 CFR 325.2(a)(3)). If the comments received by the Corps indicate that the views of the applicant are necessary in order to conduct a public interest determination on a particular issue, then the applicant will be given the opportunity to furnish his views on the issue.

Conflict Resolution. At the earliest practicable time, The Corps will supply copies of substantive comments to the applicant for his information. These comments may be supplied in the form of summaries, the actual letters (or portions of them), or representative comments. The applicant then must be given the opportunity to contact the objectors if he chooses in an attempt to resolve conflicts regarding the application, although the final decision on issuance of the permit rests with the Corps.

The importance of conflict resolution was underscored recently in *Mall Properties, Inc. v. Marsh* (672 F.Supp. 561 (D.Mass. 1987), *rejecting remand as appealable order,* 841 F.2d 440 (1st Cir. 1988)). In that case, Mall Properties, Inc. sued to reverse denial of a § 404 permit when it was discovered that it had not been informed of a meeting between the Corps and the governor of Connecticut, who had recommended rejection of the application. The federal district court agreed that Corps regulations require that an applicant be given an opportunity to rebut negative comments made of the application, and the case was remanded to the Corps so this requirement could be completed.

Our experience has been that the conflict resolution phase of a permit application is often of critical importance. It is frequently the first time that parties on both sides of the issue have the opportunity to sit at the same table and attempt to resolve differences through proper discussion and negotiation. Timely, thoughtful, and responsive conflict resolution often leads to equitable agreements without the need for unpleasant confrontation and expensive litigation.

7.3 CORPS RESPONSES TO § 404 PERMIT APPLICATIONS

The Corps can respond in several ways to a § 404 permit application. First, a determination is made whether the Corps has jurisdiction over the wetland in question. The issue of Corps jurisdiction is discussed in Sections 7.1.2 - 7.1.5, and related issues of wetland determination and delineation are discussed in Chapter 5. Second, the Corps may determine that a public hearing is required, during which time any party wishing to make an oral statement regarding the permit application has an opportunity to do so. Third, the Corps may determine that the project will not degrade wetlands in a manner that would prohibit a § 404 permit and issue the permit. A fourth alternative response is the issuance of a General or "Nationwide" permit, or determination that the project meets any of several statutory exemptions to the requirement for a § 404 permit. Fifth, the Corps may issue the permit while requiring project modifications and/or mitigation. Finally, the Corps may determine that the project is unacceptable under § 404, and reject the application. However, to make the issue even more complex, § 404(c) of the Clean Water Act gives EPA veto authority over a permit even after it has been issued.

The regulations instruct the district engineer to have decided on all applications within 60 days after receipt, but this time limit is rarely met because of the time required for compliance with other regulations and statutes (33 CFR 325.2(d)(3)), understaffing in Corps offices, and other reasons.[6]

7.3.1 The Public Hearing

Once the Corps has determined that it has jurisdiction over a wetland such that the requirement for a § 404 permit review is triggered, the district engineer *may* determine that a public hearing is to be held. Although the Corps may announce the public hearing in its public notice, it is more common for the announcement of the hearing to follow the receipt of comments (see Section 7.2.2).

The regulations specify that "any person" may request, in writing, a public hearing on a permit application within the time period specified in the public notice for comments (33 CFR 327.4(b)). The written request must specify with particularity the reasons for holding

---

[6] In Dufau v. United States, 21 Envtl. L. Rep. (Envtl. L. Inst.) 20814, 22 Cl.Ct. 156 (1990), the U.S. Claims Court rejected an applicant's claim that a 16-month delay on a permit amounted to a temporary taking. See Section 7.4.2 for a discussion of regulatory takings challenges.

a public hearing. Following such a request, the district engineer may attempt to resolve the issues informally (usually by meeting with the applicant and the person requesting the hearing). Otherwise, the district engineer must grant the request for the public hearing, or make a determination in writing that "the issues raised are insubstantial or there is otherwise no valid interest to be served by a hearing" (33 CFR 327.4(b)). In practice, public hearings are seldom granted except for large or very controversial projects.

Notice of Hearing. Once a decision has been made that a public hearing will be held, the local Corps office issues a public notice of the hearing, which should normally provide for a period of at least 30 days until the hearing takes place (33 CFR 327.11(a)). The notice contains information on the time, place, and nature of the hearing, along with information on draft environmental impact statements or environmental assessments as may be required under NEPA, if applicable (see Section 7.6.2). The notice is given to all "interested parties," including anyone who requested a hearing, all affected federal agencies, state and local agencies, and is posted in appropriate governmental buildings and provided to newspapers of general circulation (33 CFR 327.11(a) and (b)).

Form and Conduct of the Hearing. The presiding officer of the hearing is the district engineer, deputy district engineer, or other qualified person (33 CFR 327.5(a)). The district counsel (or a designee) is usually present as advisor to the presiding officer. Any other person may appear at the hearing on his own behalf, or may be represented by counsel or other representative (33 CFR 327.7).

At the hearing, the presiding officer opens the proceedings with an opening statement outlining the purposes for the hearing, and describing the procedures to be followed (33 CFR 327.8(a)). On some occasions, the applicant may be given an opportunity to describe the project prior to statements by other witnesses. Oral and/or written statements may be made by "any person," although the presiding officer normally sets time limitations on oral presentations (e.g., five minutes per person). Since the hearing is a non-adversarial proceeding, no cross-examination is allowed (33 CFR 327.8(d)). However, any person (including the applicant) may be given an opportunity for rebuttal of other statements, although rebuttal may be limited by the presiding officer.

All statements at public hearings are on the record, and verbatim transcripts of the proceedings are made available for public inspection, and are available for purchase by any person. All written statements and exhibits offered in evidence at the hearing become part of the record, subject to exclusion by the presiding officer for redundancy (33 CFR 327.8(f)). After the close of the hearing, the presiding officer must allow a period of at least ten days for submission of additional written comments (33 CFR 327.8(g)).

7.3.2 Issuing the Permit

Once the district engineer has considered all the information on the permit application (including all public comments and hearings), he is required to issue a statement of findings (SOF), or issue a record of decision (ROD) in those cases where an environmental impact statement has been prepared pursuant to the requirements of NEPA (33 CFR 325.2(a)(6)). The SOF or ROD must include the district engineer's views on the effects of the proposed project on "the public interest," including compliance with all aspects of the § 404(b)(1) guidelines (40 CFR Part 230). The district engineer must normally accept the decisions of local and state governing bodies on zoning and land use issues (33 CFR 320.4(j)(2)), or he must include in the SOF or ROD an explanation of why national issues should override local or state decisions (33 CFR 325.2(a)(6)).

Corps regulations specify a careful "public interest review," which involves a cost-benefit balance that is nearly unique among environmental regulations:

> "The decision whether to issue a permit will be based on an evaluation of the probable impacts, including cumulative impacts, of the proposed activity and its intended use on the public interest. Evaluation of the probable impact ... on the public interest requires a careful weighing of all those factors which become relevant in each particular case. The benefits which reasonably may be expected to accrue from the proposal must be balanced against its reasonably foreseeable detriments." [33 CFR 320.4(a)]

This general balancing process determines whether a permit will be issued, and what kinds of conditions it will require. Among the factors to be considered are conservation, economics, esthetics, general environmental concerns, fish and wildlife values, water quality, and public welfare (see 33 CFR 320.4(a) for additional guidance on the balancing process).

If a § 404 permit is warranted, the district engineer decides what special conditions will be required, and what the duration of the permit should be. Once the district engineer's name and signature are affixed to the permit, it normally becomes a final action unless he must forward it to his superiors in the Corps or other agencies for final action (33 CFR 325.8(b)).

Corps regulations require the district engineer to consult with the FWS, NMFS, and state agencies responsible for fish and wildlife, and give full consideration to their views (33 CFR 320.4(c)). However, courts have held that the Corps need not defer to recommendations of these agencies (see *Sierra Club v. Alexander,* 484 F.Supp. 455 (N.D.N.Y. 1980), *aff'd,* 633 F.2d 206 (2d Cir. 1981)).

The duration of a § 404 permit generally depends on the nature of the permitted activity. Permits for permanent structures are usually indefinite with no expiration date. However, where a structure or project is temporary in nature, the permit will be of limited duration with a definite expiration date (33 CFR 325.6(b)). Permits automatically expire on the expiration date specified on the permit unless they are modified, suspended, or revoked (33 CFR 325.6(a)). Guidelines on permit modification, suspension, and revocation are given at 33 CFR 325.7.

7.3.3 General and Nationwide Permits

Section 404(e) of the Clean Water Act authorizes the Corps to issue "general" permits on a state, regional, and nationwide basis for any category of activities which are similar in nature and will have only minimal individual and cumulative environmental impacts (33 USC § 1344(e)(1)). The apparent intention of general permits is that unnecessary time and paperwork can be avoided for those activities that will have similar impacts in each case, such that complete, individual § 404 permit applications would be redundant.

In general, an activity that is covered by a general permit does not require a § 404 permit application, so long as the person complies with the conditions incorporated in the general permit (33 CFR 320.1(c)). Some nationwide permits require a pre-notification procedure prior to beginning construction (33 CFR 320.1(c) and 330.6(a)). To be on the safe side, it is probably prudent to contact the local Corps office prior to construction under any circumstance.

Nationwide Permits. The most significant general permits are called "nationwide" permits. These are issued by Corps headquarters in Washington, D.C., and apply equally in all parts of the nation. Because of their broad reach, nationwide permits must undergo a review process prior to implementation separate from that for more localized general permits (see 33 CFR 330.1).

There have been 26 nationwide permits issued to date, and these are listed and described in Corps regulations at Appendix A to 33 CFR Part 330. Nationwide permits cover such activities as construction of aids to navigation and regulatory markers (permit no. 1); scientific measurement devices (permit no. 5); utility line backfill and bedding (permit no. 12); minor discharges that do not exceed 25 cubic yards (permit no. 18); oil spill cleanup activities (permit no. 20); and maintenance dredging of existing marinas, canals, and boat slips (permit no. 35).

Nationwide permits incorporate a series of general conditions that apply to all nationwide permits. These are listed at Appendix A(C) to 33 CFR Part 330. Many of these pertain to environmental controls and maintenance associated with the permitted activities. It should be noted that nationwide permits still require compliance with the federal Endangered Species Act (see Section 7.6.6), as well as coastal zone management consistency requirements (see 33 CFR 330.4(d)).

Two or more nationwide permits can be combined to authorize a "single and complete project" (33 CFR 330.2(i)), but the same nationwide permit may *not* be used more than once for a single and complete project (33 CFR 330.6(c)). For projects that combine individual permits with nationwide permits, the applicant may proceed with the parts of the project covered by a nationwide permit while awaiting review of the individual permit (33 CFR 330.6(d)). Where an unauthorized fill occurred unintentionally, a nationwide permit may be applied as an after-the-fact permit.

It should be noted that district engineers have override authority over nationwide permits, which allows them to suspend, modify, or revoke nationwide permits if individual or cumulative impacts are more than minimal, or when the activity is contrary to the public interest (33 CFR 330.1(d), 330.4(e), and 330.5). It has been held that the Corps may suspend, modify, or revoke a nationwide permit without providing a hearing (*O'Connor v. Corps of Engineers*, 801 F.Supp. 185 (N.D.Ind. 1992)).

Nationwide Permit 26. Without question the most controversial nationwide permit is Nationwide Permit 26, which exempts from the permit process any discharges of dredged or fill materials into "headwaters and isolated waters," provided: (1) the discharge does not cause the loss of more than ten acres of waters; (2) the Corps district is notified of the discharge and loss; and (3) the discharge is part of a single and complete project. Nationwide Permit 26 applies to wetlands above the headwaters, as well as the water bodies to which they are adjacent.

Nationwide permit 26 is the result of a 1984 settlement agreement between the Corps and 16 environmental groups which had challenged a predecessor of the permit (see *National Wildlife Federation v. Marsh*, 14 Envtl. L. Rep. (Envtl. L. Inst.) 29262 (D.D.C. 1984)). Much of the controversy surrounding the Nationwide Permit 26 is that the potential for destruction of wetlands is enormous. Reviews of the issues surrounding the Nationwide Permit 26 debate are in Blumm and Zaleha (1989), Goode (1989), Goldman-Carter (1989), and Want (1993).

Other Exemptions. Section 404(f)(1) of the Clean Water Act (33 USC § 1344(f)(1)) lists several kinds of activities that are exempt from the requirement for a § 404 dredge and fill permit. No permit is required for discharges:

> (A) from normal farming, silviculture, and ranching activities such as plowing, seeding, cultivating, minor drainage, harvesting for the production of food, fiber, and forest products, or upland soil and water conservation practices;

(B) for the purpose of maintenance ... of currently serviceable structures such as dikes, dams, levees, groins, riprap, breakwaters, causeways, ...

(C) for the purpose of construction or maintenance of farm or stock ponds or irrigation ditches, or the maintenance of drainage ditches;

(D) for the purpose of construction of temporary sedimentation basins on a construction site ...;

(E) for the purpose of construction or maintenance of farm roads or forest roads, or temporary roads for moving mining equipment;

(F) resulting from any activity with respect to which a State has an approved [statewide water quality plan under 33 USC § 1288(b)(4)].

Several of these exemptions have been tested in court. For example, a federal district court in *United States v. Zanger* (767 F.Supp. 1030 (N.D. Cal. 1991) refused to allow a landowner to claim several different exemptions for an alleged stream bank erosion repair which was really a redirected stream channel and fill of the former streambed. The same court held in *Leslie Salt Co. v. United States* (22 Envtl. L. Rep. (Envtl. L. Inst.) 20361 (N.D. Cal. 1992)) that blockage of water flow through a culvert was not maintenance of a drainage ditch or existing structure.

The 404(f)(1) exemptions are limited by the § 404(f)(2) "recapture clause," which makes the exemptions inapplicable if the purpose of the activity is to bring navigable waters into a use to which they were not previously subject, or where the flow of water is impaired. Courts have construed narrowly the reach of § 404(f)(1) exemptions based on the § 404(f)(2) recapture clause (see review of cases in Want, 1993).

7.3.4 The EPA Veto

Section 404(c) of the Clean Water Act authorizes the Administrator of the U.S. EPA to veto a § 404 dredge and fill permit if, after notice and opportunity for public hearings and consultation with the Corps, she determines that "the discharge [of dredged or fill materials] ... will have an unacceptable adverse effect on municipal water supplies, shellfish beds and fishery areas (including spawning and breeding areas), wildlife, or recreational areas" (33 USC § 1344(c)). While EPA has actually used its veto only rarely, the *threat* of a possible veto has led the Corps to place modifications and conditions on many § 404 permits.

The scope of the EPA veto has been clarified by several major court cases. The famous *Sweedens Swamp* litigation, which is discussed in section 8.2.4 in the context of mitigation, addressed the issue in several separate decisions (see review in Bosselman, 1989). In *Newport Galleria v. Deland* (618 F.Supp. 1179 (D.D.C. 1985), the court held that the EPA veto was not reviewable in court because it was not a final agency action.[7] Later, a different court in *Bersani v. Deland* (640 F.Supp. 716 (D. Mass. 1986) rejected the developer's argument that EPA veto authority had expired because no decision regarding a veto was reached within the 30-day limit for a public hearing. In another decision, yet a third court held in *Bersani v. U.S. EPA* (674 F.Supp. 405 (N.D.N.Y. 1987), aff'd, 850 F.2d 36 (2d Cir. 1988)) that EPA could use the same criteria in its veto decision (i.e., availability of alternative sites) that the Corps had used in issuing the permit.

The EPA veto has been upheld by most courts.[8] In fact, the only case to date to reverse an EPA veto under § 404(c) is *James City County, Va. v. U.S. Environmental Protection Agency* (758 F.Supp. 348 (E.D. Va. 1990), aff'd in part and remanded, 955 F.2d 254 (4th Cir. 1992)). In that case, EPA had exercised its § 404(c) veto over a Corps permit for fill to construct a dam for a proposed reservoir. The court held that the evidence did not support EPA's conclusion that the county had practicable water supply alternatives, and remanded the case to EPA to see if the project's environmental effects alone justified a veto.

---

[7] In an unrelated case, Reid v. Marsh, 20 Env't Rep. Cas. (BNA) 1337 (N.D. Ohio 1984), the court held that the EPA veto is discretionary, and not subject to judicial review.

[8] For example, see Creppel v. United States, 19 Envtl. L. Rep. (Envtl. L. Inst.) 20134 (E.D. La. 1988); Russo Development Corp. v. Reilly, 20 Envtl. L. Rep. (Envtl. L. Inst.) 20938 (D.N.J. 1990); and City of Alma v. United States, 744 F.Supp. 1546 (S.D. Ga. 1990).

EPA subsequently vetoed the permit for a second time on the basis that it would have unacceptable adverse impacts, and the court again reversed the decision, this time because EPA considered environmental concerns but not human needs (*James City County v. U.S. Environmental Protection Agency*, 23 Envtl. L. Rep. (Envtl. L. Inst.) 20228 (E.D. Va. 1992)).

Despite the adverse decision in *James City County*, the EPA veto of dredge and fill permits under § 404(c) remains a powerful force demanding environmental sensitivity in § 404 permit decisions.

7.4 REMEDIES FOR § 404 PERMIT DENIALS

7.4.1 Lack of Administrative Appeals Mechanism

No mechanism exists for administrative appeal to the Corps, nor is there any right to an adjudicatory hearing for § 404 permit denials like there is for denial of other types of environmental permits. Still, a permittee or other party who is adversely impacted by a Corps wetland permit decision is not without legal remedy. After the Corps rules on a federal wetland permit application, a final decision may be subject to judicial review under the Administrative Procedure Act (5 USC §§ 701-706). In some cases, parties opposing the approval of a permit may have standing to institute a "citizen suit" under the Clean Water Act, National Environmental Policy Act or other environmental law (see Section 7.5).

However, permit applicants must be aware that judicial review is usually an uphill battle because in reviewing the Corps decision, the court is bound by a "substantial evidence" standard. Under this standard, the court may set aside the permit decision only if it finds that the Corps action is arbitrary, capricious or not supported by substantial evidence in the record (see *Avoyelles Sportsmen's League, Inc. v. Marsh*, 715 F.2d 897, 904 (5th Cir. 1983)). The courts are bound by the administrative record and may not admit new evidence unless the record is so scant that it makes judicial review virtually impossible (see *Friends of the Earth v. Hintz*, 800 F.2d 822, 828 (9th Cir. 1986)). In other cases, applicants who have been denied a permit may challenge the Corps decision in federal court on constitutional grounds. For example, regulatory takings actions have frequently been brought in federal court to challenge permit denials.

7.4.2 Regulatory Takings Challenges

A regulatory taking results when a governmental regulation places such a burdensome restriction on a landowner's use of property that the government has for all intents and purposes "taken" the landowner's property. The debatable issue is exactly how much the regulation must interfere with private property rights before it is deemed a "taking." Justice Holmes opted for a balancing test, stating that "Government could hardly go on if to some extent values incident to property could not be diminished without paying for every such change in the general law" (*Pennsylvania Coal Co. v. Mahon*, 260 U.S. 393 (1922)). The courts have increasingly grappled with this issue and no clear test has emerged. Although the U.S. Supreme Court has decided several land use takings cases since 1978,[9] the general consensus has been that no bright-line test can be followed in the regulatory takings context. Although several factors have been utilized to determine when a land use regulation results in a "taking" of property, "... the Supreme Court has characterized the analytic process as one relying instead on ad hoc, factual inquiries into the circumstances of each particular case..." (*Loveladies Harbor, Inc. v. U.S.*, 21 Cl. Ct. 153, 155, 20 Envt'l Rep. 21,207 (1990)).

Regulatory takings, just like physical governmental takings, such as condemnation, are governed by either the due process clause of the Fourteenth Amendment or the "taking" clause of the Fifth Amendment to the U.S. Constitution.[10] The Fifth Amendment is held

---

[9] Penn Central Transportation Co. v. New York City, 438 U.S. 104, 124 (1978); Agins v. City of Tiburon, 447 U.S. 255 (1980); Williamson County Regional Planning Comm'n v. Hamilton Bank of Johnson City, 473 U.S. 172 (1985); MacDonald, Sommer & Frates v. County of Yolo, 477 U.S. 340 (1986); Keystone Bituminous Coal Ass'n v. De Benedictis, 480 U.S. 470 (1987); Nollan v. California Coastal Commission, 483 U.S. 825 (1987); First English Evangelical Lutheran Church v. County of Los Angeles, 482 U.S. 324 (1987).

[10] The Fourteenth Amendment to the Constitution states: "Nor shall any state deprive any person of life, liberty, or property without due process of law; nor deny to any person within its jurisdiction the equal protection of the laws." U.S. Const. Amendment XIV.

The Fifth Amendment states: "No person shall be ... deprived of life, liberty or property without due process of law; nor shall private property be taken for public use, without just compensation." U.S. Const. Amendment V.

applicable to the states by incorporation into the "due process" clause of the Fourteenth Amendment (*Chicago, Burlington & Quincy R.R. v. Chicago*, 166 U.S. 226 (1897)). Whether or not the taking is caused by physical interference or regulatory interference with a landowner's property rights, just compensation is required under the Constitution whenever a taking occurs. State constitutions also contain due process and taking clauses, and form the basis for state claims for just compensation.

Clearly, whenever a government takes a landowner's property for a public purpose pursuant to its power of eminent domain, just compensation is due. The owner is generally entitled to just compensation calculated according to the property's fair market value based on the property's highest and best use unaffected by the intended condemnation (*U.S. v. Miller*, 317 U.S. 369 (1943); see also *Olson v. U.S.*, 292 U.S. 246 (1934)). However, in the regulatory taking area, possible just compensation is dependent on a court's assessment of individual facts and circumstances. The courts rely on a variety of factors in determining whether a government regulation has resulted in a taking of property for which just compensation is owed.

Government denial of a wetlands permit to promote environmental protectionist goals may serve a legitimate public purpose while depriving a landowner of all beneficial use of his property. In such an instance, courts have employed a balancing of interests test focusing on the nature and extent of the benefit derived for the public and the nature and extent of the loss occasioned on the landowner. This balancing test is the most frequently relied on method for determining whether a regulatory taking has occurred. Under this type analysis, the courts look to three factors in making a taking determination: (1) the character of the government action; (2) the economic impact of the regulation; and (3) interference with the landowner's reasonable investment-backed expectations. This test was originally established by the landmark case of *Pennsylvania Coal Co. v. Mahon* (260 U.S. 393 (1922)). No case requires that each of these factors be met before a taking can be found. They are used as guidelines in deciding whether a taking has occurred (*Loveladies Harbor, Inc. v. U.S.*, 21 Cl. Ct. 153, 160 n. 9, 20 Envt'l Rep. 21,207 (1990)).

7.4.3 Harm-Prevention Analysis

In the context of wetland regulation, the courts often take an additional factor into account in ruling on whether a particular land use regulation constitutes a regulatory taking. Many courts will employ a harm-prevention analysis and uphold a regulation even when all economically viable use of the landowner's property has been taken so long as the regulation is aimed at preventing a serious public harm.[11] Although this harm prevention analysis is grounded in longstanding precedent that economic loss to a private property owner is irrelevant when government regulation furthers the legitimate purpose of preventing harm to the public (*Mugler v. Kansas*, 123 U.S. 623, 668-69 (1887)), it is questionable how valid this line of cases may be in light of a 1992 U.S. Supreme Court decision in *Lucas v. South Carolina Coastal Council* (112 U.S. 2886, 34 Env't Rep. Cas. (BNA) 1897 (1992)).

Under Lucas, if a government regulation deprives a landowner of all beneficial and productive use of his property, just compensation is due even if the regulation is aimed at preventing a serious public harm. The only exception to this categorical rule is when in the context of a total regulatory taking, the government can show that the landowner's expectations regarding use of the property are unreasonable in light of state common law nuisance and property law principles. Courts have generally found that dredging or filling a wetland does not constitute a nuisance.[12]

---

[11] See, e.g., McNulty v. Town of Indialantic, 727 F. Supp. 604 (M.D. Fla. 1989), rejecting a landowner's contention that a setback requirement on oceanfront property worked a taking because "the government can destroy all economic use if necessary to avoid a public nuisance or nuisance-like use." 727 F. Supp. at 609; Presbytery v. King County, 114 Wash.2d 320, 787 P.2d 907 (1990), cert. denied, 111 S.Ct 284 (1990), holding that if a regulation protects the public from harm, and does not infringe on the landowner's right to possess, exclude others and dispose of his property, no taking has occurred; Claridge v. New Hampshire Wetlands Board, 125 N.H. 745, 485 A.2d 287 (1984), where the court upheld the denial of permit to fill tidal wetlands, thereby rendering the property of negligible economic value, based on a harm prevention rationale that the regulation prevented destruction the coastal habitat that would posing a further risk to the public welfare.

[12] *See, e.g.*, Ciampitti v. United States, 18 Cl. Ct. 548 (1989). Florida Rock Industries, Inc. v. United States, 21 Cl. Ct. 161, 31 Env't Rep. Cas. (BNA) 1835 (1990) (holding that dredging associated with limestone mining is not a noxious use);

7.4.4 Regulatory Takings in the Wetlands Context

In deciding regulatory takings challenges to wetland regulations, courts generally consider the "parcel as a whole," meaning the whole parcel as it was originally purchased. Thus, in *Deltona Corp. v. United States* (657 F.2d 1184 (Cl. Ct. 1981), *cert. denied*, 455 U.S. 1017 (1982)), the 404 permit denial did not preclude all economically viable use of property because accessible uplands interspersed with wetlands could still be developed.

In more recent wetlands taking cases, however, the U.S. Claims Court has declined to define the "parcel as a whole" as either the parcel which was originally purchased or the parcel that remained at the time of the alleged taking. Instead, the court has only considered the acreage to which a § 404 permit applies and to any other acreage directly affected by the denial, such as upland acreage entirely surrounded by wetlands.[13]

By narrowing the scope of the property interest at stake, the Claims Court has increased the economic impact of the permit denial and, thus, the likelihood that it will constitute a "taking." The court in *Loveladies Harbor v. United States* found that although denial of a permit under § 404 does not interfere with the property owner's right to possess the land, exclude others from it, or to transfer it, the result may nonetheless be confiscatory (21 Cl. Ct. 153, 31 Env't Rep. (BNA) 1847, 1853 n. 9 (1990)). Thus, the nature of the government action in wetlands cases is less determinative in establishing a taking. Economic factors are given greater weight in wetland cases.[14]

For example, in *Florida Rock Industries, Inc. v. United States* (21 Cl. Ct. 161, 31 Env't Rep. Cas. (BNA) 1835 (1990)), the court

---

Loveladies Harbor, Inc. v. United States, 21 Cl. Ct. 153, 31 Env't Rep. (BNA) 1847 (1990) (holding that filling wetlands for development is not a nuisance where it is conducted in a manner that will not violate applicable water quality standards).

[13] Florida Rock Industries, Inc. v. United States, 21 Cl. Ct. 161, 31 Env't Rep. Cas. (BNA) 1835 (1990); Loveladies Harbor, Inc. v. United States, 21 Cl. Ct. 153, 31 Env't Rep. (BNA) 1847 (1990).

[14] *See, e.g.*, Formanek v. United States, 18 Cl. Ct. 785 (1989) (an offer to purchase at a fraction of the value of the property before permit denial made by a conservation group not a sufficient "economically viable use" to prevent a court from finding that a taking has occurred.)

measured the value of the land based on its highest and most profitable use before permit denial and compared that to its fair market value for the highest and best use after permit denial. The court found that takings had occurred because substantially all the value of the properties had been lost as a result of the government action.

Where the property includes accessible and developable uplands that are beyond the jurisdiction of the Corps and may therefore be developed without a permit, courts have noted that all economically viable use may not have been eliminated by the regulation (see *Deltona Corp. v. United States*, 657 F.2d 1184 (Cl. Ct. 1981), cert. denied, 455 U.S. 1017 (1982)). However, where uplands are completely surrounded by wetlands, and thus permit denial effectively precludes use of the uplands, the existence of such uplands may support a takings claim. In fact, such uplands surrounded by wetlands may be considered taken as a result of the wetland permit denial (*Loveladies Harbor*, 21 Cl. Ct. 153, 31 Env't Rep. (BNA) 1847, 1853 (1990)).

In *Ciampitti v. United States* (22 Cl.Ct. 310, 32 Env't Rep. Cas.(BNA) 1608, 21 Envt'l L. Rep. (ELI) 20,866 (Cl. Ct. 1991)), the Claims Court ruled that the denial of a § 404 permit did not constitute a compensable taking where plaintiffs purchased their property in 1983 aware of state and federal restrictions on development in wetlands. The Claims Court reasoned that because plaintiffs had ample warning of the likelihood that the wetlands could not be developed, the permit denials did not interfere with reasonable, distinct investment-backed expectations.

7.5 CHALLENGING ISSUANCE OF A § 404 PERMIT

It is not uncommon for a person other than the applicant or permittee of a § 404 permit to wish to challenge the issuance of the permit. Many such challenges are initiated by environmental organizations or other groups of concerned citizens, but it is not uncommon for challenges to come from local or state governments, or from other agencies. Because there is no direct appeal mechanism to the Corps (see Section 7.4.1), the person wishing to challenge issuance of the permit must file a lawsuit in federal district court, although there are some shortcuts available under § 505 of the Clean Water Act (33 USC § 1365), the "citizen suit" provision.

7.5.1 Challenges in Federal District Court

Private parties have no right to an administrative appeal of a Corps permit decision.[15] As a result, a person wishing to challenge a permit issuance or denial, or a person wishing to challenge a wetlands determination, must do so by filing a lawsuit in federal district court according to federal question jurisdiction and the Administrative Procedures Act (5 USC § 551 *et seq.*).

A party wishing to challenge issuance of a § 404 permit must follow carefully the procedures for initiating and pursuing a lawsuit in federal court. The following Sections discuss some of the major issues involved in such challenges, but it should be cautioned that the process in more complex than is represented here, and the prudent litigant will seek professional legal advice before proceeding.

Ripeness. Courts will dismiss a lawsuit which has not "ripened" into a case or controversy suitable for adjudication. In the context of § 404 permit challenges, this usually means that there has been final agency action on the permit. Most lawsuits on the ripeness issue have been brought by disappointed landowners who challenge negative permit decisions, but it is clear that courts will not hear a case on wetland determination issues until the Corps and EPA have made final decisions on the wetlands (see *Avella v. U.S. Army Corps of Engineers*, 20 Envtl. L. Rep. (Envtl. L. Inst.) 20920 (S.D. Fla. 1990), *aff'd per curiam*, 916 F.2d 721 (11th Cir. 1991)), or on permit issuance issues until a permit has actually been issued or denied (see *Route 26 Land De. Assoc. v. United States*, 753 F.Supp. 532 (D.Del. 1990)).

In a related issue, an EPA decision to veto a permit under § 404(c) (see Section 7.3.4) does not become ripe until the veto is finalized. In *Newport Galleria Group v. Deland* (618 F.Supp. 1179 (D.D.C. 1985), the first *Sweedens Swamp* case) the court held that an EPA decision to consider a § 404(c) veto was not final agency action, and denied review.

---

[15] The EPA, USFWS, and National Marine Fisheries Service have a limited power of administrative review resulting from a series of MOAs from 1982 to 1992, and § 404(q) of the Clean Water Act, 33 USC § 1344(q). The MOAs are discussed in National Wetlands Newsletter, 1986; and Want, 1989.

Statute of Limitations. For most kinds of lawsuits, there exist various legal limitations on the time in which the lawsuit must be brought, known as "statutes of limitations." Interestingly, there are no specific statutes of limitations in the Clean Water Act or elsewhere that serve to limit the time in which an action against the Corps, EPA or other agencies may be brought in a wetland determination or permit case.

A case which specifically addressed the issue of a statute of limitations limit on filing § 404 enforcement actions is *North Carolina Wildlife Federation v. Woodbury* (19 Envtl. L. Rep. (Envtl. L. Inst.) 21308 (E.D.N.C. 1989)). In that case, the court held that the general five-year limit for civil penalty actions (but not injunctions; 28 USC § 2462) applied to a citizens suit challenge. Apparently, the only significant limitation to a lawsuit seeking an injunction is the doctrine of "laches," or extreme lack of diligence in pursuing the suit (see *United States v. Hobbs*, 736 F.Supp. 1406 (E.D. Va. 1990)), although it is possible that the United States is not subject to laches (see *United States v. Schmitt*, 734 F.Supp. 1035 (E.D.N.Y. 1990)).

Right to Jury Trial. In a significant wetlands decision, the United States Supreme Court in *Tull v. United States* (481 U.S. 412 (1987)) held that there is a right to a jury trial when the government seeks civil penalties, but *not* when it seeks an injunction or other equitable relief. Even so, the jury decides only the substantive issue of whether the property owner violated the provisions of the Clean Water Act, while the judge decides the amount of the penalty (see review in Want, 1993). Another court held that a jury trial is not required under § 10 of the Rivers and Harbors Act (*United States v. M.C.C. of Florida*, 863 F.2d 802 (11th Cir. 1989)).

Attorney's Fees. An important aspect of wetlands litigation is the circumstances under which a prevailing party may be awarded attorney's fees and expert witness' fees. Under § 505(d) of the Clean Water Act, such fees are awarded at the court's discretion to a "prevailing or substantially prevailing party" (Water Quality Act of 1987 § 505(c), amending 33 USC § 1365(d)).

In *National Wildlife Federation v. Hanson* (18 Envtl. L. Rep. (Envtl. L. Inst.) 20008 (E.D.N.C. 1987, *aff'd*, 859 F.2d 313 (4th Cir. 1988)), the court upheld an award of over $398,000 despite the government's claim that wetland determinations are discretionary actions, and only non-discretionary actions are subject to citizen suits. However, the court in *Golden Gate Audubon Society v. U.S. Army Corps of Engineers* (738 F.Supp. 339 (N.D. Cal. 1988) refused to grant attorney's fees because wetland jurisdictional determination is a discretionary matter (see Want, 1993).

Standard of Review. The typical standard that must be met by a person wishing to challenge a § 404 permit action is to prove that the agency's action was "arbitrary and capricious, an abuse of discretion, or otherwise not in accordance with law," which is the standard under § 10(e) of the Administrative Procedures Act (5 USC § 701). Some courts have argued that the Corps should receive special deference due to the technical expertise required in wetland determinations (*Avoyelles Sportsmen's League, Inc. v. Marsh*, 715 F.2d 897, 904 (5th Cir. 1983)).

Challenges to Corps permit issuance have met with some success. For example, in *Sierra Club v. Sigler* (695 F.2d 957 (5th Cir. 1983)) an environmental group successfully challenged a Corps permit based on the lack of an Environmental Impact Statement under NEPA. *Hough v. Marsh* (557 F.Supp. 74 (D. Mass. 1982) was a successful challenge based on failure to follow EPA's § 404(b)(1) guidelines. Several suits have been successful based on the Corps' failure properly to conduct a public interest review (e.g., *North Carolina v. Hudson*, 665 F.Supp. 428 (E.D.N.C. 1987)).

7.5.2 The Administrative Record and *De Novo* Review

An important issue in many lawsuits challenging either the issuance of a Corps permit or a wetland determination is whether the court is limited to reviewing the record as previously developed by the agencies, or whether it can gather additional evidence during the litigation. The latter is called a *de novo* review. Unfortunately, courts have been inconsistent in dealing with the issue.

In general, principles of administrative law prohibit a reviewing court from engaging in a *de novo* examination of evidence. In reviews of agency decisions, courts are limited to a review of the record as developed by the agency, because the court would otherwise be substituting its judgment for the expertise of the agency. Fortunately, this principle doesn't apply to situations in which the agency has an inadequate record, considers the wrong evidence, or fails to produce a record at all.

The leading wetlands case on *de novo* review of agency decisions is *Avoyelles Sportsmen's League v. Marsh* (715 F.2d 897 (5th Cir. 1983)). There, the district court had allowed expert witnesses for the plaintiff to present evidence on the quantity of wetlands involved in the dispute, but the appellate court held this to be error since the district court should have limited its review to the administrative record which was extensive. Several subsequent courts have followed *Avoyelles* (see review in Want, 1993).

However, several other courts have held that in enforcement actions the Corps bears the burden of persuasion in wetland jurisdiction issues, and that a *de novo* review of evidence is appropriate. In *Stoeco Dev., Ltd. v. Dep't of the Army Corps of Engineers* (792 F.2d 339 (D.N.J. 1992)), the court added that the Corps lacks an administrative procedure in its wetland determination process to generate a sufficient record for appeal, although the procedures exist in its permitting decisions. In *Leslie Salt Co. v. United States* (896 F.2d 354 (9th Cir. 1990), *cert. denied*, 111 S. Ct. 1089 (1991)) the court distinguished between citizen challenges to wetland determinations which are subject to the "arbitrary and capricious" standard when *de novo* review is not allowed, and Corps enforcement actions against a property owner when the Corps bears the burden of persuasion and *de novo* review is allowed. The question of *de novo* review of wetland determination actions remains unanswered at this point, and it is likely that additional evolution and refinement will take place in future cases.

In cases where a person has challenged the issuance of a § 404 permit, the majority of courts continue to adhere to the view that review is limited to the administrative record. However, some plaintiffs have been successful in arguing that a violation of NEPA occurred, and most courts are more receptive to allowing *de novo* evidence to be considered in NEPA cases which usually involve failure by the Corps to prepare an EIS (see Section 7.6.2). While some courts have allowed additional evidence in such cases (see *Louisiana Wildlife Fed'n v. York*, 603 F.Supp. 518 (W.D. La. 1984), *aff'd in part, vacated in part and remanded*, 761 F.2d 1044 (5th Cir. 1985)), others have rejected it (see *Town of Norfolk v. U.S. Army Corps of Eng'rs*, 22 Envtl. L. Rep. (Envtl. L. Inst.) 20105 (D. Mass. 1991), *aff'd*, 968 F.2d 1438 (1st Cir. 1992)).

7.6 IMPACT OF OTHER FEDERAL ENVIRONMENTAL LAWS

7.6.1 Overview of Other Environmental Laws That May Affect Wetland Activities

An applicant seeking Corps approval for a § 404 permit must keep in mind that final issuance of a permit requires compliance with other state and federal laws. The most important state approvals include water quality certification under § 401 of the Clean Water Act (33 USC § 1341) and a determination of consistency with state Coastal Zone Management Programs (see Section 7.6.5). Crucial federal law approvals include compliance with the National Environmental Policy Act and Endangered Species Act (see Sections 7.6.2 and 7.6.6). The following list outlines many of the important federal environmental laws that may impose additional permitting and regulatory compliance requirements on activities undertaken in wetland areas:

* Coastal Zone Management Act, 16 USC 1451 et seq.
* Endangered Species Act, 16 USC 1531 et seq.
* National Historic Preservation Act, 16 USC 470 et seq.
* Preservation of Historical and Archaeological Data Act, 16 USC 469 et seq.
* Marine Mammal Protection Act, 16 USC 1361 et seq.
* National Environmental Policy Act, 42 USC 4321 et seq.
* Rivers and Harbors Act, 33 USC 401 et seq.
* Clean Water Act, 33 U.S.C. 1251 et seq.
* Clean Air Act, 42 USC 7401 et seq.
* Coastal Barriers Resources Act, 16 USC 3501 et seq.
* Marine Sanctuaries Act, 16 USC 1431 et seq.

Obviously, discussion of each of these federal environmental laws is beyond the scope of this volume. A brief discussion of some of these laws is provided merely to alert the reader to the extensive federal system of environmental laws that may affect wetland activities. Of course the type and magnitude of the particular activity will determine which federal laws will impose additional regulatory requirements. It is again important to stress that further regulatory requirements may come from the state and local level (see Section 7.9 for information on state and local requirements).

7.6.2 National Environmental Policy Act (NEPA)

In 1969 Congress enacted NEPA to address the growing concern over the environment and the need for environmental protection (National Environmental Policy Act of 1969, 42 USC § 4331). Most states have enacted their own "little NEPA" environmental quality review acts modeled after the federal statute (see Robinson, 1982; Mandelker, 1985). NEPA sets goals and provides a broad scheme to protect the natural environment, including wetland areas (42 USC § 4321). NEPA requires that all federal agencies participate in achieving environmental protection goals (42 USC § 4332(2); see e.g., *Zabel v. Tabb*, 430 F.2d 199 (5th Cir. 1970), *cert. denied* 401 U.S. 910 (1971)).

Section 102 sets forth procedures for federal agencies to incorporate environmental considerations in their decisionmaking processes (42 USC § 4332(2)(C); 40 CFR 1502.3). The most important procedural requirements are contained in NEPA's action-forcing provisions, which are designed to ensure consideration of environmental factors and public participation in major federal agency decisionmaking (42 USC § 4332(2)(C)). To satisfy NEPA's procedural requirements, agencies must prepare environmental assessments (EAs) for proposed activities (40 CFR 1501.4(b), 1508.9), unless the action falls within a categorical exclusion (40 CFR 1501.4(a)(2)). Based on determinations made in the EA, the agency is then required to decide whether a full environmental impact statement ("EIS;" 40 CFR 1501.4(c), 1508.11) is needed or whether it may issue a finding of no significant impact ("FONSI;" 40 CFR 1501.4(e), 1508.13) The federal agency may also choose to bypass the EA step and prepare an EIS for the proposed activity if it is one for which agency regulations usually require an EIS (40 CFR 1501.4(a)(1)).

NEPA requires that an EIS to be prepared and included "in every recommendation or report on proposals for legislation and other major federal actions significantly affecting the quality of the human environment..." (42 USC § 4332(C); see also 40 CFR 1502.3, 1508.18; *Upper Snake River Chapter of Trout Unlimited v. Hodel*, 921 F.2d 232 (9th Cir. 1990)).

Once an agency determines that an action requires the preparation of an EIS, it must initiate the "scoping" process, "an early and open process for determining the scope of issues to be addressed and for identifying the significant issues related to a proposed action" (40 CFR 1501.7, 1508.25) during which the agency notifies and invites all concerned agencies and individuals to participate in the decisionmaking process (40 CFR 1501.7(a)(1)). The agency identifies the issues to be analyzed and determines the scope and depth of the environmental analysis. Then the agency prepares the EIS, which concisely describes how NEPA's policies will be achieved and analyzes a range of alternatives to the proposed action. The environmental analysis in the EIS must take a "hard look" at the environmental consequences of the proposed action (*Kleppe v. Sierra Club*, 427 U.S. 390, 410 n.21 (1975); see also *Marble Mountain Audubon Society v. Rice*, 914 F.2d 179 (9th Cir. 1990)).

An agency must also take the cumulative impact of a project into consideration when deciding whether to prepare an EIS (see *City of Tenakee Springs v. Block*, 915 F.2d 1308 (9th Cir. 1990); *Sierra Club v. Penfold*, 857 F.2d 1307, 1320-21 (9th Cir. 1988)). Cumulative impacts can result from individually minor but collectively significant actions taking place over a period of time (40 CFR 1508.7). Developers proposing projects in wetland areas should, therefore, be careful to consider impacts from other development projects in the vicinity of the proposed site when reviewing the environmental impacts of the development activity.

7.6.3 River and Harbors Act of 1899 (RHA)

Under § 10 of the Rivers and Harbors Act of 1899(RHA), a developer or landowner may need to apply for an Obstruction and Alteration Permit from the Army Corps of Engineers (33 USC § 403). Section 10 of the RHA prohibits the "creation of any obstruction not affirmatively authorized by Congress, to the navigable capacity of any of the waters of the United States...." Generally, the RHA requires that a landowner secure a § 10 permit from the Corps before building any wharf, pier, or other structure in any water of the United States outside established harbor lines (see *State of New York v. DeLyser*, 759 F. Supp. 982 (W.D.N.Y. 1991)).

7.6.4 Clean Water Act (CWA)

Besides the 404 program that applies primarily to dredge and fill activities in wetland areas, other provisions of the Clean Water Act may come into play depending on the type of proposed activity. The CWA has a detailed regulatory scheme concerning discharges of pollutants into U.S. waters. All point source pollution is prohibited unless authorized by a National or State Pollutant Discharge Elimination System permit (33 USC § 1342). Whether the developer or landowner applies for a state or federal permit will depend on whether the state has an approved permit program. If the state has an approved program in place, no federal permit is required (33 USC § 1342(c)). The effluent limitations specified in the permit are based on two separate regulatory standards. First, the water quality standards for receiving waters require that discharges not degrade water quality below certain levels.[16] Second, the discharges must meet certain technological treatment performance standards. The technologically based effluent limitations are known as best practicable technology, best conventional technology, and best available technology ("BAT;" 33 USC § 1311(b)). These technology-based requirements are contained in "effluent guidelines" for each specific industry or category (33 USC § 1314(b) and 40 CFR Part 401).

Recognizing the environmental problems stemming from non-point source pollution, Congress also implemented a national program authorizing federal funding to states for the control of nonpoint source pollution in 1987 Amendments to the Clean Water Act (Clean Water Act § 319, 33 USC § 1329). In 1990, EPA developed guidelines on how to ensure effective application of water quality standards to wetlands. State nonpoint source assessments completed for EPA showed that nonpoint source pollution can have a significant adverse impacts on wetlands. EPA concluded in its final report to Congress on § 319 of the Clean Water Act that "all state nonpoint source programs need to include provisions for preserving and protecting wetlands" (U.S. Environmental Protection Agency, 1992).

---

[16] The standards are set by the federal government in the case of a National Pollutant Discharge Elimination System (NPDES) permit or by the state in accordance with federally established criteria in the case of a State Pollutant Discharge Elimination System (SPDES) permit.

Nonpoint source pollution control is therefore certain to be a new regulatory focus in maintaining environmental integrity of wetlands.

7.6.5 Coastal Zone Management Act

In 1972, Congress passed the Coastal Zone Management Act ("CZMA;" 16 USC § 1451 *et seq.*) aimed specifically at management of coastal environmental problems. With passage of the CZMA, the federal government set up a complex coastal zone management scheme, implemented primarily at the state and local government levels, which places stringent controls on activities that adversely impact on the coastal environment, including coastal wetlands.

The program implementation stage began in 1976 when the State of Washington submitted the first program to be approved by the Secretary of Commerce. The State of Oregon followed in 1977. The majority of state programs were approved during the period 1978-1982. By the end of 1992, 29 of 35 potentially eligible coastal states and U.S. territories had received federal approval for coastal management programs (CMPs). Virginia's CMP received federal program approval in 1986. Ohio is currently awaiting federal approval of its program. Texas, Georgia, Illinois, Indiana, and Minnesota are have yet to submit coastal management programs to the Secretary of Commerce.

The nature and structure of CMPs vary widely from state to state. Some states, like North Carolina and Connecticut, passed comprehensive legislation as a framework for coastal zone management. Other states, like Oregon, use existing land use legislation as the foundation for their programs. Finally, states like Florida and Massachusetts linked existing, single-purpose laws into a comprehensive umbrella for coastal zone management.

Once a state has a federally-approved CMP, it possesses a powerful tool for regulating activities affecting its coastal zone. Section 307 of the CZMA provides that all federal activities and projects affecting the state's coastal zone, as well as activities carried out by private parties that require a federal permit or license, be consistent with the state's approved CMP (16 USC § 1456(c)). Applicants for a federal permit must demonstrate to the state coastal zone authority that the proposed activity complies with the policies of the state's CMP (16 USC § 1456(c)(3)(A)). If the state authority finds that the proposed activity is inconsistent with the CMP, the federal agency may not issue the necessary permits unless the Secretary of Commerce overrides the state's objection on appeal or by his own initiative.

Applicants for a federal wetland permit should first consult with the designated state agency to obtain the views and assistance of that agency regarding the means for ensuring that the proposed activity will be conducted in a manner consistent with the State's management program. This is the most common sense method of assuring that the project will not later be contested for failure to comply with the state CMP. The applicant should also consult the coastal management program document to determine it policies and goals. Included in the CMP is a list of federal license and permit activities which the state deems likely to affect the coastal zone and which the state will expect to review for consistency with its program. This list is mandatory as part of the approved program and must describe the types of federal licenses and permits involved.

The issuance of state and local permits also depends on whether the activity for which the permit is sought will be carried out in a manner that is consistent with the state CMP. The procedures for certifying consistency vary from state to state.

7.6.6 Endangered Species Act of 1973

The ultimate goal of the Endangered Species Act ("ESA;" The Endangered Species Act of 1973, 16 USC § 1531 *et seq.*) is "to provide a means whereby the ecosystems upon which endangered species and threatened species depend may be conserved, [and] to provide a program for the conservation of such endangered species and threatened species . . . ." (16 USC § 1531(b)). Congress enacted the statute in 1973 in response to its finding that "various species of fish, wildlife, and plants in the United States have been rendered extinct as a consequence of economic growth and development untempered by adequate concern and conservation" (16 USC § 1531(a)(1)). The ESA reflects congressional recognition of the benefits of species preservation and diversity.[17]

The ESA protects the ecosystems that support endangered

---

[17] 16 USC § 1531(a)(3). The preservation of biological diversity as a subsidiary goal of the ESA was noted in S. Rep. No. 307, 93d Cong., 1st Sess. 1-2, reprinted in 1973 U.S. Code Cong. & Admin. News 2989, 2990.

species by: (1) providing for the designation of "critical habitat" (16 USC § 1533(a)(3)); (2) requiring federal interagency cooperation to "insure that [agency action] is not likely to . . . result in the destruction or adverse modification of habitat" (16 USC § 1536(a)(2)); and (3) forbidding significant habitat modification, as a form of "taking," prohibited under § 9. Critical habitat is defined as an area inhabited by a threatened or endangered species "on which are found those physical or biological features (I) essential to the conservation of the species and (II) which may require special management considerations or protection" (16 USC § 1532(5)(A)(i)). FWS regulations suggest that when designating critical habitat, the Secretary should consider a wide range of habitat requirements such as space, food, water, light, nutrients, cover or shelter, sites for reproduction, and the historic distribution of the species (50 CFR 424.12(b)).

Sections 7 and 9 of the ESA are the two primary provisions for implementing endangered species preservation. Section 7 protects against federal agency actions, either undertaken by the federal government or subject to federal approvals, which jeopardize the continued existence of endangered or threatened species or that destroy or adversely affect critical habitats (16 USC § 1536(a)(2)). With the large number of projects involving federal permits, approvals, funding, or participation, the potential impact of § 7 is quite extensive. In the landmark case of *Tennessee Valley Authority v. Hill* (437 U.S. 153 (1978)), an endangered species of fish, the snail darter, temporarily blocked construction of the Tellico Dam.

Section 9 of the ESA applies to much a larger groups of people, entities, and projects than those covered by § 7 restrictions on federal actions. Section 9 prohibits all "takings" of endangered species by any "person" under the jurisdiction of the United States (16 USC § 1538(a)(1)). This broad authorization is meant to include private individuals and entities, as well as federal, state, and local governments and officials (16 USC § 1532(13)). Property owners and developers should always give consideration to whether an endangered species occupies a wetland area.

Section 9 prohibits persons from "taking," within the United States and its territory or on the high seas, any species of wildlife or fish that is on the endangered species list (16 USC § 1538(a)(1)(B)-(C)). Still, § 10(a) authorizes the granting of incidental taking permits, which allow for some harm to individual members of a species if certain stringent mitigation measures are met (16 USC § 1539(a)). To apply for relief under this provision, landowners and developers must prepare Habitat Conservation Plans (HCPs), which describe the permitted development and any required mitigation measures (16 USC § 1539(a)(2)(A)).

The provisions of § 9 that protect wildlife and fish differ from those that protect plant species. In contrast to wildlife and fish, § 9 does not outlaw takings of plants (16 USC § 1538(a)(2)(B)). The FWS has specifically limited § 9's applicability for plants to acts of vandalism and interstate or international trade in endangered plants. It does not cover normal land development activities, including mining, grazing, logging, and other activities that might displace plants (see H.R. Rep. No. 467, 100th Cong., 1st Sess. 15 (1988)).

As defined in the ESA, the term "take" means: "to harass, harm, pursue, hunt, shoot, wound, kill, trap, capture, or collect, or to attempt to engage in any such conduct" (16 USC § 1532(19)). The FWS has defined "harm" to include habitat modification or degradation that actually kills or injures wildlife by significantly impairing essential behavioral patterns, such as breeding, feeding or sheltering" (50 CFR 17.3). This ban on actions that cause death or injury extends to actions that cause a decline in the overall population of the endangered species. For example, in *Palila v. Hawaii Department of Land & Natural Resources (Palila I)* (471 F.Supp. 985 (D. Haw. 1979), aff'd, 639 F.2d 495 (9th Cir. 1981)), the Sierra Club moved to prevent the destruction of the vegetation on which a bird, the Palila, exclusively survived. However, in order to prevail on § 9 taking grounds, plaintiffs must show that the harm is relatively certain or imminent (see *North Slope Borough v. Andrus*, 486 F.Supp. 326 (D.D.C. 1979), aff'd in part & rev'd in part, 642 F.2d 610 (D.C. Cir. 1980)).

The FWS interprets the term "to harass," as included in the definition of "take," to mean "an intentional or negligent act or omission" creating a likelihood of injury to wildlife by disrupting it to such a degree as to disturb normal behavioral patterns such as breeding, feeding or sheltering (50 CFR 17.3). This definition of "harass" has an even greater potential for impact than the term "harm" because the act or omission need not be intentional or knowing; negligence is sufficient to constitute a taking. Moreover, the act or omission need not result in actual injury; likelihood of injury suffices to trigger § 9. Thus, a developer's mere failure to plan a wetland project with the effects of traffic, noise, human presence, and such in mind for an area inhabited by an endangered species might constitute a taking under this definition of harass, provided injury was likely and the developer could have reasonably foreseen the disturbance caused by the project.

7.6.7 Marine Mammal Protection Act of 1972

In 1972, Congress enacted the Marine Mammal Protection Act ("MMPA")[18] to "maintain the health and stability of the marine ecosystem" and to protect all marine mammals (16 USC § 1361(6)). In the MMPA, "marine mammal" is defined as "any mammal which (A) is morphologically adapted to the marine environment (including sea otters and members of the orders Sirenia, Pinnipedia and Cetacea), or (B) primarily inhabits the marine environment (such as the polar bear)" (16 USC § 1362(5)). The primary purpose of the MMPA is aimed at "a complete cessation of the taking of marine mammals and a complete ban on the importation into the United States of marine mammals and marine mammal products" (16 USC § 1362(7)). "Take" is defined in the MMPA as "to harass, hunt, capture, or kill, or attempt to harass, hunt, capture, or kill any marine mammal" (16 USC § 1362(13)). Neither the MMPA nor its implementing regulations require that a taking be intentional, and the legislative history of the Act supports this stance (H.R. Rep. No. 707, 92d Cong., 1st Sess. 23 (1971); see also 50 CFR 183). The MMPA would generally have only a limited impact on activities in coastal wetlands, however, because the coastal zone is home to many marine mammals, landowners and developers must remember to carry out their land use activities in a manner that would not result in a "taking" of marine mammals.

7.6.8 Marine Sanctuaries Act

The Marine Sanctuaries Act (16 USC §§ 1431-1434), provides for regulation and monitoring of the use of selected areas of the marine

---

[18] 16 USC §§ 1361-1407. The MMPA divides jurisdiction over marine mammals between the Departments of Commerce and the Interior. The Secretary of Commerce has responsibility for cetacea and pinnipedia, with the exception of walruses. The Secretary of Interior has jurisdiction over all other marine mammals. 16 USC § 1362(12). MMPA responsibilities are delegated by DOC and DOI to the National Marine Fisheries Service and the Fish and Wildlife Service, as well as the Marine Mammal Commission.

environment valued for their uniqueness, beauty, and historical significance (16 USC § 1431(b)). The Department of the Interior designates certain marine areas as sanctuaries in which land use activities are prohibited unless a special use permit is issued by the Secretary (16 USC § 1441(b)). If a developer or owner has property within a coastal wetland sanctuary, a special use permit would become an additional hurdle in the development process. These permits are only good for five years, unless renewed upon expiration (16 USC § 1441(b)(2)).

7.7 ADVANCE IDENTIFICATION OF WETLANDS (ADID)

One of the newest and most promising mechanisms available for governmental protection of wetlands is the "Advance Identification of Wetlands" (ADID or AVID) process which is authorized by the EPA's § 404(b)(1) regulations.

Under the regulations, EPA and the Corps, on their own initiative or at the request of any other party (and after consultation with any affected state), "may identify sites which will be considered as: (1) possible future disposal sites, including existing disposal sites and non-sensitive areas; or (2) areas generally unsuitable for disposal site specification ..." (40 CFR 230.80(a)).

The designation of an area as acceptable for § 404 dredge and fill disposal activities does not, however, constitute a § 404 permit, nor does identification of an area as not available necessary preclude a § 404 permit in the future (40 CFR 230.80(b)). On the other hand, it operates as a form of notice to property owners as well as local, state, and federal agencies of the likely acceptability of an individual or general § 404 permit in an affected area, and may facilitate permit approval. As such, the ADID is a valuable and powerful planning tool that is increasing in popularity across the nation. As of June, 1991 20 ADID projects had been completed and another 39 were ongoing (U.S. Environmental Protection Agency, 1991b). Over 20 others have been subsequently initiated, approved, or completed.

7.7.1 The ADID Process

As a general matter, the purpose of an ADID is to help protect and manage the nation's remaining wetlands by determining which wetlands are of high ecological value and should be protected from dredge and fill activities. By making these determinations *before* an application for a permit takes place, the intent is to prevent inadvertent (or intentional) unpermitted wetland losses.

The ADID process begins when the EPA initiates ADID procedures. Under the § 404(b)(1) guidelines, EPA (and the Corps) must consider the likelihood that future dredge and fill activities within the ADID boundaries will result in compliance with the guidelines. To facilitate this analysis, EPA and the Corps review available water resources management data, including data from the public, other federal and state agencies, and information from approved coastal zone management plans and river basin plans (40 CFR 230.80(d)).

As stated by the regulations, however, the process can also be initiated by "any person" (40 CFR 230.80(a)). Although initiation by a person other than the EPA can take place in a variety of ways, it is common for the person to apply directly to the nearest EPA Wetlands Division office. There is no standard form which the application must follow, but it is prudent for the following information to be included: (1) the purpose and scope of the proposed ADID; (2) the proposed ADID study area; (3) objectives of the ADID (including effects on ecological resources, public interest, fish and wildlife values, recreation, storm and flooding protection, health and welfare, and opportunities for interagency cooperation); (4) other potential uses of ADID study data and findings to other laws and regulations; and (5) identification of existing data resources.

Once the ADID process commences, an appropriate public notice is issued (40 CFR 230.80(c)). The Corps (or other permitting authority) maintains a public record of the identified areas, and a written statement of the basis for identification (40 CFR 230.80(e)).

7.7.2 The Rookery Bay ADID

Collier County in the southwest corner of the state of Florida is one of the fastest growing areas in the nation. It is also one of the most environmentally sensitive areas, containing vast expanses of wetlands, including much of the Everglades (see Section 3.2.3), the Big Cypress Swamp, and a variety of others, and several endangered species, such as the Florida panther, West Indian Manatee, and red-cockaded woodpecker. Of particular importance was the presence within the area of the Rookery Bay National Estuarine Research Reserve, which contains 8,400 acres of a pristine estuarine ecosystem.

In response to the threats to these valuable wetland resources, an ADID application was submitted to EPA in July, 1990 on behalf of the Florida Audubon Society, Collier County Audubon Society, and The Conservancy, Inc. of Naples, FL (see Figure 7.1). Within months, EPA began investigating the possibility of conducting an ADID within Collier County, and the decision to pursue the ADID followed soon after.

The final Rookery Bay ADID project encompasses 108,000 acres in Collier County, FL. EPA has been and remains the lead agency for the ADID, although other agencies in the "study team" include the Corps, FWS, Florida Department of Natural Resources, Florida Game and Freshwater Fish Commission, South Florida Water Management District, and Collier County. A series of preliminary workshops and public meetings have taken place, and public comments have been received. The ADID project is scheduled to be completed in 1994 (U.S. Environmental Protection Agency, 1992e).

7.8 SPECIAL AREA MANAGEMENT PLANS (SAMPS)

7.8.1 Purpose and Development of SAMPs

The Special Area Management Plan (SAMP) process is a comprehensive plan providing for natural resource protection and reasonable economic growth, which contains detailed statement of policies and criteria to guide land and water uses in specific geographic areas. The Coastal Zone Management Act (16 USC § 1451 *et seq.*) defines "special area master plan" as:

> "a comprehensive plan providing for natural resource protection and reasonable coastal-dependent economic growth containing a detailed and comprehensive statement of policies; standards and criteria to guide public and private land uses of lands and waters; and mechanisms for timely implementation in specific geographic areas within the coastal zone" (16 USC § 1453(17).

A SAMP provides predictability to developmental interests by

The Regulatory Framework 265

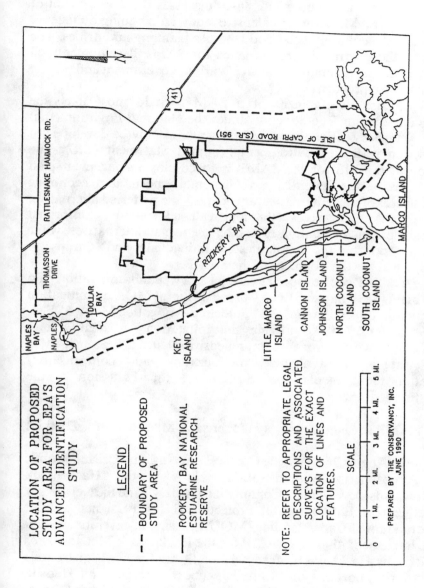

Figure 7.1: Map of the proposed Rookery Bay ADID, which accompanied the July, 1990 ADID application. The final ADID study area was over twice the area of the proposal.

establishing an area-wide basis for regulatory actions founded on cumulative effects of changes in the environment. The Corps Regulatory Guidance Letter No. 86-10 discusses the development of SAMPs. The SAMP requires extensive study an planning by federal, state and local environmental and land use planning authorities. The nature of the geographic area targeted for SAMP development will determine the degree of involvement of various government and private interests in the process.

Because development of a SAMP would most likely be considered a major federal action under the National Environmental Policy Act, (See Section 7.6) a SAMP would usually be developed in conjunction with an Environmental Impact Statement (EIS). The function of the EIS for the SAMP is to develop management plan alternatives, assess potential environmental, social, and economic consequences of each alternative, and identify the preferred alternative. A benefit of the EIS process is that it provides a forum for the informed identification and evaluation of management plan alternatives, while allowing opportunity for interested individuals and groups to participate in the development of the SAMP.

A SAMP can be especially useful as a wetlands mitigation plan for an area. (See Chapter 8) The Corps/EPA joint MOA on mitigation determinations (Section 404(b)(1) Guidelines Mitigation MOA, 55 Fed. Reg. 9210 (Feb. 7, 1990) provides that mitigation consistent with an EPA- and Corps-approved comprehensive plan, such as a SAMP, would satisfy the avoidance, minimization, and compensatory mitigation requirements. (See Section 8.1 for discussion of the mitigation sequence).

7.8.2 SAMP for New Jersey's Hackensack Meadowlands District

On August 26, 1988, the Corps, EPA, National Oceanographic and Atmospheric Administration (NOAA), the Hackensack Meadowlands Development Commission (HMDC) and the New Jersey Department of Environmental Protection (NJDEP) signed a Joint Memorandum of Understanding (MOU) agreeing to development and future implementation of a SAMP for the Hackensack Meadowlands District. The Hackensack Meadowlands District is a 32 square mile area covering portions of 14 municipalities in Bergen and Hudson Counties, New Jersey. Remaining undeveloped areas within the District are primarily wetlands, which are under substantial development pressure.

The SAMP will be a comprehensive plan for natural resource protection and reasonable economic growth within the District. It will contain a comprehensive statement of policies and criteria to guide future uses, including restoration and enhancement of land. water, and wetlands in the District. The SAMP will also foster compliance of future development activities with environmental laws and regulations, including the EPA/Corps Section 404(b)(1) guidelines. The SAMP will also provide assistance to the HMDC in ongoing efforts to revise its land use master plan for the District.

Because of the potential far-reaching environmental consequences of the SAMP, the Corps and EPA have agreed to prepare an EIS for the SAMP. EPA has also initiated an Advance Identification (ADID) study of the Hackensack Meadowlands wetlands as a first step toward identifying which wetland areas of the District are the most valuable, and therefore, in need of the highest levels of protection. (See Section 7.7).

Few SAMPs have been completed to-date. The SAMP under development for the Hackensack Meadowlands District should provide an important model for development of SAMPs of other wetland and environmentally sensitive areas of the country.[19]

7.9 STATE AND LOCAL WETLAND REGULATIONS

7.9.1 State and Local Wetland Permit Requirements

Landowners and developers who wish to carry out activities in wetland areas must find their way through a whole maze of federal, state and local laws and regulations before finally securing the necessary approvals for a specific activity or project. Federal wetland approvals can not be obtained unless necessary a state wetland permit approvals have been secured (33 CFR 320.4(j)). A certification of water quality may be necessary under federal and state law. For some projects, an environmental impact assessment may be required pursuant to a state's environmental quality review act. Local zoning regulations, such as

---

[19] For further information on the New Jersey Hackensack Meadowlands District SAMP, contact HMDC at (201) 460-1700.

wetland ordinances, may impose additional requirements. In addition to all these laws, the landowner or developer may need to certify that coastal wetland activities will be carried out in a manner consistent with a state's coastal management program.

The landowner or developer should consult state wetland maps to determine if the project is in an area subject to state wetland regulation. Federal wetland maps are also available from the Fish and Wildlife Service. The Army Corps of Engineers relies on these maps as a guide to whether certain areas will be subject to Corps jurisdiction, although not so much as a regulatory device like the states. Since each state's regulatory scheme for wetlands varies, it is necessary to determine the appropriate procedures and consult the state's wetland regulations when applying for a state wetland permit. The degree of regulation may be minimal in some states, while quite stringent in others. Some states have separate regulations for tidal and freshwater wetlands. The regulations themselves may be difficult to locate. Some states may have enacted a specific law to govern wetlands, while others have tagged wetlands regulation onto an already existing environmental law. Finally, a number of states have delegated authority to local governments to regulate wetlands. In these states, an enabling law is usually passed that contains certain standards that must be maintained when municipalities enact local wetland ordinances.[20] Thus, a landowner or developer may need to comply with three layers of regulation from federal, state, and local authorities.

By way of illustration, a description of the New York tidal wetlands law and the New Jersey wetlands law is provided in the following sections. A New York state tidal wetlands case is discussed to show how state requirements may impact a proposed wetland project. Finally, a U.S. Claims Court case arising in New Jersey is examined to demonstrate the important interplay between state and federal wetlands regulation. The importance of understanding wetlands maps is also

---

[20] For example, see Va. Code Ann. § 28.2-1302, which provides for adoption of local wetland zoning ordinances. In Connecticut, each local government is required to establish an "Inland Wetlands Agency" or else authorize an existing board or commission to carry out the state's wetlands regulations. Conn. Gen. Stat. § 22a-42(a). See also Mario v. Town of Fairfield, 585 A.2d 87 (Conn. 1991) concerning local jurisdiction of inland wetland agencies. Other state laws governing local oversight of wetland regulation include: Fla. Stat. § 403.916; Mass. Gen. L. ch. 131, § 40; N.Y. Envt'l Conserv. Law Sections 24-0501, 25-0507.

stressed in the two case studies.

7.9.2 New York State Tidal Wetland Regulation

Pursuant to the Clean Water Act § 404 wetland program administered by the Army Corps of Engineers, individual state's may adopt and administer their own wetland protection programs once approved by the Army Corps (33 USC § 1344(g), (h)). New York has separate regulatory programs for tidal and freshwater wetlands (N.Y. Envtl. Conserv. Law art. 24 (freshwater wetlands); N.Y. Envtl. Conserv. Law art. 25 (tidal wetlands).). New York State statutory authority regarding tidal wetlands is found in Article 25 of the New York Environmental Conservation Law (N.Y. Envtl. Conserv. Law art. 25 (McKinney)). The New York Department of Environmental Conservation is responsible of administering the statute. The stated policy of Article 25 is "to preserve and protect tidal wetlands, and to prevent their despoliation and destruction, giving due consideration to the reasonable economic and social development of this state" (N.Y. Envtl. Conserv. Law § 25-0102). "Tidal Wetlands" are defined as "those areas which border on or lie beneath tidal waters, such as, but not limited to, banks, bogs, salt marsh, swamps, meadows, flats or other low lands subject to tidal action, including those areas now or formerly connected to tidal waters."[21]

Under the statute, the commissioner of environmental conservation is required to inventory all tidal wetlands in New York state, setting forth boundaries using photographic and cartographic techniques to clearly and accurately map the state's tidal wetlands (N.Y. Envtl. Conserv. Law § 25-0201(1)). The commissioner is given broad discretion in amending the wetlands inventory map (see *Thompson v. Dept. of Environmental Conservation,* 130 Misc.2d 123, 495 N.Y.S.2d 107 (Sup. Ct. 1985), *aff'd,* 132 A.D.2d 665, 518 N.Y.S.2d 36 (App. Div. 1987), *appeal denied,* 71 N.Y.2d 803, 527 N.Y.S.2d 769, 522 N.E.2d 1067 (1988); *Jack Coletta, Inc. v. New York State Dept. of*

---

[21] N.Y. Envtl. Conserv. Law § 25-0103(1)(a). The definition also contains a listing of vegetation indicative of tidal wetlands. N.Y. Envtl. Conserv. Law § 25-0103(1)(b). The New York case study in Section 7.9.4 provides an excellent example of a "formerly connected" tidal wetland. See also O'Brien v. Barnes Building Co., Inc., 85 Misc.2d 424, 380 N.Y.S.2d 405, aff'd mem. sub. nom. O'Brien v. Biggane, 372 N.Y.S.2d 992 (App. Div. 1975).

Environmental Conservation, 128 A.D.2d 755, 513 N.Y.S.2d 465 (1987), *appeal denied,* 70 N.Y.2d 602, 518 N.Y.S.2d 1025, 512 N.E.2d 551 (1987)). The commissioner must file a detailed description of the technical criteria used to delineate the tidal wetlands with the secretary of state (N.Y. Envtl. Conserv. Law § 25-0201(2); 6 N.Y.C.R.R. § 661.27 contains the regulations pertaining to maintenance and amendments to the inventory map).

Under the tidal wetlands program the Department of Environmental Conservation works with local governments to protect tidal wetlands found in the locality (N.Y. Envtl. Conserv. Law § 25-0301). In addition to any permits required by the municipality in which the tidal wetland is located, the statute requires application to the commissioner for additional permits before any form of draining, dredging, excavation, dumping, filling, or erection of any structure can be undertaken (N.Y. Envtl. Conserv. Law § 25-0401). The statute requires that any land use activity in tidal wetlands be compatible with land use regulations promulgated pursuant to the statute (N.Y. Envtl. Conserv. Law § 25-0302).

7.9.3 New Jersey State Coastal Wetlands Regulation

The New Jersey coastal wetlands statute has provisions not much different from the New York tidal wetlands scheme. The New Jersey coastal wetlands law is part of Title 13, "Conservation and Development," codified as Chapter 9A, "Coastal Wetlands" (N.J. Stat. Ann. § 13:9A). Under the statute, the state legislature declares tidal wetlands one of the most vital and protective areas of that state, and "that in order to promote the public safety, health and welfare, and to protect public and private property, wildlife, marine fisheries and the natural environment, it is necessary to preserve the ecological balance of this area and to prevent its further deterioration..." (N.J. Stat. Ann. § 13:9A-1(a)). Like the New York statute, the Commissioner of Environmental Protection is required to inventory all tidal wetlands in the state and map them with boundaries showing the areas that are at or below high water (N.J. Stat. Ann. § 13:9A-1(b)). "Coastal wetlands" are defined as "any bank, marsh, swamp, meadow, flat or other low land subject to tidal action in the State of New Jersey . . . , including those areas now or formerly connected to tidal waters whose surface is at or below an elevation of 1 foot above local extreme high water, and upon which may grow or is capable of growing some, but not necessarily all, of the following: [plant species]" (N.J. Stat. Ann. § 13:9A-2).

The Commissioner is authorized to adopt, modify, or repeal orders regulating dredging, filling, removing, altering or polluting coastal wetlands. Like the New York scheme, the New Jersey statute requires that a permit be secured to conduct any of these activities (N.J. Stat. Ann. § 13:9A-4). The Department of Environmental Protection oversees determinations on all permit applications.

7.9.4 New York State Wetlands Case

In *Gazza v. New York State Department of Environmental Conservation* (139 A.D.2d 647, 527 N.Y.S.2d 285 (1988)), a landowner sought review of the state Department of Environmental Conservation's denial of his requests to demap a portion of his property designated as "formerly connected" tidal wetlands and for a setback variance. The New York Supreme Court dismissed the landowner's petition, and he appealed to the Supreme Court, Appellate Division, which affirmed the lower court ruling.

The most interesting facet of the case is whether the property in question should be classified as "tidal wetlands" within the meaning of the state Tidal Wetlands Act (N.Y. Envtl. Conserv. Law art. 25 (McKinney)). The property is designated on the state's tidal wetland map of the area as a formerly connected (FC) tidal wetland (New York State Tidal Wetlands Inventory Map No. 702-502). The map shows that approximately one-half acre has been designated as "FC," which area is substantially located on the landowner's premises. The landowner sought demapping of the formerly connected tidal wetlands in order to build a residence on his property near the FC designated area. Following the landowner's request to demap a portion of his property designated as tidal wetlands, the Department of Environmental Conservation (DEC) field staff made a visual inspection of the property to make a tidal wetland assessment. The field staff concluded that the property in question was inundated with tidal waters and vegetated by at least five of the plant species indicative of tidal wetlands (*Gazza v. New York State Department of Environmental Conservation*, 139 A.D.2d 647, 527 N.Y.S.2d 285, 286 (1988)). The New York Supreme Court concluded that the tidal wetlands delineation was supported by substantial evidence in the record and upheld DEC's refusal to demap the tidal wetland area (*Gazza v. New York State Department of Environmental Conservation,* 139 A.D.2d 647, 527 N.Y.S.2d 285 (1988)).

The basic inquiry by the landowner was whether the property in question should remain mapped as tidal wetlands as defined in the New York Tidal Wetlands Act (N.Y. Envtl. Conserv. Law art. 25 (McKinney)). The statute defines "tidal wetlands" as:

> (a) those areas which border on or lie beneath tidal waters such as, but not limited to, banks, bogs, salt marsh, swamps, meadows, flats or other low lands subject to tidal action including those areas now or formerly connected to tidal waters;
> (b) all banks, bogs, meadows, flats and tidal marsh subject to such tides, and upon which grow or may grow some or any of the following: [plant species] (N.Y. Envtl Conserv. Law § 25-0103(1)(a), (b) (McKinney)).

DEC commonly refers to these two elements as the "tidal criterion" and the "botanical criterion" (see *O'Brien v. Barnes Building Co., Inc.,* 85 Misc.2d 424, 380 N.Y.S.2d 405, 422 (Sup. Ct. 1975)). DEC has taken the position that the presence of tidal wetland plant species must be coupled with a finding that the area is subject to tidal action. The mere presence of plants named in the statute will not sustain a classification of the area as "tidal wetlands."

DEC's field study found the area in question to be vegetated by at least five species of tidal wetland plants. In *Gazza v. New York State Department of Environmental Conservation* (139 A.D.2d 647, 527 N.Y.S.2d 285, 286 (1988)), the landowner did not debate the existence of the tidal wetland vegetation, however, because under the legal definition as interpreted by the agency, the existence of plants was not enough to justify the legal classification of the property as wetlands, the main challenge was to the tidal criterion in the statute. The landowner raised the interesting question whether the formerly connected tidal wetland, which was subject to tidal action solely due to artificial, as opposed to natural, means should be deemed to have met the legal definition of "tidal wetlands" (Brief of Petitioner-Appellant at pp. 20-29)

The record in the case indicates that the area designated as "FC" on the property closely resembles the "high marsh" category (Answer Paragraphs 3-6, R86-88). "High marsh" is described by DEC's regulations as a zone "that receives only occasional tidal flooding coincident with extreme lunar tides and occasional storms" (6 N.Y.C.R.R. § 661.2). The term "high marsh" is also defined as "the normal uppermost tidal wetland zone, designated HM on an inventory map, usually dominated by salt meadow grass, Spartina patens; and spike grass, Distichlis spicata. This zone is periodically flooded by spring and storm tides ..." (6 N.Y.C.R.R. § 661.4(hh)(5)).

DEC interprets the statutory phrase "areas now or formerly connected to tidal waters" as meaning only "those areas naturally subject to tidal action in the recent past" (*O'Brien v. Barnes Building Co., Inc.*, 85 Misc.2d 424, 380 N.Y.S.2d 405, 424 (Sup. Ct. 1975)(emphasis added)). The sole source of tidal water on the property is a man-made mosquito ditch that was constructed in the 1930s. The landowner thus contended that the property had not been naturally subject to tidal action in the recent past, and therefore, failed to meet the tidal waters criterion of the statute (Brief of Petitioner-Appellant at p. 28).

The New York Supreme Court has held that "[t]he whole thrust and purpose of the Tidal Wetlands Act is to preserve lands subject to the natural action of the tides and prevent their despoliation and destruction ..." (*State v. Lang*, 84 Misc.2d 106, 375 N.Y.S.2d 941, 944, aff'd, 52 A.D.2d 921, 383 N.Y.S.2d 400 (1976)), where the court held that "[t]o conclude that the Tidal Wetlands Act was intended to include such artificially created wetlands unreasonably strains the statutory language," 375 N.Y.S.2d at 944)). The landowner relied on this language in contending that when the influx of tidal waters is artificially induced, the land is not a tidal wetland within the meaning of the statute (Brief of Petitioner-Appellant at p. 28). However, even though no evidence existed that the mapped area was ever a naturally occurring tidal wetland but for the construction of the drainage pipe, the Appellate Division of the New York Supreme Court held that denial of the landowner's request for demapping had a rational basis and was supported by substantial evidence in the record (*Gazza v. New York State Department of Environmental Conservation*, 139 A.D.2d 647, 527 N.Y.S.2d 285, 286 (1988)). The environmental expert for the landowner contended that but for the artificial, man-made drainage ditch, the wetlands on the property would disappear entirely, however, this evidence did not sway the opinion of the court (Haje Affidavit, Paragraph 7, R50).

In this case, the scientific evidence on wetland delineation proved more important than the legal definition of tidal wetlands contained in the statute. The court was not persuaded by the legal reasoning of the landowner, which was quite strongly presented and supported by legal precedent and DEC's own interpretations of tidal wetlands. The demapping was denied because field studies by both the DEC staff and the landowner's own environmental expert showed the presence of wetland plants and tidal waters. Thus, the scientific evidence lead the court to conclude that the area was properly mapped as a tidal wetland, despite strong legal evidence that the area did not meet the legal definition in the New York Tidal Wetlands Act.

Clearly, an attorney representing a landowner in such a case must understand the maps and faces a strong burden to overcome the validity of the mapping. Even if the landowner had been successful in demapping the area on the state inventory map, he might later have been faced with a challenge to the Army Corps of Engineers delineation if a permit had been denied. The next case study demonstrates the importance of checking both the state and federal maps.

7.9.5 United States Claims Court Wetlands Case

Ciampitti v. United States (22 Cl.Ct. 310, 32 Env't Rep. Cas. (BNA) 1608, 21 Envt'l L. Rep. (ELI) 20,866 (1991)), was an action that was heard before the United States Claims Court to decide whether a permit denial to fill wetlands on Robert Ciampitti's property constituted a taking without just compensation under the Fifth Amendment to the U.S. Constitution. The Claims Court found no taking (22 Cl.Ct. at 322; see Section 7.4). The relevant holding in the case for purposes of this case study is the court's conclusion that there was no taking because the wetland permit denials did not interfere with distinct, investment-backed expectations, in that Ciampitti knew about the difficulty attendant upon developing the wetlands well before he purchased the property (22 Cl.Ct. at 320-321). He should have known the property was located in tidal wetland areas by consulting state and federal wetland inventory maps.

Wetlands maps play an important evidentiary role in the case.

The property for which Ciampitti sought development permits was designated as a wetland on state and federal wetland maps.

The facts are lengthy, therefore only the most relevant facts are summarized here. Robert Ciampitti owned property in Cape May County, New Jersey, which he had acquired over a period of years in a series of acquisitions. The property at issue is located in Lower Township in an area known as Diamond Beach. Lower Township straddles a barrier island on the South New Jersey coast. It includes the beach, upland and marsh area of a portion of the island between Cape May City on the south and Wildwood Crest on the north. The western extremity of Diamond Beach adjoins an area of marsh and open water known as Jarvis Sound which separates the barrier island from the mainland.

As early as 1980, the engineering firm of George E. Speitel & Associates (Speitel) had notified Ciampitti that there was possible state and federal jurisdiction over wetlands in Diamond Beach (22 Cl.Ct. at 314). Speitel requested that a Corps representative conduct a field study of the property. Shortly thereafter, Speitel was notified that it was unlikely that a federal permit to fill the area would be granted. Ciampitti conceded that he was told by Speitel that portions of the western end of Diamond Beach might be subject to federal wetlands regulations, and that if they were, a permit would have to be obtained from the Corps. Ciampitti was also advised of the Corps' suggestion that if he wanted to know the extent of federal wetlands jurisdiction, he should submit a map of the property and request clarification on designation. Ciampitti declined to follow that suggestion.

7.9.6 Importance of Wetland Maps in the Case

Despite his knowledge that the federal government might have an interest in regulating Diamond Beach wetlands, Ciampitti was not unduly concerned about the need for a federal permit. Two things justified his thinking. First, he supposed that if a permit was necessary, Speitel would take care of it. Second, he knew that in 1981 there had been fill activity, road building, and utility installation going on in the area. In part, Ciampitti's attitude was prompted by the fact that the Corps, unlike the New Jersey Department of Environmental Protection (DEP), does not perform advance designation in an authoritative way of lands determined to be subject to regulation. Further, unlike the state, the Corps does not file maps in local courthouses. Michael Claffey, a biologist with the Corps, testified that the landowner or developer is expected to ascertain by inquiry whether a permit to fill will be required. He did point out, however, that because New Jersey regulations are somewhat less comprehensive in scope, a state designation of wetlands conclusively means that the site will be protected as federal wetlands (22 Cl.Ct. at 315). In addition, the Fish and Wildlife Service has made National Wetlands Inventory Maps available, which are developed from aerial photographs that reflect vegetation and habitat. The Corps relies on the maps, not as a regulatory device, but as a way to give a quick estimate of whether lands will be regulated. The map reflects that the western end of Diamond Beach is estuarine wetlands.

During the 1970's the state began developing maps, based on infrared aerial photography, showing wetlands. This method is based on the presence of certain salt-water dependent vegetation. A map was adopted in 1975 which included the Diamond Beach property. This map was filed in the county courthouse (22 Cl.Ct. at 315). Claffey testified that the map is very reliable, and that the wetlands indications were in fact borne out during his three visits to the site in 1987 and 1990.

As it turns out, Speitel applied to the Corps on Ciampitti's behalf for a permit, pursuant to § 404 of the Clean Water Act, to maintain existing unauthorized fill, to place substantial additional fill within 11 western blocks within Diamond Beach, and to dredge an area for a marina. As part of the approval process, the Corps instructed Ciampitti that the project needed the approval of the DEP. The DEP wrote the Corps on February 6, 1986, advising that the proposed activities would affect state-regulated wetlands, and would be inconsistent with the state's Coastal Zone Management Program. In the Corps' June 5, 1986 denial of a § 404 permit, it recited that the project would contravene federal regulations found at 33 CFR 320, as well as federal guidelines regulating the discharge of fill or dredged material, and that the project had been determined by the state to be inconsistent with its Coastal Zone Management Program. In addition, the Corps observed in a separate Statement of Findings that much of the land in question was subject to state wetlands regulation, yet Ciampitti had not yet applied for a state permit (22 Cl.Ct. at 316). Corps regulations specifically require denial where independently required state permits are not obtained (33 CFR 320.4(j)).

In this case, the landowner clearly failed to recognize the importance of consulting wetland maps for his property. The role of the mapping of Ciampitti's property as a state and federal wetland was a critical issue that the landowner disregarded. The case indicates how the state and federal wetland designations often overlap and how the two regulatory schemes mesh. The developer needs to be aware of both state and federal wetland delineation. Approval of a federal permit cannot occur if the developer has not obtained any necessary state permit approval on the property first. Only when the property at issue is solely under federal jurisdiction as a wetland will the federal permit alone be required. Still, the state may object that the developer's activity contravenes the state coastal zone management program even when the property is not specifically designated as a state wetland. Finally, the Army Corps does not map wetlands. The Fish and Wildlife Service produces federal wetland inventory maps. These need to be consulted when the property shows signs of being a wetland. If the property is mapped as such on the FWS map, the Corps will probably require a permit. Finally, even if the area is not designated on any state or federal map as a wetland, if the area has wetland characteristics, it is always best to seek a determination from the Corps or state wetland agency before proceeding with any activity.

8

Wetland Mitigation

James F. Berry
and
Mark S. Dennison

INTRODUCTION

As discussed in Chapter 7, one of the requirements for many wetlands development projects is some form of "mitigation" to restore or replace lost wetland values. Generally speaking, mitigation is the attempted replacement of the functions and values of wetlands proposed for filling through creation of new wetlands or enhancement of existing wetlands; that is, "compensating" for lost functions (see Kruczynski, 1990).[1] The logic behind this requirement is that some forms of damage to wetlands may be unavoidable, even though a particular project is environmentally sound. If mitigation can create a situation where the *overall* effect is a net *gain* in wetland values (or, at least, no net loss), then the project should be allowed.

The federal Clean Water Act allows for mitigation when wetland values are adversely affected. It is not the only federal law to allow (or require) mitigation for environmental damage resulting from an action. In fact, the majority of federal environmental laws allow

---

[1] The U.S. Army Corps of Engineers and the USEPA specifically refer to "compensatory mitigation" as restoration of existing degraded wetlands, or creation of man-made wetlands. See Section 8.3.3.

some degree of mitigation.[2] A case in point is the Clean Air Act which allows a transfer of rights to pollute the air from one polluter to another in the same area, which led speculators with the Chicago Mercantile Exchange to invest recently in "pollution futures."

For many people mitigation represents a valuable compromise between a desire to protect wetland resources, while at the same time allowing property development for a variety of human uses. Developers are often required to readjust lot lines, redirect stormwater or other runoff, or completely relocate a development project in order to protect important wetland values. Many developers have argued that there is often a net *improvement* of the affected wetlands as a result (Salvesen, 1990). Nevertheless, the concept of mitigation as applied to wetlands has been controversial since its inception (see Kusler, 1986; Redmond, 1992).

The first major exercise of the mitigation requirement was in 1973 (the year after passage of the Clean Water Act) when the Corps and other federal agencies permitted a land swap in which Marco Island, Florida was allowed limited development in exchange for the preservation of sensitive marine and estuarine wetlands (Phillips, 1987). Since then, there has been considerable debate as to what exactly constitutes "mitigation" for wetland losses, what the ratio of wetlands lost versus replacement must be, and under what circumstances mitigation should be required. Many property owners see the mitigation requirement as an unfair burden on private property rights (see Chapter 1). Disagreement among agencies on the precise meaning and application of the mitigation requirement has been a source of confusion for all sides (see U.S. Environmental Protection Agency, 1990; and Sections 8.2 and 8.3).

Adding fuel to the controversy is the failure of wetland scientists (or other people) to reach consensus on the effectiveness of wetland mitigation (see Houck, 1989). Many wetland scientists have worked to develop the technology necessary to create and restore wetlands for

---

[2] For example, regulations under NEPA require preparation of mitigation plans as part of an Environmental Impact Statement (see Park City Resource Council v. U.S. Dept. of Agriculture, 817 F.2d 609 (10th Cir. 1987)). The U.S. Supreme Court said that the EIS must simply "discuss possible mitigation measures" in Robertson v. Methow Valley Citizens Council, 490 U.S. 332 (1989), although it is unclear what effect the Methow Valley decision will have on wetland mitigation requirements outside the EIS process.

mitigation purposes, and there have been some important successes (see Chapter 9). But other wetland scientists have argued that mitigation is impractical (or impossible), and that wetland values once lost can never adequately be replaced (see Kruczynski, 1990a, 1990b; Salvesen, 1990). There is evidence that mitigation projects that appear successful over the short term may prove to be ineffective over the long term, and that certain types of wetlands are better candidates for mitigation than others (e.g., marsh mitigation projects are more successful than forested wetlands; Kusler and Kentula, 1990).

Information on the effectiveness of mitigation projects remains scant, largely because monitoring efforts of wetland mitigation projects have been inadequate or unavailable (Kusler and Kentula, 1990). One of the few nationwide reports available was for the period from 1980-81 (when wetland mitigation was still in its infancy), when the Office of Technology Assessment reported that of the 50,000 acres of wetlands lost by permitted projects, only 5,000 acres of mitigation were required. Thus, 90% of permitted losses were unmitigated (Conservation Foundation, 1988; Leslie, 1989).

Several states have reported on the effectiveness of mitigation under state wetland laws. For example, the Florida Department of Environmental Regulation, which administers Florida's wetlands mitigation programs (Fla. Stat. § 403.913 *et seq.*), reported that from 1985 through 1990, 1,262 permits requiring mitigation were issued authorizing 3,305.42 acres of wetlands to be destroyed, and requiring 3,344.9 acres to be created, 7,300.9 acres to be enhanced (or restored), and 7,587.54 acres to be preserved. Of 63 permits that were reviewed in 1991 (34 freshwater and 29 saltwater), only four were found to be in full compliance with mitigation requirements and were likely to be ecologically "successful." In about 34% of cases, no required mitigation had been attempted even though wetland losses had occurred (Florida Department of Environmental Regulation, 1991; see Redmond, 1992).

Unfortunately, with national wetland policy in a state of flux, and disagreement among policy makers, the scientific community, and the regulated community, it is unlikely that these controversies will dissipate in the near future (see Zallen, 1992).

8.1 THE REGULATORY FRAMEWORK

The requirement for mitigation as compensation for damages

sustained by wetlands is an important part of the federal government's "no net loss" policy (see U.S. Fish and Wildlife Service, 1990). However, mitigation is not specifically required under § 404 of the Clean Water Act, the statute that requires a permit for dredging or filling a wetland. The mitigation requirement is found in other federal statutes, notably the National Environmental Policy Act (NEPA; 42 USC § 4321 *et seq.*) and the Fish and Wildlife Coordination Act (16 USC § 661), which require mitigation for *all* federal actions which adversely affect the environment. The issuance of a § 404 dredge and fill permit is the kind of federal agency action which triggers NEPA and its mitigation requirement (see Chapter 7).

8.1.1 The Origin of Wetland Mitigation Policy

The Council on Environmental Quality (CEQ) adopted the regulations implementing NEPA in 1978 (40 CFR 1508.20), and the U.S. Fish and Wildlife Service (FWS) did likewise in their 1981 comprehensive policy which attempted to clarify the objectives of mitigation as well as providing guidelines for its implementation (46 Fed Reg. 7644-63 (Jan. 23, 1981)). The U.S. Army Corps of Engineers (Corps), which issues dredge and fill permits under § 404 of the Clean Water Act, took a somewhat more restrictive view of mitigation than did the CEQ, regarding mitigation as a last step after attempts at avoidance, minimizing impact, and repair of damage had failed. This conflict eventually led to the famous *"Sweedens Swamp"* controversy, and the EPA/Corps Memorandum of Agreement discussed in Section 8.2.

In its 1978 regulations, the CEQ defined the various mitigation alternatives under NEPA that continue to form the backbone of federal mitigation policy (Leslie, 1990):

* Avoid environmental impacts completely by avoiding an action (or part of an action) that might lead to environmental degradation;

* Minimize environmental impacts by limiting the degree or magnitude of an action and its implementation;

* Rectify environmental impacts by repairing, rehabilitating, or restoring the environmental impact;

* Reduce or eliminate environmental impact over time by preservation and maintenance operations over the life of the project; and

* Compensate for the environmental impact of the action by replacing or providing substitute resources or environments.

Unfortunately, the CEQ failed to spell out details as to how these policies were to be implemented. As a result, both the policies and their implementation remain controversial. Among the more controversial unresolved policy issues are: (1) the types of mitigation that are acceptable for a particular project or wetland; (2) the methodology used to determine appropriate mitigation techniques; (3) the amount of mitigation that should be required; (4) the wetland values that are to be preserved or replaced, and appropriate methods of measurement and valuation; (5) the appropriate parties to undertake and pay for mitigation; and (6) the methods for monitoring the project and assuring compliance (see Leslie, 1990).

Current mitigation policies administered by the Corps under its § 404 regulations state that:

> "Mitigation is an important aspect of the review and balancing process .... Consideration of mitigation will occur throughout the permit application review process and includes avoiding, minimizing, rectifying, reducing, or compensating for resource losses. Losses will be avoided to the extent practicable. Compensation may occur on-site or at an off-site location." [33 CFR 320.4(r)(1)]

Despite the relatively straightforward language of the regulations, the Corps is currently modifying its methods of applying its wetland mitigation policies.[3] Many specific applications of Corps policy are regulated by the provisions of the 1990 Memorandum of Agreement ("MOA") between the Corps and EPA, which will be discussed further in Sections 8.3 and 8.4.

Current Corps policy permits mitigation to occur on-site (i.e., restoring those wetlands located on the project site which are degraded by the project or by previous actions), or off-site (either the purchase of wetlands at another site, or creation of new wetlands at a site where none exist), although there is a strong preference for on-site mitigation (see Section 8.3.1). The "on-site vs. off-site" question has received considerable attention, particularly from environmentalists, scientists, and some federal agencies who argue that off-site mitigation allows "the sacrifice of valuable natural wetlands for the ill-considered promise of some future, potentially less desirable, wetland replication" (U.S. Environmental Protection Agency, 1990; see also Berry, 1992b, and Chapter 9). Opponents of mitigation plans have argued successfully that neither a rock quarry[4] nor a marsh fed by urban runoff[5] was sufficient mitigation for destruction of valuable wetlands (see Houck, 1989).

8.1.2 Kinds of Mitigation Measures

The Corps' regulations place mitigation measures into three categories (33 CFR 320.4(r)). First, minor project modifications (those considered feasible by the applicant) can include reductions in scope and size; changes in construction methods, materials, or timing; and operation and maintenance practices that reflect a sensitivity to environmental quality. For example, erosion control features could be required on a fill project to reduce sedimentation impacts (33 CFR

---

[3] In the July 1, 1992 edition of its regulations, section 320.4(r) contains a statement noting that the Corps' regulatory policy statement is *not* a substitute for the 404(b)(1) guidelines, and that an interagency Working Group is currently working on mitigation requirements. July 1, 1992 Edition of 33 CFR 320.4(r), footnote 1.

[4] Bersani v. U.S. Envtl. Protection Agency 674 F.Supp. 405 (N.D.N.Y. 1987), aff'd sub nom. Robichaud v. U.S. Envtl. Protection Agency, 850 F.2d 36 (2d Cir. 1988), cert. denied, 109 S.Ct. 1556 (1989) [the famous "Sweedens Swamp" case discussed in Section 8.3].

[5] Nat'l Audubon Soc. v. Hartz Mountain Development Corp., 14 Envtl. L. Rep. (Envtl. Law Inst.) 20,724 (D.N.J. 1983).

320.4(r)(i)). Second, additional mitigation measures may be required to satisfy legal requirements under § 404(b)(1), which provides for the EPA review process of potential environmental impacts under § 404 dredge and fill permits (33 CFR 320.4(r)(1)(ii)). EPA's regulatory guidelines under § 404(b)(1) contain a list of some such measures (40 CFR 230.70-230.77). Third, still other mitigation measures may be required as a result of the Corps' public interest review process (see Section 7.2) if found to be reasonable and justified by the District Engineer, but only if the mitigation measures are required to ensure that "the project is not contrary to the public interest" (33 CFR 320(r)(1)(iii)).

The third requirement mandates that the Corps need not require mitigation measures unless it can be shown that failure to require the measures would be "contrary to the public interest." Therefore, a party wishing to attack the Corps' mitigation requirements (or, conversely, its failure to require mitigation) has the burden of demonstrating what constitutes "the public interest," which is a vague and often frustrating concept at best.

The Corps' § 404 regulations are reasonably explicit with respect to the kinds of wetlands losses for which mitigation is required:

> "All compensatory mitigation will be for significant resource losses which are specifically identifiable, reasonably likely to occur, and of importance to the human or aquatic environment. Also, all mitigation will be directly related to the impacts of the proposal, appropriate to the scope and degree of those impacts, and reasonably enforceable." [33 CFR 320(r)(2)]

In other words, wetland degradation judged by the District Engineer to be minor, insignificant, unimportant, or speculative does *not* trigger the requirement for compensatory mitigation. Likewise, the Corps cannot require mitigation that is unrelated to the impacts that will result from the project itself (i.e., a developer *cannot* be forced to provide mitigation for a project unrelated to the one for which the § 404 permit is granted).

8.1.3 The Role of Other Federal Agencies

In addition to the Corps and the EPA, the U.S. Fish and

Wildlife Service also has commenting authority on § 404 permits and compensatory mitigation under the Fish and Wildlife Coordination Act (16 USC §§ 661-666), NEPA (42 USC §§ 4321, 4363), and the Endangered Species Act (16 USC § 1531 *et seq.*), among others. The FWS focuses its policy specifically on the "value" of the affected wetlands, and recommends: (1) that damage to the most valued resources be avoided; and (2) that the degree of mitigation correspond to the value and scarcity of the habitat at risk (46 Fed. Reg. 7644 (Nov. 13, 1986)). In general, the Corps is deferential to FWS recommendations which tend to focus more precisely on fish and wildlife issues than does the Corps. In 1986, the FWS established four wetlands resource categories associated with its mitigation planning goals (see Table 8.1; see also, Leslie, 1990).

The FWS (as well as the Corps and EPA) uses several procedures in order to evaluate the functional values of a particular wetland for wildlife habitat. The two most common are the Habitat Evaluation Procedures (HEP) developed by the FWS, and the Wetland Evaluation Technique (WET), developed jointly by the Corps and the Federal Highway Administration (see Adamus et al., 1987; Adamus, 1988; and Chapter 10). The most frequently used procedure is HEP, which utilizes standard computer models to relate biological requirements and tolerances for certain indicator species to environmental variables as they occur on the subject property, such as water depth and quality, flooding periodicity, vegetation density and type, and soil type. The HEP procedure then provides numerical values for habitat suitability, which can be used as objective measures of the relative "quality" of wetland functional values (see Adamus, 1988).

Proper use of HEP procedures requires a relatively high scientific skill level, and formal training from the FWS. As a result, there is a danger that the HEP procedure may be misused. For example, a consultant to a developer in southeastern Florida recently attempted to use the HEP procedure to show that extensive wetlands on a site proposed for development had low habitat value as it existed. This was done by carefully choosing indicator species which no longer occurred with regularity on the site due to various hydrological modifications in the past. The consultant then used the HEP to show that "improved" habitat would result from mitigation projects such as borrow lakes, with a net ecological benefit for the selected species (see Kruczinski, 1990b). A proper HEP would have shown that the development would have

286   Wetlands

dramatically decreased habitat value for the wetland species that occurred there.[6]

The National Marine Fisheries Service (NMFS) also comments on potential damage to fisheries. The NMFS typically becomes involved early in the planning process in order to resolve potential conflicts and minimize adverse effects on marine resources and habitats. NMFS will recommend mitigation measures for "essential public interest projects" when practical alternatives are unavailable, and recommend habitat enhancement measures (48 Fed. Reg. 53147 (Nov. 13, 1986)).

8.2 CORPS/EPA 404(b)(1) MITIGATION GUIDELINES AND JOINT MOA

8.2.1 Overview

In reviewing an application for dredge and fill activities in wetlands areas, the Corps is required to consider 404(b)(1) Guidelines issued by EPA (See Section 8.1).[7] The guidelines protect wetlands by prohibiting discharges that have significant adverse effects on human health or welfare, recreation, aesthetics, economics, aquatic ecosystems, and wildlife dependent on aquatic ecosystems (40 CFR Section 230.10(c)). The Section 404(b)(1) guidelines require applicants to take all appropriate and practicable steps to minimize the adverse impacts of proposed filling activities. The regulations do not define what will be considered appropriate and practicable. In the case of non-water-dependent projects, "no discharge of dredged or fill material shall be permitted if there is a practicable alternative to the proposed discharge which would have less adverse impact on the aquatic ecosystem ...." (40 CFR Section 230.10(a)). Several controversial

---

[6] The developer in this case attempted to use the HEP results to convince Florida state agencies that the habitat value was low. It is unlikely that federal Corps or EPA officials would have been convinced by this use of HEP. The case remains unresolved at this writing.

[7] Section 404(b)(1), 33 USC 1344(b) authorizes the Secretary of the Army to specify disposal sites "through the application of guidelines developed by the Administrator" of the EPA "in conjunction with" the Corps. The guidelines are found at 40 CFR Part 230.

wetlands cases have considered the meaning of this "practicable alternatives" language (see Section 8.2.4).

Because various conflicts arose between EPA and Corps over interpretation of the 404(b)(1) guidelines, including the "practicable alternatives" provision (see Section 8.2.3), the two agencies issued a joint Memorandum of Agreement (MOA) concerning the determination of mitigation measures under Section 404(b)(1). (Section 404(b)(1) Guidelines Mitigation MOA, 55 Fed. Reg. 9210 (Feb. 7, 1990)). Under the joint MOA, before a permit may be issued, the applicant must first attempt to avoid wetlands impacts, then minimize impacts, and finally as a last resort, compensate for unavoidable impacts.

8.2.2 EPA/Corps Mitigation MOA

The EPA/Corps Joint MOA adopts the goal of no overall net loss of wetland values and functions. By emphasizing wetland "values and functions," the MOA authorizes a less than one to one replacement of wetland acreage where the wetlands lost are significantly degraded.

Although the MOA recognizes that the "no net loss" goal may not be achieved with respect to every permit application, it does not address whether such losses may be "made up" on other permit applications, thus meeting the "no overall net loss" requirement.

The MOA sets forth a strict sequence of regulatory considerations: avoidance, minimization, and compensatory mitigation. First, wetland impacts must be avoided to the maximum extent practicable. Cost, existing technology, and "logistics in light of overall project purposes" may be considered in determining what is "practicable." In evaluating alternative upland sites, the adverse environmental impacts at the upland site must be considered before concluding that a "practicable alternative" exists.

If unavoidable impacts remain, they must be minimized to the extent appropriate and practicable. Only as a last resort, where unavoidable adverse impacts remain after minimization procedures, may the Corps consider compensatory mitigation proposals. Thus, the Corps may not consider a compensatory mitigation proposal if it determines that adverse impacts may be avoided.

The sequencing requirement does not apply where the project will result in environmental gain or insignificant environmental loss. The sequencing requirement is considered satisfied if the project is consistent with a comprehensive plan, such as a Special Area Management Plan (see Section 7.8), which has been approved by the Corps and EPA, and which assures that the compensatory mitigation requirements of the Section 404(b)(1) Guidelines will be met.

The MOA specifies that "functional values" of wetlands should be determined by applying aquatic site assessment techniques generally recognized by experts in the field. These resource assessments should be tailored to the particular site and performed by qualified professionals. Despite this clarification in the MOA, scientific uncertainty remains concerning assessment results.

The MOA states a preference for in-kind, onsite compensatory mitigation over out-of-kind, offsite mitigation. Where off-site measures are necessary, they should be in close physical proximity to the affected site and preferably in the same watershed. Finally, restoration of degraded wetlands is preferred over creation of new wetlands, because of scientific uncertainty with respect to the success of wetland creation (See Chapter 9).

The MOA requires at least a "one for one functional replacement (i.e., no net loss of values), with an adequate margin of safety to reflect the expected degree of success associated with the mitigation plan." However, the MOA recognizes that one-to-one replacement of functional values may not be appropriate and practicable in all cases, such as where hydrological conditions make restoration or replacement impracticable. Thus, where replacement is practicable and the wetland filled and wetland "created" have the same value, a one-to-one ratio of replacement acreage is required. However, where the wetland being impacted has high functional values and the replacement acreage has lower functional values, the permittee may have to replace more than one acre of wetlands for every acre lost. Similarly, the MOA authorizes a less than one-to-one replacement of wetland acreage where the wetlands lost are significantly degraded.

8.2.3 "Practicable Alternatives"

Perhaps the most important (and most controversial) aspect of the § 404(b)(1) Guidelines is the "practicable alternatives" provision. The guidelines create a presumption against filling wetlands by prohibiting the discharge of dredged or fill material into waters where "a practicable alternative to the proposed discharge [exists] which would have less adverse impact on the aquatic ecosystem, so long as the alternative does not have other significant adverse environmental consequences" (40 CFR 230.10(a)). It also provides that a practicable alternative may include "an area not presently owned by the applicant which could reasonably be obtained, utilized, expanded or managed in order to fulfill the basic purpose of the proposed activity" (40 CFR Section 230.10(a)(2)). It further provides that, "unless clearly demonstrated otherwise", practicable alternatives are (1) "presumed to be available" and (2) "presumed to have less adverse impact on the aquatic ecosystem" (40 CFR Section 230.10(a)(3)). Thus, an applicant must rebut both of these presumptions in order to obtain a permit.

Sections 230.10(c) and (d) require that the Corps not permit any discharge that would contribute to significant degradation of the nation's wetlands and that any adverse impacts must be mitigated through practicable measures. Under Section 404(c) of the Clean Water Act (33 USC Section 1344(c)), EPA has veto power over any decision of the Corps to issue a permit. Specifically, Section 404(c) provides that the Administrator of EPA may prohibit the specification of a disposal site "whenever he determines, after notice and opportunity for public hearings, that the discharge of materials into such area will have an unacceptable adverse effect" on, among other things, wildlife. An "unacceptable adverse effect" is defined in 40 CFR Section 231.2(e) as an effect that is likely to result in, among other things, "significant loss of or damage to ... wildlife habitat." The veto procedure under Section 404(c) begins with the Regional Administrator (RA) who, under Section 231.3(a), must notify the Corps and the applicant when it is possible he will find an "unacceptable adverse effect." If within 15 days the applicant fails to satisfy the RA that no such effect will occur, the RA must publish his proposed determination to veto the grant of a permit. A period for public comment and an optional public hearing follows, after which the RA either withdraws the determination or submits a recommended determination to the national Administrator, whose decision to affirm, modify or rescind the RA's recommendation is the final determination of EPA for purposes of judicial review. The burden of proving that the discharge will have an "unacceptable adverse effect" is on EPA. (45 Fed.Reg. 85,336, 85,338 (1980); 44 Fed.Reg. 58,076, 58,080 (1979)). The Corps processes about 11,000 permit applications each year. EPA has vetoed only a handful of Corps decisions to grant permits (for example, see *James City County v. EPA*, 955 F.2d 254 (4th

Cir. 1992); and Section 7.3.4).

8.2.4 The "Sweedens Swamp" Case

The leading case concerning the practicable alternatives provision is *Bersani v. EPA*, commonly referred to as the "Sweedens Swamp" decision (674 F. Supp. 405 (N.D.N.Y. 1987), *aff'd*, 850 F.2d 36 (2nd Cir. 1988), *cert. denied*, 489 U.S. 1089 (1989)). Bersani involved the EPA's veto of the grant of a permit to fill "or alter" most of a wetland area known as Sweedens Swamp for the purpose of building a mall. The court's review of the veto pertained to disagreement between the two agencies over whether the developer had adequately rebutted the presumption of the existence of practicable alternatives to the Sweedens Swamp site (850 F.2d at 42).[8]

The Corps' Director of Civil Works found that the proposed upland alternative, called "the North Attleboro site," was unavailable to the applicant "because it ha[d] been optioned to another developer." In vetoing the grant of the permit, EPA argued that the Corps' finding was inaccurate because the North Attleboro site "could have been available to [the applicant] at the time [the applicant] investigated the area to search for a site," since the other developer did not buy the options for the site until after the applicant had entered the market. EPA also gave other reasons for the veto, including its finding that the filling of the Swamp would adversely affect wildlife.

Review of EPA's veto was sought on the ground that EPA's "market entry theory," as the developer called it, was "inconsistent with both the language of the 404(b)(1) guidelines and the past practice of the Corps and the EPA." The developer argued that both the language of the Guidelines and past practice indicated that the availability of an

---

[8] A second point of disagreement, not discussed on appeal, concerned the applicant's mitigation proposal. The Director of Civil Works at the national headquarters of the Corps granted the permit despite the recommendation of the Northeast Regional Corps to deny it. The N.E. Corps had recommended denial because it found that the North Attleboro alternative was feasible for mall development.

The Director of Civil Works decided to grant the permit "after finding that [the applicant's] offsite mitigation proposal would reduce the adverse impacts sufficiently to allow the 'practicable alternative' test to be deemed satisfied". However, the Corp's argument about the mitigation proposal was rejected by the EPA because of "scientific uncertainty of success."

alternative site should be analyzed as of the time when the applicant applied for a permit.

These arguments were rejected by both the federal District Court and the Second Circuit Court of Appeals. The developer claimed that the "most natural" reading of the alternatives provision required a time-of-application rule because the provision was written in the present tense. Both courts found that no conclusion could be drawn from the tense of the provision as to intent on timing and that intent would have to be discovered from examining the objective of the Guidelines. The Second Circuit stated that "the purpose is to create an incentive for developers to avoid choosing wetlands when they could choose an alternative upland site," and that "[i]f the practicable alternatives analysis were applied to the time of the application for a permit, the developer would have little incentive to search for alternatives, especially if it were confident that alternatives soon would disappear." Thus, the court concluded that "a common-sense reading of the statute can lead only to the use of the market entry approach used by EPA."

8.3 FORMS OF MITIGATION

A wide range of mitigation measures have been allowed by the Corps. Dial and Deis (1986) summarized several of these approaches available at the time: (1) increased public access to the area; (2) acquisition of other wetlands to provide enhanced protection, or acquisition with a management commitment; (3) restoration or creation of wetlands, either as general compensation or as replacement for a specific habitat type; (4) indemnification or direct monetary payment for lost wetland values; and (5) mitigation banking (compensatory off-site wetlands restoration or creation). Most federal agencies and most states no longer permit approaches (1) or (2) unless the goal of increased public access is compensation for lost public recreational opportunities, or the acquisition includes enhancement or assurance of proper management to compensate for lost wetland values (Leslie, 1989).

8.3.1 Mitigation Alternatives

As discussed in Section 8.1, mitigation may occur "on-site" or "off-site," which has created a controversy as to which type of mitigation is best for a particular project. At present, the Corps and other federal agencies (as well as most state agencies) prefer on-site mitigation.

Alternatives for compensatory mitigation fall generally into five categories (see Kruczynski, 1990b), listed in relative order of their invasiveness:

(1) *Preservation.* Purchase of a parcel of land containing a valuable wetland, which is then placed in public ownership with provisions for long-term protection and/or management (this may take the form of "conservation easements" or other devices).

(2) *Exchange.* The exchange for a wetland area which will be damaged by a project for another wetland (typically of larger size and higher wetland values), which is placed in long-term protection as in (1).

(3) *Enhancement.* A wetland in which some functions have been degraded (or lost) is "repaired," such that the degraded (or lost) wetland functions are again available.

(4) *Restoration.* A former wetland with few (or no) remaining wetland functions is restored to a form in which specific functions (perhaps all) are available.

(5) *Creation.* A wetland is created where none previously existed. The goal is usually to create specific wetland functions.

Of course, the particular type of mitigation required for a specific project will depend on the nature of the project, the area in which the project is located, the quality of the wetlands to be damaged or destroyed (i.e., the relative "values" of the wetland), and the amount of wetlands to be damaged or destroyed.

In those parts of the country where damage to wetlands is widespread and few natural wetlands remain (e.g., northeastern Illinois), wetland enhancement, restoration, or creation is likely to be perceived by the agencies as proper mitigation for a project. The technology required to enhance, restore, and create wetlands has improved dramatically in the past decade (see Chapter 9). It has been argued that these techniques are the only ones that will promote a "no net loss" policy regarding wetlands (Kruczynski, 1990b).

On the other hand, in areas where high-quality natural wetlands are common, a project may argue in favor of preservation or exchange as mitigation. Of course, there is a net loss of wetlands under this scenario, even though one may argue that the valuable long-term protection to the preserved wetlands makes this an attractive option in specific instances, as does the possibility of preserving large tracts of contiguous natural wetlands with high diversity and wetland values. Nevertheless, current Corps and EPA practices strongly favor on-site mitigation.

Another important issue is the *amount* of land involved in mitigation. Typically, the only time a 1:1 (mitigated:damaged wetlands) "swap" is permitted is when the mitigation is offered up front, with no risk that there will be lost wetland values. For wetland restoration mitigation the exchange is typically 1.5:1, for creation 2:1, and for enhancement 3:1 (Kruczynski, 1990b). Some states require greater than 1:1 mitigation (e.g., Florida requires 2:1 in-kind mitigation with 25-year monitoring).

One of the newest mitigation alternatives is "wetland banking," which is gaining in popularity in several parts of the country. Wetland banking will be discussed in Section 8.4.

8.3.2 Sequencing

One of the most important aspects of wetlands mitigation is the proper "sequencing" of the various alternative kinds of mitigation. Current sequencing methodology by the regulatory agencies is controlled by the 1990 MOA between the Corps and EPA discussed in Section 8.2.

Under the MOA, the Corps has adopted the definition of mitigation used by the Council on Environmental Quality ("CEQ") and EPA (40 CFR 1508.20). Under this definition, "mitigation" includes: (1) avoiding impacts, (2) minimizing impacts, (3) rectifying impacts, (4) reducing impacts over time, and (5) compensating for impacts. The MOA stated further that "as a practical matter, [the types of mitigation enumerated by CEQ] can be combined to form three general types: avoidance, minimization, and compensatory mitigation" (1990 MOA at 2).

The MOA actually went further than either Corps guidelines or CEQ regulations by mandating that mitigation measures follow the proper "sequence" as spelled out in the § 404(b)(1) guidelines. During the evaluation process of a § 404 permit application:

> "The Corps ... first makes a determination that potential impacts have been avoided to the maximum extent practicable; remaining unavoidable impacts will then be mitigated to the extent appropriate and practicable by requiring steps to minimize impacts and, finally, compensate for aquatic resource values" (MOA at 2).

In determining if proposed mitigation is "appropriate and practicable," the MOA states that "such [mitigation] measures should be appropriate to the scope and degree of those impacts and practicable in terms of cost, existing technology, and logistics in light of overall project purposes" (MOA at 3).

Avoidance. The MOA interprets the § 404(b)(1) guidelines to mean that the thrust in mitigation alternatives is on "avoidance of impacts."[9] Unlike its position in *Bersani v. EPA* (see Section 8.2.4), the Corps must now determine that impacts have been avoided "to the maximum extent practicable" *before* it issues a permit. As discussed in Section 8.2.3, permits are not to be issued if there is a "practicable alternative ... which would have less adverse impact on the aquatic ecosystem, so long as the alternative does not have other significant adverse environmental consequences" (40 CFR 230.10(a)).

Minimization. The MOA next requires that any unavoidable impacts on the aquatic ecosystem must be minimized. "Appropriate and practicable" steps must be taken to minimize potential adverse impacts through project modifications and permit conditions (MOA at 3; 40

---

[9] 40 CFR 230.10(a) states: "Except as provided under section 404(b)(2), no discharge of dredged or fill material shall be permitted if there is a practicable alternative to the proposed discharge which would have less adverse impact on the aquatic ecosystem, so long as the alternative does not have other significant adverse environmental consequences."

CFR 230.10)). Subpart H of the § 404(b)(1) guidelines describes several (but not all) means for minimizing impacts. These include: (1) locating and confining the discharge to minimize smothering of organisms (40 CFR 230.70(a)); (2) designing the discharge to avoid a disruption of periodic water inundation patterns (40 CFR 230.70(b)); (3) selecting a site that has been used previously for dredged material discharge (40 CFR 230.70(c)); (4) selecting a site at which the substrate is composed of material similar to that being discharged, such as sand on sand or mud on mud (40 CFR 230.70(d)); and (5) designing to prevent the creation of standing water in an area of normally fluctuating water levels, and preventing the drainage of areas subject to such fluctuation (40 CFR 230.70(f)). The thrust of the "minimization" requirement is that an applicant must take precautions in the design and implementation of any mitigation project to minimize adverse environmental impacts to the maximum extent practicable.

8.3.3 Compensatory Mitigation.

If unavoidable adverse environmental impacts still remain after all appropriate and practicable avoidance and minimization has been implemented, then compensatory mitigation is required. The MOA regards as examples of compensatory actions "restoration of existing degraded wetlands or creation of man-made wetlands" (MOA at 3). In other words, wetland restoration and creation are to be considered only *after* all possibilities for appropriate and practicable avoidance and minimization has been exhausted. The MOA indicates a preference for restoration over compensation:

> "There is continued uncertainty regarding the success of wetland creation or other habitat development. Therefore, in determining the nature and extent of habitat development of this type, careful consideration should be given to its likelihood of success. Because the likelihood of success is greater and the impacts to potentially valuable uplands are reduced, restoration should be the first option considered." [MOA at 4]

The MOA shows a strong preference for on-site mitigation. Compensatory actions should be undertaken, when practicable, in areas adjacent or contiguous to the discharge site (on-site compensatory mitigation). If on-site compensatory mitigation is not practicable, off-site compensatory mitigation should be undertaken "in the same geographical area if practicable (i.e., in close physical proximity and, to the extent possible, the same watershed)" (MOA at 3). The preference for on-site compensatory mitigation is consistent with Corps and EPA policies, but it creates difficulties for applicants who wish to participate in off-site mitigation such as preservation or exchange.

In deciding if compensatory mitigation is acceptable, the Corps and EPA consider those functional values (see Section 2.3) which will be lost as a result of the project. Scientific techniques such as HEP and WET (discussed in Section 8.1.3) are used to determine the value of wetlands that will be impacted, as well as those which replace them. Compensatory mitigation must be designed to replace the specific functional values which will be lost, rather than a surface area or "acre for acre" tradeoff (see Zallen, 1992). In circumstances where there is inadequate scientific information on functions and values for a specific wetland a one-for-one acreage replacement may be used, but the ratio will be greater where the impacted wetland has demonstrably higher functional values than the replacement, or where the likelihood of success of the proposed replacement is low.

The MOA also shows a preference for "in-kind" compensatory mitigation to "out-of-kind." In other words, compensatory mitigation should be designed to the extent possible in a manner such that the exact wetland values that will be impacted are replaced. In most instances, this will mean replacing a wetland with the same wetland type that is impacted (i.e., replacing a forested wetland with a forested wetland, or a bog with a bog). Mitigation must provide "at a minimum, one for one functional replacement (i.e., no net loss of values)" (MOA at 4). The MOA specifically authorizes the use of "approved" mitigation banks as acceptable compensatory mitigation).

Exceptions to the Sequencing Requirement. The MOA specifically recognizes two limited exceptions to the mitigation sequence requirement. First, it may be necessary to deviate from the normal sequence when the Corps and EPA agree that a proposed discharge is necessary to avoid environmental harm. The MOA gives the examples of such deviations "to protect a natural aquatic community from saltwater intrusion, chemical contamination, or other deleterious physical or chemical impacts" (MOA at 3). A second exception to sequencing occurs when the Corps and EPA agree "that the proposed discharge can reasonably be expected to result in environmental gain or insignificant environmental losses" (MOA at 3).

There is an additional implied exception. In discussing one-for-one replacement of functional values, the MOA recognizes that this minimum requirement "may not be appropriate and practicable, and thus may not be relevant in all cases" (MOA at 4-5). In a controversial footnote, the MOA further explains:

> "For example, there are certain areas where, due to hydrological conditions, the technology for restoration or creation of wetlands may not be available at present, or may otherwise be impracticable. In addition, avoidance, minimization, and compensatory mitigation may not be practicable where there is a high proportion of land which is wetlands." [MOA at 5, footnote 7].

For the latter statement, EPA has given the example that the creation of tundra or permafrost in Alaska is not technically feasible at this time, such that mitigation may not be practicable (Zallen, 1992). The degree to which this exception applies to other situations is the subject of current discussions within the Corps and EPA.

Other Considerations. The MOA acknowledges that mitigation compliance is a complex and difficult process, and encourages all applicants for § 404 permits to arrange preapplication meetings with the Corps, as well as other federal agencies (e.g., EPA and the FWS) and state and local governmental authorities. Such meetings are crucial to applicants because they provide a background in specific application procedures which may vary from site to site, and often indicate to the applicant what specific mitigation procedures, monitoring requirements, etc. may be considered acceptable. Failure to arrange a preapplication meeting often results in unnecessary expenditures of time, effort, and finances.

Virtually all mitigation actions require some form of monitoring by the applicant and the Corps during the construction phase as well as once the mitigation is in place. Compensatory mitigation techniques with high levels of scientific uncertainty will require long-term monitoring, reporting, and possible remediation actions. Monitoring is to be directed toward determining whether permit conditions are being met, and whether the purposes of the mitigation are actually being achieved. If permittees are found to be in non-compliance, the Corps will notify them, request a corrected plan, attempt to resolve the violation, then issue an order demanding compliance. If permittees fail to comply with the order, then the Corps may suspend or revoke the permit (see 33 CFR 325.7(c) for procedures), and/or recommend legal action, which includes both civil and criminal penalties (33 CFR 326.5).

Finally, the Corps and EPA have been involved in additional rulemaking regarding mitigation since 1991. The Bush administration directed the agencies to reconsider such issues as a market-oriented mitigation banking system, and to consider mitigation sequencing only for "high value" wetlands that would be defined with a new rule (Zallen, 1992). A new final rule is scheduled to be completed by September, 1993, but it is not possible at this writing to determine precisely what form the new rule will take given the change in administration and other intervening factors.

8.4 WETLAND BANKING

Mitigation "banking" is a concept developed originally in the early 1980's by the U.S. Fish and Wildlife Service in an attempt to increase the effectiveness of wetlands mitigation while reducing the costs to the regulated community. The practice remains controversial, but it is increasing in popularity and was described by Sokolove and Huang (1992) as "the most promising solution to the loss of wetlands during development."

Mitigation banking is considered an acceptable form of compensatory mitigation under the 1990 Corps/EPA MOA discussed in section 8.2 and 8.3 above. Where a mitigation bank has been approved by the Corps and EPA as mitigation for a specific identified project, use of the bank for those projects is considered as meeting the requirements for compensatory mitigation (see Section 8.3.2) *regardless of the practicability of other forms of compensatory mitigation* (MOA at 4). It should be cautioned, however, that simple purchase or "preservation" of existing wetlands will only be accepted as compensatory mitigation in exceptional circumstances.

8.4.1 The Concept of Mitigation Banking

Wetland banking is quite simple in principle. In typical situations, either a deteriorated (or deteriorating) wetland is enhanced or restored, or a completely new wetland is created where none existed before (see Chapter 9). The restored or created wetlands are owned by private or public entities (including federal, state, and local agencies) who receive wetland "credits" which can be applied at a later time for "debits" resulting from unavoidable impacts to natural wetlands. This approach has the appeal of apparently conforming with the federal governments "no net loss of wetlands" policies.

The number of credits received from a particular wetland bank is judged on the basis of specific wetland values by the FWS, using techniques such as the HEP procedure discussed in Section 8.1.3. The accumulated credits in the wetland bank are then used to offset specific wetland values when a subsequent development project causes unavoidable wetland losses.

Anderson and Rockel (1991) identified four objectives which should increase the cost effectiveness of mitigation banks. First, permit applicants should experience savings in both time and money because mitigation costs and procedures are identifiable at the outset. Second, most mitigation banking projects are for large projects, and they experience the cost savings associated with economies of scale. Third, larger mitigation banking projects should tend to favor fish and wildlife because entire ecosystems are protected. Finally, wetland restoration and enhancement (as opposed to creation) involves already-functioning wetlands, such that uncertainty about success of mitigation actions should be reduced.

8.4.2 The Tenneco LaTerre Wetland Mitigation Bank

Most wetland banking projects have, indeed, involved large-scale projects. Perhaps the best known mitigation banking project is the Tenneco LaTerre wetland mitigation bank in Terrebonne Parish, Louisiana (see Zagata, 1985). This project was undertaken in an area within the coastal Mississippi River delta where subsidence and intrusion of salt water was gradually destroying a large freshwater marsh. The Tenneco Company (and its predecessor the LaTerre Co.) invested over $20 million in constructing over 100 weirs, 200 bulkheads and mud dams, and 30 miles of levees whose primary function was protection of the integrity of the freshwater marshes.

The Tenneco LaTerre mitigation bank was established by a 1984 Memorandum of Agreement (MOA) between the Tenneco Co. and five government agencies: the FWS, Soil Conservation Service, National Marine Fisheries Service, the Louisiana Department of Natural Resources, and the Louisiana Department of Wildlife and Fisheries. Under the terms of the MOA, Tenneco is required to spend $3 million per year over a 25 year period to preserve and enhance fish and wildlife habitat for approximately 5,000 acres of marsh owned by Tenneco, and another 2,200 acres owned by other parties but which lie within the system of dikes. Tenneco is responsible for maintaining and monitoring the wetlands during this period. In November, 1988, American Petrofina acquired mineral rights to the LaTerre site, and assumed management responsibilities of the LaTerre mitigation bank (see Anderson and DeCaprio, 1992).

In return for its investment in the LaTerre mitigation bank, Tenneco (and Petrofina) have earned nearly 8.5 million Habitat Units (HUs), which are calculated by multiplying the habitat suitability for a particular species of interest to wildlife managers (scaled from zero to one) times the number of acres protected. These credits may be used to offset mitigation requirements either off-site or on-site, although on-site projects require a 2:1 ratio of credits to debits. As of 1992, relatively few of the available credits had been redeemed, most of these for dredging of canals to allow drilling equipment to be brought to a site by barge (Anderson and DeCaprio, 1992).

While the Tenneco LaTerre mitigation bank has been viewed by many as a success, there are also critics. For example, it has been argued that the system of dikes and weirs on the site serves to retard the decline of the marshes within the bank, but it hastens the decline of wetlands elsewhere, and also inhibits natural movements of wildlife within the area (Anderson and DeCaprio, 1992). It is also possible that this mitigation bank offers little more than a delaying tactic, since even Tenneco's projections suggest that loss of freshwater wetlands will merely be postponed for 25 years (see Zagata, 1985). Finally, the Tenneco LaTerre mitigation bank was designed to *restore* wetlands which were perceived as declining but not to create any new ones, such that lost wetlands arguably resulted in a net loss of wetlands.

Wetland mitigation banks have been established in several other states, and the list is growing yearly (see review in Anderson and Rockel, 1991).

8.4.3 Advantages and Disadvantages of Mitigation Banks

Advantages. As noted above, mitigation banking has a high degree of intellectual and practical appeal at many levels. To developers and the regulated community in general, mitigation banks offer much greater flexibility than traditional mitigation procedures. By buying or using existing mitigation credits, developers save the time and expense of designing a specific mitigation plan for each project. In addition, the higher success of mitigation banks due to their larger size, preferable placement, and superior design, combined with the fact that mitigation credits are only issued after the bank has been certified a success, should improve generally the success of mitigation. Finally, because mitigation banks can store credits, the process should save time, money, and create greater certainty for the developer during the permit process (Kusler, 1992).

Another potential advantage of wetland mitigation banking is that it encourages the development of large-scale, cost-effective mitigation sites which include a larger portion of available ecosystems for fish and wildlife. In addition, single, large mitigation banks will reduce the number of mitigation sites, and will also allow more efficient use of limited agency compliance staff (Lewis, 1992).

Disadvantages. Despite the appeal and potential utility of wetland mitigation banks, there have been critics and problems with the process from the beginning. Many of these are criticisms and problems of mitigation procedures in general, but others are specific to mitigation banking. One such problem is that wetland functions and values tend to be specific to a particular site, since a wetland's relationship to other wetlands, sources of ground water and surface water, and adjacent upland areas usually determines its values. The "off-site" nature of wetland banks is contrary to the on-site mitigation preference by federal agencies.

Another problem with mitigation banking is that it tends to result in the creation of those wetland types that are the easiest and cheapest to create, namely shrub wetlands and marshes (Kusler and Kentula, 1990; Kusler, 1992). As a result, the habitat type received in exchange may be quite different from the wetlands that were damaged by the project.

A problem shared by all mitigation projects which has a particular impact on mitigation banking is the lack of technical expertise by many individuals involved in both the planning and monitoring of wetland banking projects. No guidelines currently exist to identify the qualifications necessary for an effective designer of mitigation projects, and agencies responsible for monitoring the mitigation banks are usually understaffed and incompletely trained. As a result, mitigation banks are often poorly designed, poorly monitored, and destined for failure from the outset. To make matters worse, mitigation banks are often creatures of complex agreements between private and public interests, and are subject to a variety of federal, state, and local laws such that design and monitoring require an advanced level of expertise that is often unavailable and always expensive.

Recommendations. Despite the problems and criticisms, several recent writers have suggested ways in which the most positive aspects of wetland mitigation banks can be retained. Noted wetland expert and attorney Jon Kusler has advocated the use of "joint projects" as an alternative to mitigation banks. These joint projects are composed of groups of developers who allocate funds for a specific mitigation project to compensate for specific wetland losses. The resulting level of specificity facilitates cooperative actions between the developers and agencies, and reduces the problems associated with governments holding private money. Because of the problems associated with mitigation banks, some private and public environmental organizations (e.g., the California State Coastal Conservancy) prefer joint projects to mitigation banks (Kusler, 1992).

Prominent environmental attorney Lindell Marsh of Irvine, California has suggested a planning-intensive system which would impose "impact fees" on a development as it proceeds. Money collected by the agencies as a result would be invested in conserving wetlands or habitat. Such a system would have the advantages of creating area-wide planning schemes that would effectively assess values, and identify those wetlands that should be preserved or can be lost with minimal environmental cost (L. Marsh, personal communication; see Marsh and Acker, 1992). The basis for this approach already exists in the Special Area Management Plan system (SAMP, see Section 7.8), Habitat

Conservation Plans under the Endangered Species Act, and Advance Identification of Wetlands under the § 404(b)(1) guidelines (see Section 7.7).

Sokolove and Huang (1992) recommended increasing incentives to industry and development interests by increasing opportunities for private investment in mitigation banks. They argue that developers (particularly in a slow economy) would invest in a private mitigation bank with low costs and little risk, rather than investing in highly speculative traditional development projects. Not only would developers acquire mitigation credits to offset future losses, but they would also experience a windfall of public support for their environmental sensitivity.

9

Assessing Risks to Ecological Resources in Wetlands

Thomas P. Burns

INTRODUCTION

The fate of wetlands and other ecological resources on or near hazardous waste sites in the U.S. depends on three key acts of Congress. These are the National Environmental Policy Act (NEPA, 42 USC § 4321 *et seq.*), the Resource Conservation and Recovery Act (RCRA, 42 USC § 6901 *et seq.*), and the Comprehensive Environmental Response, Compensation, and Liability Act (CERCLA), as amended in the Superfund Amendment and Reauthorization Act of 1986 (SARA, 42 USC § 9601 *et seq.).* Were it not for these laws, populations and ecosystems would face the threat of continuing exposure to hazardous wastes or the often destructive consequences of remediation technologies chosen without regard to their ecological impacts. RCRA and CERCLA protect ecological resources in the environs of active and inactive hazardous waste sites, respectively. These acts require the U.S. Environmental Protection Agency (EPA) to consider the risks to ecological resources, including wetlands, when evaluating proposed remedial actions. They go beyond protecting human health to protecting the health of the environment, because NEPA mandates that impacts to the environment be considered in any proposed project that receives federal support. Other legislation and executive orders, such as the Clean Water Act, and Executive Order 11990 protecting wetlands, specify additional requirements for the remediation and ecological risk assessment processes.

An ecological risk assessment (ERA) considers if and where there is imminent and substantial danger to animal and plant populations and their habitats. More generally, risk assessment is the comprehensive evaluation of the potential consequences of an action and the uncertainties associated with that analysis. Like an intelligent gambler, risk assessment tries to answer the questions: "What is the payoff or loss?" and "What are the odds of winning or losing?" The EPA now uniformly requires ERAs for federal cleanup projects because the cost of putting our ecological resources at zero risk of adverse effects from contamination or remedial technologies is prohibitive and because there is insufficient knowledge to make remedial decisions with complete certainty of their consequences (Barnthouse et al., 1986).

In addition to the EPA, ERAs typically involve the lead agency, natural resource trustees, and the ecological risk assessor. The lead agency is responsible for implementing the remedial investigation, which includes the ERA. An example of a lead agency is the U.S. Department of Energy (DOE), which has responsibility for many hazardous waste sites associated with its nuclear energy and weapons production activities. Natural resource trustees are designated by the President, State governors, and American Indian tribal chairmen as required by CERCLA. The Federal trustees are the Secretaries of the Departments of Interior, Commerce, Defense, Energy, and Agriculture and their delegates at subordinate administrative agencies (e.g., the National Park Service). The trustees' mandate is to "protect and restore [living and nonliving] resources under their jurisdiction," and they are given wide latitude to define what biota and "their supporting ecosystems" are to be held in the public trust (Environmental Protection Agency, 1992a). Their primary role in the ERA is to provide review and advice to the ecological risk assessors and, if requested to do so, to conduct a preliminary natural resource survey to identify trust resources at the site.

Ideally, an ERA at RCRA or CERCLA sites is conducted by qualified environmental scientists, including biologists, zoologists, botanists, and ecologists, as part of a larger remedial investigation team. They must characterize the ecological resources at the site, the exposure of animals and plants to contaminants, the potential effects on ecological resources, and the resulting risks of significant adverse effects on those resources from the given exposures. In the remainder of this Chapter there will be discussion of the general features of the ERA process in the context of remedial activities at hazardous waste sites and on particular problems of assessing ecological risks to wetlands.

9.1 ENDPOINTS AND UNDERSTANDING

The overall goals of an ERA are to identify the ecological resources that are present at the site (habitat characterization) and to evaluate if and how exposure to hazardous substances has impacted, is presently impacting, and will potentially impact those resources. This is foremost a scientific endeavor, but the need for a timely and cost effective result limit the scope and extent of the research. At a minimum, the ERA describes the types of habitats and their location and size, documents the presence and habitat use of as many species as possible, considers how organisms are exposed to contaminants, identifies the possible effects following exposure, and characterizes the risks to the ecological resources. The presence of wetlands usually prompts a formal delineation. There is seldom sufficient time and money to do extensive sampling of biota for estimates of population size, analysis of gut contents, or determination of contaminant body burdens. Only rarely are site-specific toxicology studies conducted using local biota and contaminated media from the site. Despite these limitations, a serious assessment of the risk to biota at a site is possible by extrapolating from published ecological and toxicological data (Barnthouse et al., 1986; U.S. Environmental Protection Agency, 1989, 1992b). The result of these practical constraints is greater uncertainty to the final characterization of risk.

The specific goals of any given ERA are often stated as assessment endpoints (U.S. Environmental Protection Agency, 1989; Suter, 1993). Assessment endpoints are the site-specific questions the ERA as a scientific endeavor is designed to answer (e.g., "Is the amphibian community at a wetland site being impacted by the nearby landfill?"). Two types of assessment endpoints can be recognized: those with ecological relevance and those with social and political relevance.

Ecologically relevant assessment endpoints are defined by the scientific issues of the ERA, i.e., the nature of the contaminants and the ecosystem at risk. The number and types of insect larvae inhabiting streams, for example, may mean little to the general public, but insect larvae are critical to the normal functioning of stream ecosystems. Likewise, the detritus-eating organisms in coastal salt marshes, myriad microscopic invertebrates, flies, and fiddler crabs, are critical links in the food web (e.g., Teal, 1962). Detailed knowledge of the plants and animals inhabiting an ecosystem, as well as the ways in which ecosystems are put together, how they function and change through time, and how they are integrated into a regional or landscape context are crucial to choosing ecologically relevant assessment and measurement endpoints.

Although the ERA primarily addresses scientific issues, social and political concerns may nevertheless help focus the investigation. Specific ecological resources are obvious subjects of assessment endpoints if they are protected by existing legislation (e.g., Executive Order 11990 and the Endangered Species Act). For example, our concern for migratory waterfowl, which is reflected in the Migratory Bird Treaty Act (16 USC § 703-711), might define the health of overwintering migrants as an assessment endpoint for an ERA at a threatened wetland. The measurement endpoint might be the abundance and contaminant body burdens of their prey, or the number of visiting birds and their body burdens upon arrival and departure. Even in the unlikely event that migrating waterfowl play no important role in the functioning of a given wetland ecosystem, nor constitute more than an insignificant part of the wetland biota, they have social, political, and economic value to the human community. All ecological assessment endpoints, in fact, reflect the values of our society because baseline risk assessments must, by law, consider the health of the environment.

Traditionally valued, expected, and implicit in scientific studies, quality is now an explicit goal of the ERA process. The Environmental Protection Agency (1989) has promulgated quality assurance (QA) guidelines, and lead agencies, e.g., DOE, have responded by instituting their own QA programs and by requiring contracting risk assessors to submit QA plans. The amount of resources invested to assure quality is probably reasonable given the liabilities, but it can appear unbalanced when financial constraints limit the amount of data that can be collected for the ERA, which then becomes one of the bases for costly decisions. Although wetlands can be relatively small in area, especially in urban industrial centers, collecting data in wetlands has additional costs, and undoubtedly, most ecologists know less about the biotic inhabitants of wetlands than other types of ecosystems. For the foreseeable future at least, maximizing quality and minimizing uncertainty in ERAs involving wetlands will be difficult.

9.2 THE ECOLOGICAL RISK ASSESSMENT

Over the past several years, the EPA, in consultation with ecologists from government, industry, and academia, has provided guidance on the conduct of ERAs (Environmental Protection Agency, 1989, 1991, 1992b). Below, a working model of an ERA constructed around EPA's (1992b) suggested general framework is summarized. Particular ERAs will, of course, have specific goals, endpoints, and scope in addition to different contaminants and ecological resources at risk. Thus, each ERA will require individualized adjustments to the model. The general framework should, however, be equally useful to most ERAs including those involving wetlands.

The *Framework for Ecological Risk Assessment* (Environmental Protection Agency, 1992b) identifies three "phases" comprising four interrelated activities (Figure 9.1). The three phases are problem formulation, analysis, and risk characterization. The second phase comprises two activities: Exposure Assessment and Effects Assessment. These four interrelated activities are considered individually below. Each involves important considerations to ensure a scientifically sound and high quality ERA.

9.2.1 Problem Formulation

The first phase of an ERA has as its primary objective a concise and accurate statement of the problem being addressed. It "establishes the goals, breadth, and focus" (Environmental Protection Agency, 1992b) of the scientific investigation. As discussed above, the specific goals of an ERA are stated as assessment endpoints, and these are developed by the risk assessor or the lead agency in cooperation with the natural resource trustees. The "breadth" or scope of the ERA is broadly defined by the lead agency but, more often than not, is determined by limits on time, money, or both. Sampling animal populations, analyzing tissues for body burdens, conducting toxicity tests, and building simulation models to predict effects on populations and ecosystems are desirable but expensive activities. Finally, the focus of the ERA is partially specified by the assessment endpoints, practical constraints defining what can be accomplished and, most importantly, a conceptual model of the ecosystem.

The conceptual model, whether or not it is explicitly identified

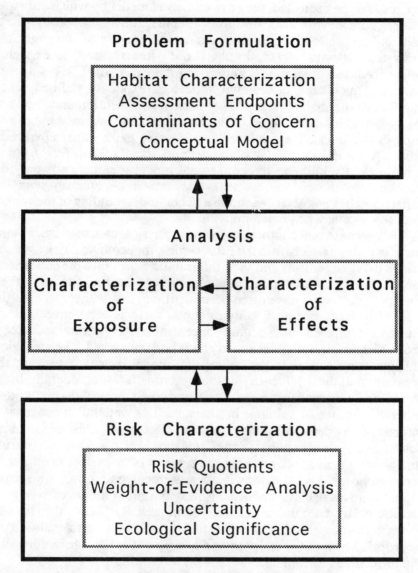

Figure 9.1: The three phases of ecological risk assessment: problem formulation, analysis, and risk characterization.

as such, embodies the risk assessors' current understanding of the relationships between the sources of risk (stressors), the ecological resources to be protected, the environmental media by which stressors come in contact with those resources, and the pathways by which direct and indirect effects on those resources propagate through the ecosystem. It is usually desirable, if not often realized, to explicitly define the conceptual model underlying the ERA. An ERA is a team project. Unless the conceptual model is explicitly defined, each individual potentially carries a unique and incomplete understanding of the system in his or her mind. This can only lead to inconsistencies and gaps in the finished product or considerable extra effort correcting them. A botanist's model of a wetland inevitably differs from a zoologist's. By building the conceptual model together, not only do members of the risk assessment team reach a common point of departure, they ineluctably acquire a richer understanding of the system than the one they possessed individually.

Because much information about physical characteristics at the site invariably exists before the ERA begins, the ecological risk assessor brings living organisms and their interrelationships into the conceptual model. The potential contaminants present at the site and their transport pathways have often been identified in advance. If, as is the case with many chemical contaminants, there is a potential risk to human health, then contaminant concentrations in various media (e.g., groundwater, air, soils) will already have been measured. The ecological risk assessors complete the conceptual model by characterizing the ecosystems at the site, identifying the presence of state or federally-listed threatened and endangered species and their critical habitats, and otherwise unique or valuable habitats, such as wetlands. Because it is impossible to deal with every species, a set of organisms for more detailed analysis of exposure and effects should also be identified during this phase. These species may be the subjects of assessment endpoints, species chosen as indicators in measurement endpoints, or simply representatives of classes of organisms with similar modes of exposure, sensitivity to contaminants, and responses. In sum, the habitat characterization should provide sufficient information to evaluate the appropriateness of the endpoints and scope of the ERA for addressing the problem as formulated.

A second major objective of the problem formulation stage is screening of contaminants for further consideration as ecological *contaminants of concern* (COCs). Whether or not a contaminant present in a particular medium (e.g., the sediments in a wetland or the soils of a hazardous waste landfill) is scrutinized in greater detail during subsequent phases of the ERA often depends on whether or not the ratio of the contaminant's measured concentration in a medium to a benchmark or 'threshold' toxicity concentration exceeds unity. This threshold is typically either a state or federal criteria [e.g., EPA's ambient water quality criteria (AWQC)], or a contaminant concentration from published toxicological studies [e.g., the no observed adverse effect level (NOAEL)]. More complex screening methods that also consider physical and chemical characteristics, which influence the likelihood of a contaminant posing an ecological threat, are currently being developed.[1]

The simple logic of so-called "quotient methods" is that if the amount of contaminant to which an organism is exposed is less than the critical threshold concentration, the population will not be at serious risk (Barnthouse et al., 1986). Thus, unless it can be shown that a contaminant at a concentration below the established criteria causes directly or indirectly adverse effects on the ecosystem as a result of some subtle nonlethal effect on the organism, there is little reason to further consider the contaminant in the ERA. Unfortunately, there are seldom resources to make this determination because it requires detailed simulation models of the network of interactions among organisms in the ecosystem (Bartell et al., 1992; Burns et al., unpublished ms). Instead, the comparison is made while acknowledging that there is uncertainty about the meaning of the COC screening threshold for protecting ecosystem health.

9.2.2 Exposure Assessment

The exposure assessment, one of two activities in the analysis phase of the ERA (Environmental Protection Agency, 1992b), seeks to quantify or otherwise characterize the magnitude of the exposure to contaminants that is actually experienced by the organisms at a site. The difference between the concentration of contaminant measured in an environmental media, such as water, soil or air, and the

---

[1] See Science Applications International Corporation, *Baseline Risk Assessment for the Wayne Site, Wayne, NJ.* DOE/OR/21950-012. (Predecisional Draft, January 1993).

concentration that organisms experience will vary depending on the organisms' behaviors and modes of existence, the contaminants' physical and chemical characteristics, and the media of exposure. For example, PCBs at high concentrations in a wetland sediment might be unavailable for uptake by crayfish living on the sediment surface because of their tendency to bind to particulate organic matter. Sediment dwellers (e.g., *Tubifex* worms) and especially those organisms ingesting sediments will, on the other hand, be exposed to a higher concentration of PCBs. The complexities introduced by the diversity of organisms and contaminants that are often present at hazardous waste sites require well reasoned conceptual models and choices for indicator organisms and, whenever possible, well designed sampling programs.

Data limitations frequently constrain the exposure assessment to be little more than a reiteration of the results of the remedial investigation, which quantifies the concentration of contaminants in the various media, supplemented with evidence that certain types of organisms would or would not actually be exposed to the measured concentrations. Actual exposures to organisms could be greater or less than those measured in environmental media. Organisms in direct contact with contaminated media may be exposed by multiple modes (e.g., inhalation and ingestion of soil contaminants). Organisms not directly exposed to contaminants can have greater exposures than those exposed directly, if contaminants become concentrated in the tissues of their prey.

The organisms considered most valuable by modern society are frequently the "top" predators, which are most at risk from ingesting contaminated prey. For example, a bald eagle hunting over a wetland with contaminated sediment is unlikely to come in direct contact with or ingest those sediments. Its exposure to contaminants will depend on the proportion of its diet that is contaminated, how much food of what kinds it takes from the site, and the contaminant concentrations in the wetland organisms it eats. The concentration of contaminant in the prey is a function of the prey species' diets and the bioconcentration factors for those organisms that are in direct contact with contaminated sediments or sediment porewater. An organism's bioconcentration factor (BCF) for a given chemical in a given medium, usually water, is the contaminant body burden achieved after some specified period of exposure to the contaminated medium divided by the concentration of the chemical in the exposure medium.[2] The BCF for an organism-contaminant pair is thus an important bit of information in evaluating the risks to the overall health of the environment as indicated by the abundance and health of top predators permanently residing and foraging in the system. In general, ecological risk assessors face a difficult challenge evaluating the direct and indirect exposure to contaminants of organisms in the networks of feeding relationships and other material transfers that are ecosystems.

9.2.3 Effects Assessment

An effects assessment quantitatively links concentrations of contaminants to adverse effects (Environmental Protection Agency, 1991). This second activity in the analysis phase requires the kind of data that comes from experiments where an organism or population is exposed to a known concentration of a contaminant for a specified period of time (dose) and the effects of that exposure recorded (Timbrell, 1982; Rand and Petrocelli, 1985). Ideally, the effects assessment should be based on dose-response data from studies at the site using sensitive organisms and contaminated media from the site (Environmental Protection Agency, 1992c; Suter, 1993). This ideal is seldom attained. Ecological risk assessors usually rely heavily upon published laboratory studies, which vary in design, quality, and appropriateness for the site, and as a result, the effects assessment introduces additional uncertainties into the ERA.

The effects assessment faces the additional problem of evaluating effects to populations and ecosystems, as required by law. There are no established methods for extrapolating from organismal toxicity to population and ecosystem level effects. This does not mean, however, that ecological science does not have a theoretical and practical basis for evaluating these effects. Sensitivity analysis, uncertainty analysis and indirect effects analysis of simulation models can be useful tools in this regard (Tomovic, 1963; Gardner et al., 1981; Bartell et al., 1992; Burns et al., *unpublished ms.*). Building simulation

---

[2] An organism's bioconcentration factor (BCF) for a given chemical in a given medium, usually water, is the contaminant body burden achieved after some specified period of exposure to the contaminated medium, divided by the concentration of the chemical in the exposure medium.

models is difficult where the data are abundant, so the short-term prospects of using models of wetland ecosystems to assess direct and indirect effects are not good. Nevertheless, further support for studies of how population size is regulated and how perturbations propagate through ecosystems will do much to advance the application of ecological knowledge to current problems.

9.2.4 Risk Characterization

The climax of the ERA is the characterization of risks to ecological resources and the evaluation of the significance of potential adverse effects on the environment. The actual risk characterization also usually includes a discussion of the uncertainties associated with the assessment, many of which have been identified in the preceding discussions of the problem formulation and analysis phases. Given the results of the exposure and effects assessments and the recognized uncertainties, the risk assessor analyzes and interprets the evidence to answer the questions posed in the problem formulation phase of the ERA. The results of the risk assessment subsequently feed into the risk management process, which weighs the various options' costs and benefits to the interested parties.

For example, consider a hypothetical case where the surface water innundating a wetland near a hazardous waste site is contaminated with mercury at concentrations above EPA water quality criteria. The risk characterization evaluates the risk to the biota and concludes that the vegetation is at little risk because plants do not take up mercury. Raccoons, which are the indicator species for mammalian predators — the subject of an assessment endpoint — are judged to be at significant risk. This is because raccoons feed predominately on crayfish which bioconcentrate in their tissues mercury obtained directly from the water and from their benthic invertebrate prey. The mercury concentration that raccoons experience in their diet is calculated from published methylmercury bioconcentration factors for crayfish and estimates of crayfish consumption rates for insect larvae. This exposure estimate is compared to published values for the toxicity of methylmercury fed to laboratory rats (because there are no data for raccoons), and it is found that the calculated exposure is 100 times greater than the rat NOAEL. Despite the uncertainty introduced by using rat toxicity data, laboratory BCF values, and published estimates of feeding rates from raccoons at other locations, the risk assessors conclude that the raccoons are at risk of significant adverse effects and that surface water contaminants could have serious effects on the overall health of the wetland ecosystem. All that remains is for the risk assessors to present the supporting arguments for their conclusions.

The example above uses one of several approaches to characterizing risk: the Quotient Method (Barnthouse et al., 1986). This approach was discussed earlier as a way to screen contaminants for further consideration. In the risk characterization phase, the quotient obtained by dividing the measured or calculated exposure of a given ecological resource by a predetermined concentration, which is presumably protective of that resource (e.g., AWQC, NOAEL), is taken as an estimate of the risk. The higher the quotient, the greater the risk. Shortcomings of this approach are discussed in detail elsewhere (Suter, 1993). Quotient methods are currently the most practical means of quantitatively comparing ecological risks.

Weight-of-evidence analysis, a second approach to characterizing risks, can be used alone or to complement quotient methods. In this approach, one makes a proposition concerning the scientific issues of the ERA and then presents an argument for or against that proposition using whatever site-specific or literature data, insights, and professional judgement one can muster (Suter and Loar, 1992). As a formal approach, it should be familiar to members of the legal profession and clinical diagnosticians; informally, it is what we all do in our daily lives. Additional guidance on developing weight-of-evidence arguments is available (Hill, 1965; Suter, 1993). The value of this approach is that it acknowledges that more than precise quantitative data can have a legitimate bearing on the assessment of the risks of adverse effects on ecological resources, as long as the uncertainties and shortcomings of the data are acknowledged.

Other approaches to quantifying the risk to ecological resources are being developed. Suter (1993) makes a case for an approach modeled after the medical sciences (e.g., epidemiology and pathology) in which adverse effects are demonstrated and the cause of these effects is proved using something analogous to Koch's postulates. The rigor of this approach is a worthy goal for enhancing the ERA process as a scientific endeavor, but it is unrealistic to expect this goal to be reached given the usual constraints on time and funding. Likewise, simulation models of contaminant fate and transport (Mackay, 1991) and dynamics of interacting populations (Bartell et al., 1992) offer potentially useful, if not absolutely necessary, tools for rigorously assessing exposure and effects. Yet, they too suffer from the constraints of too little data and too little time and money to learn more which make them all but impossible for most ERAs. Assessing risks to ecological resources in wetlands and other ecosystems will continue to be less a than ideal process until a serious commitment is made to relax these constraints. In the meantime, ERAs, uncertainties and all, must suffice to meet the practical needs of the user.

9.3 ERA'S AND WETLANDS

9.3.1 Examples of Wetland ERAs

A diversity of ERAs are in progress at hazardous waste sites that comprise wetlands. These range from large, relatively well-funded projects, like those at DOE weapons production and research facilities in Oak Ridge (TN), Augusta (GA), Hanford (WA), and Brookhaven (NY), to small National Priorities List or "Superfund" sites. Current reports are available from the DOE or local public information offices. Of those sites with which I am familiar, wetlands potentially impacted by contamination and the "standard engineering practices" used in remediation include small ponds with typical emergent vegetation (e.g., *Phragmites* spp.) surrounded by commercial developments, large semi-wild areas of vegetated drainage swales bordering the Niagara River, and drainage ditch/creeks snaking through suburban communities. East Fork Poplar Creek, which runs for 24 km through commercial, residential, and relatively undisturbed bottomland hardwood forest on the DOE's Oak Ridge Reservation in Oak Ridge, TN, includes 16 formally delineated wetlands within its 500-year floodplain that are at varying risk from mercury and other contaminants in sediments and surface waters.[3] In all of the above cases, where wetlands have been formally delineated, they received or are receiving explicit attention in the ERA process.

Although it is tempting to discount the value of these "urban"

---

[3] See Science Applications International Corporation, East Fork Poplar Creek/Sewer Line Beltway Remedial Investigation Report. DOE/OR/02-1119&DO. (Predecisional Draft, January 1993).

ecosystems from an economic perspective, they can have much ecological value from a regional perspective, which recognizes their importance as animal movement corridors, species refuges, hydrological buffers, water purifiers, etc. Perhaps even more significantly, the public is beginning to value these natural areas as the last "wild" places left in their local environment. Wetlands are ever more likely to have numerous advocates involved in the ERA process, including natural resource trustees. This is desirable. It wasn't too long ago that the this-isn't-a-wetland-its-a-swamp attitude was ubiquitous among regulators, elected officials, and consultants (Owens-Smith et al., In Press). Today, few of those involved in the ERA process are ignorant of the laws protecting wetlands.

9.3.2 Special Problems With Wetland ERAs

Despite the growing interest in protecting local wetlands, the problems faced in assessing the risks to them are considerable and require the expertise of wetland scientists. Some problems in wetland ERAs are special to wetlands; others are common to most sites but present extra difficulties in wetlands. Compiling a comprehensive list of species that use a site at any time during the year is difficult under the best of circumstances. It is even more difficult and yet more crucial to identify the many species using a wetland. Habitats that are relatively rare and isolated in a region, like many wetlands, serve as temporary but nevertheless critical "rest stops" for migratory animals. The short-term exposure to contaminants of these species and resulting risk should be evaluated, but this may require extensive knowledge of their ranges, diets, densities, and breeding habits, and not merely intensive knowledge of their ecology when at the wetland being assessed.

Much more needs to be done before exposure and effects assessments in wetlands are matter-of-fact. Sediment-quality criteria have only recently appeared (Long and Morgan, 1991). While these criteria were derived from data for freshwater and marine sediments, wetland sediments may be sufficiently different, or sediment dwelling species (or particular life stages) in wetlands may be sufficiently more sensitive, to require sediment-quality criteria specific to wetlands. There is scant toxicity-effects data for wildlife, in general (but see Eisler 1985-1990), and species common or restricted to wetlands, in particular. Many species from freshwater and marine wetlands [e.g. crayfish

(*Procambrus*), frogs (*Rana, Bufo*), fiddler crabs (*Uca*)] might make excellent candidates for standardized toxicity tests using a wide range of common contaminants. Other issues that must be considered are the effects of remedial activities that could increase sedimentation in wetlands downstream or alter the topology and hydrological flow patterns responsible for a wetland's existence. It is not enough that we express concern for wetlands. We must understand them better if we are going to accurately assess and effectively mitigate the risks to ecological resources in wetlands.

9.4 SUMMARY

Before the effects of a remedial activity on a wetland can be mitigated, an ERA will often be required to justify and guide the choice of remedial action. The ERA is a socially responsive scientific investigation of an ecosystem and the risks from possible anthropogenic impacts on it. It requires as much ecological knowledge as can be obtained about the organisms present, the nature and distribution of contaminants, and the relationship between proximal exposure and ultimate effects. The uncertainty in the characterization of risk that results from the assessment will be a direct function of the resources invested in acquiring accurate and precise data about the site, its inhabitants and contaminants, and the risk assessors' ability to develop a coherent and rich conceptual model of the ecosystem. That model should specify in appropriate and sufficient detail the network of pathways connecting contaminant sources to the biota and the myriad interconnections among the biota. The ERA process faces special difficulties when dealing with wetlands, which are protected by laws and executive orders, but which are still not extensively studied. Evaluating the risk to wetland resources and evaluating their ecological significance can be challenging tasks. Wetlands offer, however, special ecological and social benefits in return for the extra effort required to understand and protect them.

10
Restoration and Creation of Wetlands

Robert P. Brooks

INTRODUCTION

As discussed more fully in Section 8.3, wetland restoration and creation are forms of "mitigation" often required by federal, state, or local agencies. The goal of mitigation is usually to compensate for wetland values (see Section 2.3) which have been lost or degraded. While wetland restoration and creation activities have been somewhat controversial, they may represent the best opportunity to revive wetland values that have been degraded or lost, and to achieve a goal of "no net loss" of the nation's wetlands (see Kruczynski, 1990a, 1990b; Kusler and Kentula, 1990; and Chapter 8).

Wetland restoration and creation should be viewed as a process. The process remains essentially the same regardless of the impetus for undertaking a particular project. Project success is predicated on recognizing, from the beginning, the complexity of undertaking restoration and creation. The ecological and technical hurdles, as intimidating as they are, can pale in comparison to socio-economic and philosophical hurdles (Brooks, 1991; National Research Council, 1992). For instance, plans for restoring or creating a wetland, despite having a solid ecological foundation, are often viewed with suspicion. Can you guarantee success? Will equivalent functions be replaced? How much will it cost? There are enough success stories, there is enough need, and there is a large enough segment of society that believes that trying to restore damaged ecosystems or create lost ecosystems is a worthy and sometimes necessary goal (Brooks, 1991). The objective of this chapter is to provide an overview of wetland restoration and creation and to help you accomplish your restoration and creation projects in a satisfying and cost-efficient manner, regardless of the impetus for the project.

10.1 DEFINITIONS

A variety of terms are used interchangeably to describe restoration: reclamation, rehabilitation, regeneration, rejuvenation, recovery, and creation (Cairns, 1980). For the purposes of this publication, restoration is defined as: the process of returning from a disturbed or altered condition to a previously existing natural or altered condition (Lewis, 1990). Similarly, creation is defined as: conversion of a non-wetland area into a wetland area (Lewis, 1990). The intent is clear - to replace a wetland with another of comparable area and function or to change an undesirable environment into a more acceptable one.

The rationale for deciding to restore or create a wetland may be based on one of several reasons: 1) regulatory compliance from a legislative mandate (federal and state wetland regulations); 2) immediate threat to human health and safety (threat to a municipal water supply); 3) a grassroots effort to repair cumulative damage to an environment (a community or conservation organization restoring a degraded section of stream); or 4) an interagency initiative to increase wildlife habitat (restoration of prior-converted farmed wetlands) or protect a watershed (install low-technology wetland treatment systems to curb non-point source pollution). Regardless of the reason, one should have a plan or strategy that focuses limited energy and funds on a project of the proper scope and scale.

10.2 THE STATE-OF-THE-TECHNOLOGY

A brief review of the current status of the science and technology of wetland restoration and creation is in order before describing the process itself. Several notable books and publications describing case studies are available as models for ecological restoration projects in general. The periodical, *Restoration and Management Notes*, has been published by the University of Wisconsin Arboretum since 1981, but does not emphasize wetlands. Formed as a professional organization in 1987, the Society for Ecological Restoration and Management recently began publication of a technical journal, *Ecological Restoration*. Restoration examples from a variety of ecosystems, including wetlands, with projects ranging from reintroductions of single species to watershed-level efforts are covered in several edited volumes (Jordan et al., 1987; Cairns, 1988; Berger, 1990).

Publications focused entirely on wetland restoration and creation also are available. The creation and restoration of coastal plant communities around the world was summarized by Lewis (1982) and Thayer (1992). Gore (1992) discussed the theory and practice of restoring streams and rivers. Wolf and others (1986). produced an annotated bibliography on restoring and creating wetlands in the U.S. The Association of (State) Wetland Managers publishes proceedings from national and regional meetings, many of which cover wetland restoration. Of particular note is the volume from a 1986 symposium, *Mitigation of Impacts and Losses* (Kusler et al., 1988). The status-of-the-science in coastal and freshwater wetlands was presented in a two-volume publication edited by Kusler and Kentula (1990). Two scientific journals, *Wetlands* and *Wetlands Ecology and Management*, frequently contain articles about wetland restoration. Additional sources of information include the experience and publications of federal agencies, such as the U.S. Army Corps of Engineers (shore stabilization, wetland creation), the Soil Conservation Service (technical assistance on erosion and sedimentation control, pond and wetland construction, sources of plant materials, best management practices on agricultural lands) (U.S. Soil Conservation Service, 1992), the U.S. Forest Service (tree planting, best management practices for managing forests, habitat improvement for wildlife and fisheries)(Payne and Copes, 1988) and the U.S. Fish and Wildlife Service (wetland construction and management for waterfowl, database on published literature)(Schneller-McDonald, 1989).

A browse through the collective literature on the subject suggests that the best opportunity for success is for restoration of emergent wetlands with a component of open water (pond-like), that is, an early successional community (Hammer, 1992). Restoration, as defined previously, implies that the site was once a wetland, and hence, the essential component of adequate hydrology is present or can be restored. Emergent vegetation (e.g., grass-like and broad-leaved herbaceous plants) can grow and reproduce in one to two years (versus woody shrubs and trees), although the full extent of plant diversity may not be present for many years. There are some success stories with the more difficult ground-water based and/or forested wetlands (Best and

Erwin, 1984; Clewell, 1988; Clewell and Lea, 1990), but by far the majority of wetland projects produce, at best, marsh-like habitats. Keep in mind that emergent wetlands do not always represent an early stage of vegetative development. Stable (constant water depth) or periodic hydrologic regimes (tides, floods) can maintain dominance by herbaceous species indefinitely, with no successional progression towards shrub or forested communities.

Unfortunately, much of information about the success and failure of restoration and creation projects is based on hearsay and "windshield surveys." Few projects provide baseline data for subsequent comparisons, monitoring over multiple years is rare, and critical evaluation of wetland functions is virtually non-existent. Comparative studies between groups of wetland mitigation projects versus natural reference wetlands in Oregon, Florida, and Connecticut demonstrated the preponderance of the emergent wetland type - "a pond with a fringe of emergent marsh" (Kentula et al., 1992). These authors developed and described an approach for formulating design guidelines and setting performance criteria through a system of permit tracking, monitoring, and site evaluation. Following this approach, knowledge should expand as more and more sites are investigated.

We are left with the dilemma of deciding what types of wetlands can be restored or created. Despite the expanding supply of technical literature, several significant questions remain to be answered. What level of function is achievable for wetland projects in a given landscape setting? Do mitigation projects achieve the same level of function as compared to natural wetlands? How long does it take for projects to achieve the desired level of function (Kentula et al., 1992)? How do we measure and monitor the success and failure of restoration and creation projects? Are there certain types of wetlands (with respect to vegetative and hydrologic characteristics) that should be off limits to all but experimental attempts at restoration and creation? The remainder of this chapter examines the strategies and procedures needed to achieve success with restoration and creation projects, while at the same time attempting to answer some of these questions.

10.3 THE PROCESS

Suggested steps to follow to facilitate successful restoration and creation of wetlands are listed below and described in detail throughout this chapter. The steps are not distinct, but represent an overlapping series of tasks that must be accomplished. Note also that items 1-4 may need to be conducted before or concurrent with a permit application.

    1. Conduct functional assessment of the wetlands to be impacted, and consider functional needs for the region.
    2. Set site-specific objectives for the project in cooperation with stakeholders.
    3. Select and acquire access to a suitable site.
    4. Design conceptual plans based on site conditions and project objectives.
    5. Prepare construction plans, specifications, and budget.
    6. Implement construction and maintenance.
    7. Prepare as-built conditions for baseline information; implement monitoring protocols and prepare evaluation reports.

10.3.1 Functional Assessments

Despite the litany of functions and values espoused for wetlands, most experts agree that all wetlands are not created (or restored) equally. Stated another way, individual wetlands typically provide only one or two primary functions and wetlands vary in their functional capability. It seems logical then, to design each restoration or creation project based on one or two major functions. The choice of functions should be based on specific objectives developed for the property of interest. Assuming a decision has been made to tackle a restoration or creation project, the selection of which functions to emphasize is tied to either replacement of lost functions attributed to an encroachment, or a documented need for alternative functional replacement for that particular watershed or region.

For example, most mitigation projects undertaken as a result of regulatory action demand replacement of the wetland "in-kind" and "on-site" (i.e., replace the same type of wetland near to the property in question). However, if justified, a case could be made to construct a different kind of wetland of a type less common to the region (out-of-kind) that provides critical habitat for a plant or animal species of concern. The key to making such a trade-off, is to obtain information not only about the individual wetland in question, but also its location and relation to the surrounding landscape.

There is no standard way to assess wetland function. Although it would be convenient for the regulator and the consultant to have a universal protocol available, the geographic, vegetative, and hydrologic diversity of wetland types produces an inherent variability that defies a simple cookbook approach. What works for a riverine bottomland hardwood in the southern U.S. is not necessarily usable for a depressional wetland in the prairie pothole region. The investigator must select the most appropriate methods from an assessment tool box.

Functional assessment methods have been reviewed by Lonard and Clairain (1986). An attempt to apply the most comprehensive of these methods to the entire U.S. has been abandoned in favor or more regional approaches. However, the concepts developed by Adamus and Stockwell (1983) and others (Larson, 1976; U.S. Fish and Wildlife Service, 1980; Hollands and Magee, 1986; Marble, 1992; Ammann et al., 1986; Ammann and Stone, 1991) have continued to surface as the best way of objectively determining the relative importance of a function for a given wetland. Individual functions can be assessed independently and either kept separate or used to generate a ranking or score. See Adamus and others (1991) for an extensive review of the literature.

10.3.2 Site Objectives

Given a national wetlands policy of no net loss in area or function, projects should not be initiated without a clear statement as to what level of functional replacement is sought. Wetlands with early successional characteristics (e.g., open water interspersed with emergent vegetation, dominated by annuals and short-lived perennial species, substrates with small amounts of organic matter, moderate species diversity) are more likely to be created successfully than types with long-lived perennials, large accumulations of organic matter (e.g., peat or muck), and those represented by high taxonomic diversity. Project objectives should reflect these technological limitations.

Objectives should be adaptive and flexible, so that when more specific site information becomes available, they can be modified for realistic outcomes. Likewise, the permit conditions that often drive project objectives, must also be adjustable. For example, it is far better to specify habitat requirements based on life form of native plants and their affinity for certain hydroperiods, than it is to demand $90 \pm 5\%$

vegetative cover regardless of the species, which could end up being exotic weeds.

Site objectives should not be developed in a vacuum. Establishing a small planning committee composed of internal personnel responsible for the project is an effective way to develop a strategy. This group should formulate objectives for the project and conduct the initial screening about the ecological feasibility, economic costs, and obstacles to completion. Informal contacts with local planners, local natural resource professionals, and organizations undergoing similar projects lead to the formation of cooperative partnerships. What other related projects are underway in the community? The preliminary objectives and plan can be presented in private or public meetings to get input from agency representatives, financial sponsors, other partners, the media, and the public. Large, cooperative projects frequently require several iterations of planning, input, feedback, and modification before the final design is acceptable to all interested parties. Large projects also tend to receive more public scrutiny, whether it's desired or not.

10.3.3 Site Selection and Development of Conceptual Plans

The objectives should help screen the pool of potential sites that have the basic characteristics needed to support a successful venture. Of prime concern, of course, is ownership and access to a site. Some sites might provide optimal conditions, but may be unavailable or unaffordable. Other sites will be available and suitably located in the vicinity of the project area, but may require more topographic modification, excavation, or manipulation of the hydrologic regime to accommodate the intended wetland.

Many of the same agencies, organizations, and individuals contacted regarding site objectives also can provide input into the site selection process. Federal and state agencies and organizations may have funding available for innovative projects that lead to a net gain in wetland resources (e.g., U.S. Department of Agriculture programs, such as the Conservation Reserve and Wetlands Reserve Programs; U.S. Fish and Wildlife Service's Partners in Wildlife; Ducks Unlimited's MARSH Program). Local entities are more likely to know of areas in need of restoration where wetlands could be part of the solution (e.g., severely eroded streambanks, areas of non-point source runoff, defunct stormwater systems). Contact conservation districts, watershed associations, and local chapters of conservation and service organizations such as Trout Unlimited and Scouts for potential sites.

An initial screening of potential sites can be conducted in the office or through a brief "windshield survey." As the list of candidate sites narrows, the level of information required to make an informed decision will become increasingly more detailed (Figure 10.1). Project objectives, site selection criteria, and conceptual designs for the wetland must be gradually integrated as this process continues. For the final one or two choices, access to conduct field studies will be mandatory. A "ballpark" budget estimate should be prepared to determine if the project is economically feasible. Consulting fees, permitting costs, land acquisition costs, and approximations for construction, maintenance, and evaluation costs should be projected (see Section 10.5). At any point in the planning process, a major fiscal or ecological obstacle may shift the focus to alternative sites or solutions. The following factors should be considered during the process of selecting a site and developing conceptual plans:

1. Located within a reasonable distance of the overall project or center of operations.
2. Topography and geology must be appropriate.
3. Water quality and quantity must be sufficient to support the wetland.
4. Site morphometry and soils must be carefully considered.
5. Consider the landscape context of the project.

1. *Location.* If the project is tied to a permitted activity, there may be on-site requirements for mitigation. Most projects will have to be located in a designated drainage area (off-site, but same watershed) or management zone (county or ecoregion boundaries used by management agency) as defined by stakeholders. Exceptions to on-site replacement might include urban sites where natural lands are often absent or severely degraded, identification of a critical need or an ideal site in an adjacent drainage, and lack of properties where acquisition or permanent easements are unobtainable. Mobilization of construction equipment over long distances can add substantially to overall project costs. Thus, focusing the search for a suitable site within a radius of several miles makes good economic and ecological sense.

Restoration and Creation 327

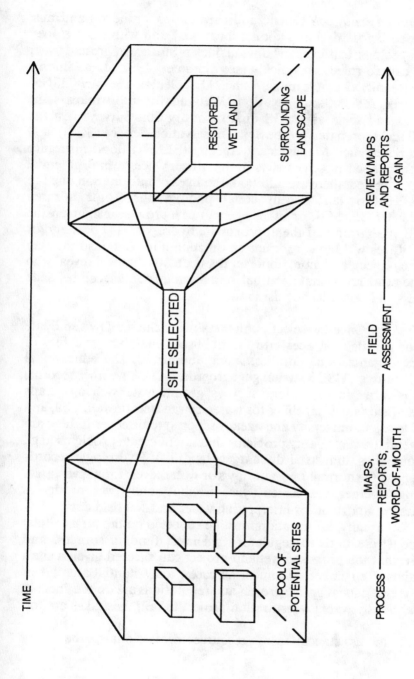

Figure 10.1: The process of selecting wetland restoration sites involves sorting through the potential pool of sites. Once a site is selected, reconsider the landscape position again.

2. *Topography and geology.* The landscape setting for the wetland must be assessed before deciding upon a location. Land with steep slopes, unstable geologic features, and obvious lack of surface or ground water sources can be ruled out quickly (see Hammer, 1992 for additional details). Readily available maps, such as U.S. Geologic Survey (USGS) quadrangles (1:24,000 scale), National Wetlands Inventory maps (same base map and scale as USGS quad), county Soil Surveys (usually 1:20,000 scale) prepared by the Soil Conservation Service (SCS)), and Federal Emergency Management Agency (FEMA) flood insurance maps can be used to screen sites in the office. Once several potential areas are targeted, there are additional sources of information one can consult. Historic and recent aerial photographs (color infrared preferred; USGS, SCS, or other agency) can provide insight on the expected vegetation and the approximate hydroperiod. Local and/or county offices will have information on zoning restrictions that may affect the project, and more importantly, tax maps that will reveal who owns the property so that an initial (and discrete) contact can be made for access and potential acquisition.

3. *Hydrology.* Once the selection process has yielded a few candidate sites, one can delve deeper into the published literature and files of agencies, organizations, utilities, water authorities, and educational institutions, (e.g., USGS stream gage records, NOAA weather records, natural resource and hydrologic surveys, ground water well logs, maps of transmission lines, pipelines for water, sewage, stormwater, oil, and natural gas, and university and agency reports) to discover if there are any potential water quality problems that might create problems (e.g., Superfund sites, industrial discharges, landfills). Hydrologic records may suggest the normal range of flows or water levels for a given area. Knowing whether the current year is abnormally wet or dry is particularly important for interpreting water and soil field data.

Eventually, information must be collected on the target site(s) to demonstrate both to regulatory agencies, funding sources, and yourself, that the proposed wetland project can succeed given a site's constraints. Remember, in almost all cases, a substantial amount of money is being spent on the project and reputations are on the line. No one consciously wants a restoration project to fail; the stakes are too high.

With assistance from resource professionals, the site's

hydrology, soils, and existing plant communities must be investigated on-site. Hydrologic investigations should include an assessment of both surface and ground water. Will point or non-point sources of surface runoff provide a significant portion of the required water supply? Are there any concerns about agricultural runoff with excess nutrients, high sediment loading from adjacent earth disturbance activities, or industrial and municipal stormwater discharges? Do ground water levels fluctuate dramatically by season? Are there current or future threats to the water supplies, such as mining, water extraction from wells, or road construction? Will the combined sources of water (precipitation, surface runoff, shallow ground water) provide sufficient hydrology during the growing season?

4. *Site morphometry and soils.* As the conceptual plan for the wetland unfolds, the site assessment team must be able to visualize how the morphometry of the proposed wetland (e.g., size, shape, slope, depth) will conform to the physical constraints of the site. A common mistake made when acquiring land is not to allow sufficient land area surrounding the intended wetland acreage for buffer zones or adequate basin slopes. Locating a wetland immediately next to a highway embankment, parking lot, livestock feedlot, or other unnatural land cover is unwise. Besides jeopardizing the wetland's integrity, post-construction functional assessments undoubtedly will take into account the landscape surroundings of the newly restored or created wetland and rank it correspondingly low. Lack of a naturally vegetated buffer will lower wildlife value, potentially allow contamination from stormwater, and reduce opportunities for volunteer colonization of native species from adjacent habitats. The latter can reduce planting costs.

It is very difficult to define precise water levels in a constructed wetland during all seasons. Thus, it is prudent to design the basin so that a sufficient amount of suitable substrate is available for the planting and colonization of plants and macroinvertebrates. The best way to accomplish this is to avoid the "swimming pool mentality" that produces vertical or very steep bank slopes (< 1v:3h). Naturally occurring wetlands usually form along hydrologic gradients that may stretch for hundreds or thousands of yards. Subtle changes in slope allow a diversity of hydrophytes to adapt to the micro-relief along that gradient. Some obligate plants may require continuous saturation throughout the year, whereas some facultative species may only tolerate saturated root zones for a few weeks. By designing and constructing the wetland with an undulating bottom and slopes that vary in elevation by as little as tenths of an inch, the wetland restorer is setting up conditions whereby the respective plant species will find their own optimal setting (Garbisch, 1986). Wherever possible, slopes should be > 1:10, although > 1:15 - 1:20 is preferred. To achieve this level of control, grading plans will need contours of no more than one-foot intervals, and possibly less (Garbisch, 1986).

If water is the defining factor of wetland formation, then the characteristics of soil are a close second. When precipitation and surface runoff are the primary sources of water supply to a wetland, the soils must be sufficiently impermeable to minimize infiltration. Conversely, if the wetland basin has been excavated to intersect shallow ground water, as is often the case, then the soil permeability must be capable of allowing the water to flow freely in both vertically and horizontal directions. By considering regional climatic factors, such as average annual temperatures, precipitation/evaporation ratios, and typical relative humidity, a water budget for the wetland can be estimated (Hammer, 1992). If the known sources of water for the project will not provide sufficient quantity, then perhaps an alternative site should be sought. As a last resort, amending the water supply with diversions from streams, lakes, or wells can be considered. Agencies or personnel with oversight for the project will expect a commitment to maintain this auxiliary source in perpetuity, a potentially expensive proposition.

Initial determinations of soil suitability usually involve coring a series of shallow holes and possibly installing simple wells. Examining the texture and color of soil to depths of four to six feet will shed light on the existing water table elevations (using characteristics similar to those used for delineating wetlands). In addition, the permeability of the soil can be determined. A cursory look at soil from a core can suggest the depth at which soils are saturated (the same process is used to evaluate sites of on-site septic systems), and thus, how much excavation may be necessary to support a wetland. However, more useful hydrologic data can be obtained by monitoring the fluctuation of subsurface water depths in simple wells (slotted plastic (PVC) pipes) for a period of months to several years, as time allows.

Many, though not all, natural wetlands have a relatively high proportion of organic matter consisting of moderately decomposed peat or highly decomposed muck. Yet designs for constructed wetlands typically call for scraping or excavating into the subsoil to an established depth. At best, a thin layer (about six inches) of topsoil is spread back over the subsoil before planting occurs. Thus, plant roots are expected to penetrate a substrate dominated by either coarse fragments (gravel and rock) or compacted clay. For most wetland types, designers should specify a much thicker layer of nutrient-rich topsoil with an organic matter content near that of natural reference wetlands of the same type.

5. *Landscape context.* Now that the screening process has focused on a single site, take another look beyond the property boundaries (Figure 10.1). What are the potential threats to a successful wetland project from adjacent land uses (e.g., urban, agriculture, transportation, mining, industry, etc.)? Will neighboring landowners be receptive to the project (e.g., negative perceptions about wetlands producing vermin and mosquitos, concern about impounded water)? Will the wetland be isolated or connected with other wetlands and bodies of water (e.g., flood storage and wildlife functions)? Given the expense of creating a wetland, is it likely to remain functional as land uses change in the foreseeable future? Although it may be convenient to squeeze a wetland into the middle of a suburban development or highway median, does it make ecological sense? Replacing lost acreage is one part of the mitigation equation, but replacing equivalent functions must also be seriously considered.

10.3.4 Prepare Construction Plans, Specifications, and Budget

At this point in the process, one assumes the project has been approved and/or endorsed by the stakeholders, whether they be regulatory personnel or organization leaders. Permanent access to the site has been secured. Knowledge gained during the site investigation can be used to prepare construction plans and specifications. A construction contract must be let to do the work, unless you are performing the work yourself. In either case, detailed drawings and specifications are needed to ensure that the wetland is installed properly. Drawings and specifications form the basis for a construction contract.

Time delays are one of the major factors that can raise the

overall cost of a restoration or creation project. Budget estimates should be based on a timetable that is tied directly to a work plan and any seasonal constraints on the construction process. Construction plans and specifications may need final approval from agency representatives before work can commence, so allow sufficient time for review. Complex projects will usually be staged through several phases. For example, a typical construction sequence for a wetland of several acres in area would involve the following stages:

1. Professional survey.
2. Earth-moving activities.
3. Soil and substrate preparation.
4. Plant establishment.

1. *Professional Survey.* The property boundaries, access points, construction staging area, actual wetland area, and critical habitats or structures that are off limits are surveyed, staked, and used to create a base map on which further planning and monitoring will be based.

2. *Earth-Moving Activities.* Many wetland projects require substantial, and expensive, excavations. Contractors considering the bid package will want to know rather precisely the amount of site preparation required (e.g., installing roads or bridges for access, grubbing of existing vegetation, stockpiling of topsoil), the cut and fill balance (e.g., grading plan, depth of excavations, off-site trucking of excavated materials, re-spreading of topsoil or hydric soil), and any specialized construction requirements, such as dams, dikes, or other water control structures. Wetland projects are atypical of work done by most firms, with the exception of a few consultants specializing in wetland restoration and creation, so interested parties should be invited to an on-site visit as part of a pre-bid conference to answer specific questions about the unique features and tasks needed for a particular project.

3. *Soil and Substrate Preparation.* Before beginning activities that disturb soils, proper erosion and sedimentation controls should be in place. During site preparation, any topsoil or hydric soil that is to be saved should be stockpiled and properly marked. Rough grading of large sites may require major earth-moving equipment (e.g., bulldozers, graders, front-end loaders, Figure 10.2). Fine grading to proper

Figure 10.2: Rough grading of a wetland basin with topsoil. (Photo courtesy of T. Rightnour, Water's Edge Technology, Inc.)

elevation should be accomplished with low-impact equipment to avoid excessive soil compaction. During this latter stage, the top layer of soil is spread throughout the site to a specified depth (6 inches minimum, 12-18 inches preferred). This can be accomplished by depositing the soil in 3-6 inch layers. On some sites, no addition of topsoil or hydric soil may be needed. Other sites may require organic amendments (e.g., composted leaf litter or food wastes) or nutrient additions (e.g., composted manure, slow-release chemical fertilizer). Analyses of soil texture, permeability, and chemistry can provide the necessary information.

4. *Plant establishment.* Planting is not required of every wetland project, but usually establishment of hydrophytes from the seed bank of hydric soils or through voluntary colonization is supplemented with

some nursery-grown stock. Timing is critical if plant survivorship is to be maximized. The proper season for germination of seeds may be quite different from that for submergent rootstocks or tree seedlings.

Nursery stock should be ordered from reputable dealers months and, sometimes, years ahead of time. Inquire about the geographic and genetic source of the stock. Is it appropriate for your area? How and when will it be delivered? Expensive planting stock delivered to a site at the wrong time (e.g., late Friday afternoon or on a hot July day) can result in a substantial waste of time and money. Herbaceous species may cost several dollars per plant, whereas woody species can cost $5-10 or more per plant. The number of plants required is usually based on a rough grid pattern related to the size of the area planted. A one-acre wetland requiring 2 x 2-foot spacing, would require over 10,000 individual plants! Although use of a geometric planting grid is convenient for computations, the actual planting can be varied to produce a more naturalistic appearance.

Even if planting stock comprises a small portion of the overall project, a plan and specifications should be prepared. Areas to be planted should be shown clearly on maps. The source, number, size, and planting requirements should be shown for each species. The selection of plant species should parallel the expected hydrology of the site. Many grasses, rushes, and sedges can tolerate moderate fluctuations in water level (Hammer, 1992). Submergents and some broad-leaved emergents may favor more stable depths of 6-18 inches, and hence, should be associated with deeper, open water areas. Relatively few woody shrubs and trees can tolerate continuous flooding, so they are usually relegated to the transition areas between saturated soils and uplands. Areas to be seeded, or where hydric soils have been spread (often called mulching), should also be mapped and described. If voluntary colonization is expected to occur in large portions of the wetland, the anticipated source and species should be noted (e.g., wind blown, water borne, or animal dispersed propagules).

10.3.5 Implementation and Maintenance

Once the plans are approved, the contract is let, and permits are secured, work can begin. DO NOT COMMENCE WORK WITHOUT THE PROPER WETLAND AND CONSTRUCTION PERMITS! Because these projects are atypical, it is imperative that a competent individual trained in wetland science and management supervise on-site activities. Some phases of construction and planting can proceed rapidly, within hours or days. To avoid mistakes and costly delays, decisions must be made promptly on-site as the work progresses. Develop good communication links with both project managers and laborers. Make sure everyone knows who is ultimately in charge of the project.

Since many aspects of construction were mentioned in the previous section, readers are encouraged to refer to publications by Garbisch and others (Garbisch, 1986, 1989; Garbisch et al., 1992) for advice on staging the construction process. A great deal of attention should be given to carefully preparing the morphometry, substrate conditions, and of course, hydrology of a site before initiating plant establishment. Depending on when the physical layout of the site is completed, it may be prudent to delay the planting sequence (except for substrate stabilization) for several months, or even until the next growing season.

The selection of plant species depends on the type of wetland being built (e.g., tidal versus nontidal, persistent emergent versus non-persistent emergent). Seeding of the desired species should be done in the Spring to maximize germination. Transplanting of potted stock or dormant underground tubers and bulbs from nurseries should occur before new growth begins. Thus, Fall through early Spring are appropriate times (Figure 10.3). Transplants tend to have greater survivorship rates than plants grown from seed. The planting requirements for species preferred for food by waterfowl are reasonably well known (Kadlec and Wentz, 1974). Recent research on a greater diversity of wetland plants is beginning to broaden the knowledge base on their site requirements and life cycles (Thunhorst, 1992).[1]

Even the best of projects rarely performs perfectly, so provisions should be made for repairs and maintenance to the site, perhaps on a time and materials basis by one of the contractors involved in construction. Unpredictable weather, abnormal water levels, wildlife depredation, and even vandalism can destroy or delay the forward

---

[1] Ed Garbisch's nonprofit corporation, Environmental Concern, Inc., produces *Wetland Journal* periodically. This newsletter covers current research, restoration, and education topics on both their own work and that of others. Address: P.O. Box P, 210 West Chew Ave., St. Michaels, MD 21663.

Figure 10.3: Plantings shrubs and tree seedlings in a newly excavated wetland creation project. (Photo courtesy of T. Rightnour, Water's Edge Technology, Inc.)

progress of the wetland.

10.3.6 Monitoring and Evaluation

1. *Purpose and Frequency of Monitoring.* At the very least, monitoring should be a requirement in permit conditions. However, monitoring serves other purposes. At the conclusion of construction and planting (if any), as-built conditions should be documented on the original base map and with a brief narrative. This serves to establish baseline conditions that may vary considerably from the original construction drawings. The reason for these potential changes is that project managers on-site must have enough flexibility during construction to modify plans and procedures to adapt to local site conditions. Soils may vary more or less than expected. Water tables may prove to be shallower or deeper than anticipated. Excessive precipitation may delay

operations for days or weeks. Since successful restoration or creation of a wetland is the major objective, changes that enhance the possibility of success should be encouraged. One cautionary note is in order. Construction contracts should contain clauses that allow for some adjustments without additional billings (perhaps 10-20% of the total amount). This should be stated clearly to potential bidders.

To make the collection of monitoring data worthwhile, there should be a succinct plan that specifies who will collect the data, who will pay for monitoring and evaluation, how the data will be collected, when it will be collected, where it will be stored permanently, and who will be responsible for evaluating the results of monitoring. Without some interpretation of results, the information simply becomes part of dead storage. The approach put forth by Kentula and others (1992) shows how these data can be put to valuable use, not only for the target project, but also as a iterative learning experience for both regulators and builders of wetlands.

Recently, there has been a trend away from requiring annual monitoring for 5-10 years. Monitoring a site for the first, second, and possibly a third year can serve to confirm that the wetland was constructed, and provide a trouble-shooting mechanism to identify problems that need timely corrections. Kentula and others (1992) suggest that a less-intensive "routine" monitoring protocol be followed during these early years. Reports generated from these monitoring activities need not be lengthy. They should build upon the as-built conditions and initial report, focusing on the major functions of interest and the direction of change, if any, among the measured parameters. Wetland projects will not likely reach their potential functional capability until 5, 10, or more years have passed. If all is well, or corrections are made early on, the site could be visited during the fifth, seventh, and tenth years to document progress. At the conclusion of the monitoring period, a "comprehensive" monitoring and evaluation report should be prepared and distributed. This report should draw some conclusions regarding the relative success of the project, both in terms of area and functions replaced. For projects not constructed under regulatory scrutiny, the level of monitoring and reporting could be less, but they should still be an essential component of the project. In the latter case, volunteers can be recruited and trained to help reduce costs (see Kentula and others (1992) for ways of organizing volunteers).

2. *Suggested Monitoring Protocol.* A basic monitoring protocol is suggested here. Readers are advised to review the following publications for details on monitoring and evaluation (Horner and Raedeke, 1989; Adamus and Brandt, 1990; Hammer, 1992; Kentula et al., 1992).

Since wetlands are defined by their hydrology, soils, and vegetation, it is logical that these parameters should be important components of the monitoring protocol. All phases of monitoring should include identifying information, such as a location map, date of construction, and ownership. As-built conditions should be documented via maps that display an overhead view and several vertical profiles. From these, the morphometry of the wetland will be documented, including area, shape, and typical slopes and depths. This morphometric data, when combined with hydrologic information, can provide estimates of basin volume and retention times that relate to flood storage and water treatment functions.

Hydrology is best recorded as a function of depth over time. That is, periodic readings are taken from staff gages (i.e., stakes with measured increments to measure depth of surface water) and shallow wells (i.e., slotted PVC pipes for measuring depth of water below the surface). If enough readings are taken, a hydrograph can be developed that shows fluctuations in depth both above and below the surface of the substrate over time (Figure 10.4). Other useful hydrologic data include flow rates and discharges from inlets and outlets, and indirect indicators, such as those used during delineation procedures.

To be deemed a success, a project must become a recognizable wetland. Therefore, soil characteristics used in jurisdictional delineations, such as the color of mottles in proportion to the soil matrix colors, are useful measures as to whether or not a site has become a wetland. More detailed investigations are necessary before the full functional capability can be assessed (e.g., changes in soil texture, organic matter, sediment accretion, and chemistry of pore water over time).

Vegetation measurements are commonly used to assess wetland success. A plant species list, developed from proper sampling procedures, will determine if hydrophytes dominate the site. When used in conjunction with the wetland indicator status of the U. S. Fish and Wildlife Service (i.e., obligate, facultative-wet, etc.), an index to site "wetness" can be developed (Wentworth et al., 1988). A common permit requirement is that the wetland must have a minimum coverage of specified vegetation after a few years (e.g., 80% cover of obligate or facultative-wet plants after two growing seasons). To arrive at this value, the site is sampled via plots or transects to derive an average measure of percent cover (see Section 2.2). Although this is an overly simplistic measure, it does reflect the productivity of the site, potential wildlife habitat, and probable direction of plant succession. An estimate of plant survivorship (another frequent permit requirement) can be obtained by tagging individual stems of planted stock and checking their condition repeatedly throughout the growing season.

Wildlife habitat is often a primary function of concern for restored and created wetlands. Meaningful direct sampling of vertebrate and invertebrate populations is time-consuming and expensive. As an alternative, surrogate measures of habitat suitability based on physical parameters and vegetation can be used. Habitat Evaluation Procedures (U.S. Fish and Wildlife Service, 1980) are the most standard techniques available. Sightings and indirect observations made while on-site should

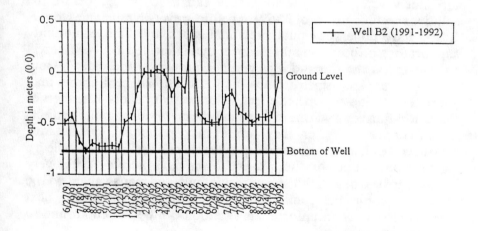

Figure 10.4: Water level fluctuations (hydrograph) from a pipe well installed in the Tipton Demonstration Wetland.

be recorded as well.

If water treatment is a designated function, then the expense of taking repeated water quality samples for substances of interest (e.g., nitrates, phosphates, heavy metals, suspended solids) may be warranted. These can be taken at inlets and outlets to obtain a crude estimate of retention or transformation of pollutants by the wetland.

Finally, a series of permanent photographic stations should be established. Color photographs, in combination with a narrative report and maps, can provide useful and valuable documentation of the wetland's progress. When made part of the permanent file on a wetland, for either formal permits or informal projects, this information can be conveniently reviewed by interested parties in the office. By coordinating your monitoring and evaluation efforts and sharing information with other active participants in the restoration field, one can achieve the maximum amount of benefit possible.

The purpose of monitoring and evaluation is not simply to add more expense to a project. When integrated early on in the restoration process, monitoring can be an effective management tool. If problems are identified quickly (e.g., wildlife depredation of newly-planted tubers, too much water, bank erosion), they can be corrected before the solutions become cost-prohibitive or before irreversible damage is done. If you have done your public relations work and built community support for a project, you may discover that interested local people will help maintain the site and repair minor problems. Individuals from birding clubs, sportsmen's clubs, and educational institutions may adopt the site and become, in effect, voluntary caretakers. Members of the restoration team will more than likely be involved in similar projects in the future. Whether they are agency personnel, consultants, attorneys, bankers, or concerned citizens, all can apply the knowledge gained from one restoration project to the next project. At the very least, awareness of wetland policy, regulations, and management will be increased within the community. The case study presented next demonstrates how an unfortunate situation can be turned around to provide some benefits.

10.4 CASE STUDY - TIPTON DEMONSTRATION WETLANDS

10.4.1 Background

In late 1989, an industrial development authority (IDA) approached the Pennsylvania State University (PSU) through an intermediary, the Ben Franklin Technology Center of Central and Northern Pennsylvania (BCT) for help on how to deal with a rather difficult and legally complex wetland encroachment in central Pennsylvania. Much of the project is documented in the public record,[2] but to avoid confusing the reader with irrelevant details, the players will be referred to by their general affiliation, not actual names, and the only the useful phases of the overall project will be discussed in detail. This is not an ideal model for future projects, but it illustrates how a seemingly intractable situation can be made a little better if everyone involved is open-minded. There were no gallant white knights or evil black knights in this case study, only people trying to do their job on a very complex field of play, namely, the arena of wetlands mitigation.

The case began in the mid-1980s with a misunderstanding. A large property acquired by the IDA for development as an industrial park was found to contain some jurisdictional wetlands. The extent of those wetlands was not fully delineated at first. A site visit several years later by agency staff identified the encroachment into wetlands after construction of the park was well under way. A cease-an-desist order was issued to halt operations. BCT had provided funding for the local IDA, and hence, sought to resolve the many problems and minimize additional legal action by involving investigators at PSU.

10.4.2 The Process

After lengthy negotiations involving at least four agencies, the

---

[2] For more information on the Tipton Demonstration Project see: Brooks, R.P., A.L. Stauffer and G.W. Petersen, Jr. 1992. Wetlands educational program and demonstration site (at the Peterson Industrial Park). Final Report. Vol. 1 - Overview. Ben Franklin Technology Center, University Park, PA. and Brooks, R. P., and A. L. Stauffer. 1992. Wetlands demonstration area and Peterson Industrial Park wetlands, Tipton, PA. Final Report Vol. 2 - Monitoring report 1991-1992. Ben Franklin Technology Center, University Park, PA.

IDA, past and current consultants, their respective attorneys, BCT, and PSU, the following plan evolved. All parties agreed, and a formal consent order demanded, that the IDA would mitigate about 30 acres of wetlands. Due to the historic disturbances to the property (it formerly served as a mud racing track and then a local air strip), the already installed infrastructure, the completed construction of facilities for four companies, and both disturbed and undisturbed wetlands that were highly interspersed with uplands, there was only room for on-site mitigation of about 15 acres within the park boundaries. An additional suitable location for 15 additional acres had to be located. A consultant was hired by the IDA to handle delineation of remaining wetlands within the park boundaries, prepare permit applications for construction of several additional facilities within the park, and to locate a suitable site for off-site mitigation of 15 acres. PSU entered into an agreement with BTC to develop an educational program on wetlands, targeted at industrial development authorities and related businesses, and to provide technical advice and perform the monitoring on the mitigation sites for the IDA.

Once the participating agencies had agreed to the total amount of mitigation required, the IDA's consultant examined several properties for the off-site mitigation. Selection criteria included lands already owned by the IDA, lands adjacent to the existing industrial park, lands with potential for additional industrial development, lands within the same drainage area, and lands with hydric soils and suitable hydrologic sources. Constraints on properties that eventually focused the search on one site ran the gamut from being too small to encompass 15 acres of wetlands to not being for sale. During this process, other parties contacted the IDA concerning possible collaboration on lands that could support 2-3 times the acreage proposed. However, no cooperative arrangements could be made.

The site eventually selected was farmland in active crops on the opposite side of the river from the industrial park. The parcel was large enough to encompass the necessary 15 acres of wetland. It also had road frontage for possible development, thereby adding potential for recovering some of the costs. Test pits and borings outfitted with wells made of PVC pipe revealed that the water table was about 3-4 feet below the surface. Unfortunately, this condition required substantial excavation across the entire site. These data were used by the consultant and agencies to reach a decision that the property was suitable, although not ideal. The IDA negotiated purchase of the 40-acre property from the landowner for about $115,000 or $3,000/acre.

All parties contributed to development of a conceptual design plan for the site. The objectives were to replace 15 acres of emergent wetland similar to that lost in the development of the industrial park. A reasonable diverse vegetative community was sought by agency personnel, including areas designed as wildlife habitat. The primary water source was to be intersection of ground water with some supplemental water from runoff and precipitation. A decision was made not to connect the basin directly with the adjacent river, except for a narrow drainage ditch that served as the outlet. Permit requirements specified the following: 1) defined acreage, 2) equivalent or better functions (none specifically mentioned), 3) saturated soils in upper soil horizon during much of the growing season, 4) timing of planting, 5) contiguous band of shrubs of at least three species, 6) use of at least four herbaceous species, 7) 75% survivorship of both shrubs (tagged) and herbaceous plants (areal coverage), and 8) monitoring for two years.

Since the mitigation area was quite large, the contract was split into separate tasks, each independently solicited for competitive bid. A new consultant was brought in to supervise the work and conduct all surveying. A major part of the project entailed excavation, earthmoving, and site preparation at a cost of about $130,000. The work area was professionally surveyed. Excavation lasted several months into the Fall and Winter of 1990. Large bulldozers and pans used by the contractor sometimes got mired in the wetter soils, particularly after heavy rains. After some debate between the agencies and the IDA about the required depth of the basin, final elevations were agreed upon, topsoil was spread to a depth of about 6 inches, and banks of the basin (but not the wetland bottom) and the upland storage piles were seeded as an erosion control measure.

Another contractor was hired to complete the planting of the wetland. A rough planting plan and list of planted herbaceous and woody species was prepared. Due to the large amount of plant material needed, geographic and genetic concerns were essentially ignored in favor of acquiring the material in time for a late Spring planting. The planting contract cost approximately $25,000 and was completed by June 1991. A substantial portion of the wetland was designed to be colonized by volunteer species, although the intended sources were not specified.

The monitoring program began before planting was completed. Sixteen slotted PVC wells were installed along four transects. In addition to water level data, information on plant coverage, species lists, hydric soil development, sediment accretion, and wildlife use was collected. Since additional wetlands were being destroyed as part of the permitting activities on the original industrial park, hydric soil was available as a soil amendment. An experiment was designed by PSU to examine the effects on wetland plants of four treatments, each replicated four times within the 15-acre wetland. Each treatment consisted of a 20 x 20 foot plot located randomly near one of the wells. The treatments were: 1) hydric soil placed on the surface, 2) composted leaf litter mixed into the soil and planted, 3) bare soil planted as in the first two treatments, and 4) bare soil with no planting acting as a control. The plots were prepared under yet another separate contract.

By the Fall of 1991, the IDA consultant was orchestrating the remaining mitigation activities at the original industrial park. Due to the placement of industrial facilities, seven separate parcels were planned for wetland creation and restoration. Some parcels were excavated to increase contact with groundwater, whereas others were simply graded to facilitate sheet flow water across the surface. Earth-moving, planting, and monitoring were handled in a similar way as in the large wetland.

Beginning in 1990, and paralleling the mitigation activities, the following educational activities were completed: 1) five half-day seminars targeted at industrial development authorities were held throughout Pennsylvania; 2) a 22-minute video on balancing wetlands and industrial development was produced; 3) the video and a 100-page manual were distributed to over 120 industrial development authorities and corporations throughout the state; 4) a large, wooden sign listing cooperators was erected on the site; 5) numerous educational and media field trips were conducted at the site, and 6) in cooperation with the IDA, an unsuccessful attempt was made to secure a state grant for the local township to fund construction of a pavilion and boardwalk at the wetland. Although the latter item did not take place, a local birding club adopted the site, installed nest boxes, and now keeps a tally of the abundant bird life using the wetland.

At this time, two years of monitoring have been completed. The 1991 drought undoubtedly hindered establishment of the wetland, but the results are generally encouraging. By the end of 1992, the 15-acre basin supported primarily wetland vegetation, hydric soils were developing in the wetter areas, water tables were higher in 1992 than in the abnormally dry year of 1991, and over 100 species of wildlife have been seen using the site. The other 15 acres of wetland divided among seven parcels are interspersed throughout the industrial park, and appear to be developing fairly well (Figure 10.5). Their presence precluded the construction of additional industrial facilities, but perhaps their long-term value will be realized by the park's corporations. The overall cost of the permitting, land acquisition, and mitigation was well in excess of one-half million dollars, or more than $35,000/acre.

10.4.3 Was the Tipton Demonstration Project a Success?

What lessons can be learned from a review of this case study?
1. The site selection process worked fairly well given the large land area required.
2. The contract bidding process worked, in that the minimum bids by qualified contractors saved the IDA a considerable amount of money on an already expensive project.
3. Despite repeated attempts to designate a single individual as a contact person to coordinate activities, communications between the various parties were less than perfect.
4. The IDA was under considerable pressure from corporate clients to secure additional permits within the park for expansions and new businesses. Thus, financial concerns drove the mitigation process more than ecological ones, although all parties were interested in completing a successful project, albeit for different reasons.
5. The morphometry of the wetland basin could have been improved by imposing more gentle slopes in some areas. Due to the limited hydrology, only the bottom of the basin can effectively become a wetland.
6. The soil amendment experiments clearly showed the ecological and economic advantages of using hydric soil if available. The drought conditions showed the obvious benefit of adding waste organic matter to help maintain soil moisture and promote plant survivorship and growth.
7. The associated educational program was well received, and appeared

Figure 10.5: Aerial view during the first year of the Tipton Demonstration Wetland adjacent to Peterson Industrial Park, Tipton, PA.

to reach a diverse clientele, however, participation by the target audience, industrial development authorities, was less than anticipated. This suggested that those that have not experienced delays due to wetland issues are not yet aware of the importance of early planning for wetlands.

8. Throughout the project, the media was kept well-informed. Although not all coverage was positive (there was an initial focus on the costs and construction delays), the cooperators were acknowledged for their constructive efforts to find a solution.

9. Finally, a jurisdictional wetland was created of about the area required. The functional assessment was necessarily vague since the permit did not specify clearly the functional objectives. Thus, wetlands of relatively low functional value were replaced by wetlands of equal area and similarly low functional value.

10.5 COST CONSIDERATIONS AND CONCLUSIONS

As stated in the beginning of this chapter, the reasons for undertaking a wetland restoration or creation project fall into one of four categories: regulatory compliance, human health and safety, grassroots project, or regional initiative. Wetlands are regulated with closer scrutiny than most other types of natural land cover or use. This makes regulatory compliance the most common reason for engaging in this process. As might be expected, few participants enter the arena altruistically. Although the intentions are often honorable, there is always a strong concern for the financial commitment involved. This is especially true for the regulated community, but cannot be ignored by voluntary participants in wetland restoration and creation. How much does it cost to restore and create wetlands? The process is complex, as is the answer to this question.

To discover the true cost of a project, one must segregate the related costs as described in the case study. A private landowner who wants to have a wetland restored on prior converted farmlands may be selected as a participant in a U.S. Fish and Wildlife Service program like Partners for Wildlife. These projects are often subsidized by the government at virtually no cost to the landowner.

A more typical scenario parallels the Tipton Case Study on a reduced scale. After a legitimate search, a developer finds no alternatives to encroaching upon a jurisdictional wetland. The permitting process is triggered, and following a public interest review, approval is given subject to a requirement to mitigate the wetland losses in area and function. Based on personal experience and a few published studies, the permitting process, including wetland restoration, typically costs anywhere from 3-14% of total project costs (Miller, 1993). Higher percentages (10-14%) generally reflect smaller projects with modest capital investment, such as establishing roads for forest harvesting operations (Miller, 1993), whereas for residential and industrial construction projects with higher levels of investment, wetland-associated costs might be 3-7% of the overall budget (J. Willenbrock, personal communication). These are real costs, and should be internalized early on in the budgeting process. Those individuals or companies that do not acknowledge wetland issues up front or that get into legal difficulties for not following the proper regulatory sequence can expect to pay considerably more money before arriving at the approval stage.

The stance of the reluctant restorer can be modified if the approaches to mitigation are kept flexible. Innovation can only be fostered if the climate is supportive rather than antagonistic. Examples include regulatory restraint for "good faith" efforts, positive public relations campaigns for public and private partnerships, and reduced risk of punishment for parties that strive to make a project successful.

10.6 GUIDE TO WETLAND PERMIT SEQUENCING AND PROJECT MANAGEMENT FOR WETLAND RESTORATION OR CREATION

To assist readers in their financial planning, Table 10.1 lists tasks and approximate costs for a typical wetland restoration or creation project stemming from a wetland encroachment permit. They are meant only to provide guidance concerning the range of the most obvious direct costs. A complete economic analysis would undoubtedly uncover additional indirect expenditures for personnel and operations. Each encroachment, each parcel of land, each restoration project will be unique. A "cookbook" approach is not possible, and not recommended.

Returning to our original questions, "Can you guarantee success? Will equivalent functions be replaced?", the answers appear to be a qualified yes. As is true of any good project, regardless of the field, careful planning, attention to detail, not promising more than you can deliver, and constructive evaluation, can increase the probability of achieving one's objectives. The simplest measure of success is that clients are satisfied. Whether it be the landowner, agency representative, or an interest group, you must ask whether they are reasonably satisfied with your performance given the uncertainties of wetland restoration technology at this point in time. The formula for successful restoration and creation of wetlands should include: 1) honesty and trust among the participants; 2) flexibility to adapt to variable conditions; 3) good communication among planners, project managers, and laborers; and 4) application of the best available knowledge and technology. Part of that knowledge base means understanding the hydrology of different wetland types in unique landscape positions. What kind of wetland are you attempting to create? Can it be created in the area chosen for the project? There is no substitute for scrutinizing the ecological

Table 10.1 Costs Associated with Wetland Restoration or Creation.

(Range of estimated costs assuming a property of about 10 acres.)

> WETLAND IDENTIFICATION ($500-1,500)
> National Wetlands Inventory Map Review
> Soils Survey and Map Review for Hydric Soils
> Field Reconnaissance
>
> JURISDICTIONAL DELINEATION ($1,500-5,000)
> Establishment of Wetland Boundaries on Proposed
>   Project Site
> Routine Method for Small Areas, Intermediate or
>   Comprehensive Method for Large, Difficult,
>   and Sensitive Areas
>
> ANALYSIS OF PRACTICABLE PROJECT
>   ALTERNATIVES ($2,000-10,000)
> Location Alternatives (upland or drained hydric soil
>   preferred)
> Design Alternatives
> Impact Minimization Alternatives for Proposed
>   Project
> Develop Preliminary Project Objectives
> Selection of Site (costs of land acquisition are not
>   included)
>
> FUNCTIONAL ASSESSMENTS OF WETLANDS TO BE
>   IMPACTED ($1,000s)
> Habitat Evaluation
> Hydrologic Assessment

Table 10.1: (Continued)

> PREAPPLICATION MEETING WITH FEDERAL AND STATE AGENCIES ($ = TIME)
>> Review of Project Need (Local Economy, Location Suitability)
>> Options Considered
>> Discussion of Mitigation Requirements
>
> PREPARATION OF PERMIT APPLICATIONS ($2,000-10,000)
>> Facility Design
>> Pollution Control Plans and Specifications
>> Conceptual Mitigation Plan
>
> DESIGN OF MITIGATION PLAN ($2,000-10,000)
>> Develop Final Project Objectives
>> Wetland Creation Plans
>> Restoration/Enhancement Designs
>> Engineering Plans and Specifications
>
> IMPLEMENTATION OF MITIGATION PLAN ($10,000s)
>> Selection of Contractor
>> Field Supervision During Construction
>> Documentation of As-Built Conditions
>
> POST-CONSTRUCTION MONITORING AND EVALUATION OF MITIGATION ($1,000s/yr, 3-10 yrs)
>> Planting Success
>> Hydrologic and Soil Monitoring

---

characteristics of a site and its surroundings.

Another part of that knowledge base is the continual use (emphasizing that the data collected be used) of a science-based monitoring and evaluation system that provides feedback both to policymakers and participants in the permitting process. Of great concern to all should be how to ensure adequate institutional memory of long-lived projects, given the usual turnover of personnel in most organizations. Kentula and others (1992) suggested a possible approach, and others certainly exist. Until we have a better, more objective system of measuring success, especially regarding functional replacement, criticism of restoration and creation may still appear legitimate.

Wetlands of many different types are being restored and created every day. Success comes easier with some types than with others. As wetland science advances as a discipline, the criteria used to select potential sites, to design and construct wetlands, and to measure their functional performance also will advance. Wetland restoration and creation projects are an integral part of our national "no net loss" policy that also promotes a long-term gain in wetland resources.

Literature Cited

Abernethy, Y. and R.E. Turner. 1987. US forested wetlands: 1940-1980. *BioScience* 37:721-727.

Adams, D.A. 1963. Factors influencing vascular plant zonation in North Carolina salt marshes. *Ecology* 44:445-456.

Adamus, P.R. 1988. Criteria for created or restored wetlands. *In:* D.D. Hook, W.H. McKee, H.K. Smith, J. Gregory, V.G. Burrell, M.R. DeVoe, R.E. Sojka, S. Gilbert, R. Banks, L.H. Stolzy, C. Brooks, T.D. Mathews, and T.H. Shear (eds.). *The Ecology and Management of Wetlands. Volume 2: Management, Use, and Value of Wetlands.* Timber Press, Portland, OR.

Adamus, P.R. and K. Brandt. 1990. *Impacts on Quality of Inland Wetlands of the US: A Survey of Indicators, Techniques, and Applications of Community-Level Biomonitoring Data.* U.S. Environmental Protection Agency, EPA/600/3-90/073, Corvallis, OR.

Adamus, P.R., E.J. Clarain, R.D. Smith, and R.E Young. 1987. *Wetland Evaluation Technique (WET).* U.S. Army Corps of Engineers, Tech. Rep. Y-87, Vicksburg, MS.

Adamus, P.R. and R.T. Stockwell. 1983. *A Method for Wetland Functional Assessment. Vol. 1: Critical Review and Evaluation Concepts.* U.S. Dept. of Transportation, Federal Highway Administration, Rept. FHWA-IO-82-23, Washington, D.C.

Adamus, P.R., L.T. Stockwell, E.J. Clairain, Jr, M.E. Morrow, L.P. Rozas, R.D. Smith. 1991. *Wetland Evaluation Technique (WET). Volume I: Literature Review and Evaluation Rationale.* U.S. Army Corps Engineers, Wetlands Research Program Tech. Rep. WRP-DE-2, Vicksburg, MS.

Albrecht, V.S. 1992. The wetlands debate. *Urban Land* (May 1992):20-23.

Albrecht, V.S. and D. Isaacs. 1992. Wetlands jurisdiction and judicial review. *Natural Resources & Environment* (Amer. Bar Assoc.) 7(1):29-31, 65-67.

Ammann, A.P., R W. Frazen, and J.L. Johnson. 1986. *Method for the Evaluation of Inland Wetlands in Connecticut.* Connecticut Dept. Environ. Prot., and U.S. Soil Conservation Serv., Bull. No. 9, Hartford, CT.

Ammann, A.P. and A.L. Stone. 1991. *Method for the Comparative Evaluation of Nontidal Wetlands in New Hampshire.* New Hampshire Dept. Environ. Services, NHDES-WRD-1991-3, Concord, NH.

Anderson, D.R., J.L. Laake, B.R. Crain, and K.P. Burnham. 1979. Guidelines for line transect sampling of biological populations. *J. Wildl. Manage.* 43:70-78.

Anderson, R. and M. Rockel. 1991. *Economic Valuation of Wetlands.* American Petroleum Institute, Discussion Paper #065, Washington, D.C.

Anderson, R. and R. DeCaprio. 1992. Banking on the bayou. *National Wetlands Newsletter* 14(1):10.

Anderson, R.C. and J. White. 1970. A cypress swamp outlier in southern Illinois. *Illinois State Acad. Sci. Trans.* 63:6-13.

Anderson, R.R., R.G. Brown, and R.D. Rappleye. 1976. Water quality and plant distribution along the upper Patuxent River, Maryland. *Chesapeake Science* (1976):145-156.

Andrewartha, H.G. 1970. *Introduction to the Study of Animal Populations.* Univ. Chicago Press, Chicago.

Andrewartha, H.G. and L.C. Birch. 1954. *The Distribution and Abundance of Animals.* Univ. Chicago Press, Chicago.

Atchue, A., F.P. Day, and H.G. Marshall. 1983. Algal dynamics and nitrogen and phosphorous cycling in a cypress stand in the seasonally flooded Great Dismal Swamp. *Hydrobiologia* 106:115-122.

Bagur, J.D., 1977. *Coastal Marsh Productivity.* U.S. Fish & Wildl. Serv., Publ. OBS-81/24, Washington, D.C

Baird, D. and R.E. Ulanowicz. 1989. The seasonal dynamics of the Chesapeake Bay ecosystem. *Ecol. Monog.* 59:329-364.

Baldwin, M.F., M. Leslie, and E.H. Clark II. 1990. Government programs inducing wetlands alteration. *In:* G. Bingham, E.H. Clark II, L.V. Haygood, and M. Leslie (eds.) *Issues in Wetlands Protection: Background Papers Prepared for the National Wetlands Policy Forum.* The Conservation Foundation, Washington, D.C.

Barclay, J.S. and W.V. White (eds.). 1981. *Playa Lakes Symposium Proceedings.* U.S. Fish Wildl. Serv., FWS/OBS-81/07, Washington, D.C.

Barnthouse, L. W., G.W. Suter, S.M. Bartell, J.J. Beauchamp, R.H. Gardner, E. Linder, R.V. O'Neill, and A.E. Rosen. 1986. *User's Manual for Ecological Risk Assessment.* Publ. No. 2679, ONL-6251. Environmental Sciences Division, Oak Ridge National Laboratory, Oak Ridge, TN.

Barrette, M. 1993. Delaware announces wetlands plan. *Environ. & Dev.* (May 1993):3-4.

Bartell, S. M., R.H. Gardner, and R.V. O'Neill. 1992. *Ecological Risk Estimation.* Lewis Publ., Boca Raton, FL.

Bayly, I.A.E. 1967. The general biological classification of aquatic environments with special reference to those of Australia. *In:* A. H. Weatherley (ed.) *Australian Inland Waters and Their Fauna.* Australian National Univ. Press, Canberra.

Beeftink, W.G. 1977. Salt marshes. *In:* R.S.K. Barnes (ed.) *The Coastline.* John Wiley & Sons, New York.

Begon, M. 1985. A general theory of life-history variation. *In* R.M. Sibly and R.H. Smith (eds.) *Behavioural Ecology.* Blackwell, Oxford.

Belt, C.B., Jr. 1975. The 1973 flood and man's constriction of the Mississippi River. *Science* 189:681-684.

Benninghoff, W.S. 1966. The Relevé method of describing vegetation. *Michigan Botanist* 5:109-114.

Berger, J.J. (ed.) 1990. *Environmental Restoration: Science and Strategies for Restoring the Earth*. Island Press, Washington, D.C.

Berry, J.F. 1975. The population effects of ecological sympatry on musk turtles in northern Florida. *Copeia* 1975:692-701.

Berry, J.F. 1992a. Wetlands: swamped in rhetoric? *Environ. & Dev.* (Jan. 1992):1-3 [reprinted: *Current Municipal Problems* 19(1):85-91].

Berry, J.F. 1992b. Wetlands: a way out of the morass? *Environ. & Dev.* (Feb. 1992):1-3 [reprinted: *Current Municipal Problems* 19(1):91-96; and *Technical Quarterly* 7(3):7-9].

Bertness, M.D. 1987. Peat accumulation and the success of marsh plants. *Ecology* 69:703-713.

Bertness, M.D. 1992. The ecology of a New England salt marsh. *American Scientist* 80:260-268.

Best, G.R. and K.L. Erwin 1984. Effects of hydroperiod on survival and growth of tree seedlings in a phosphate surface-mined reclaimed wetland. *In:* D.H. Graves (ed.) *Proceedings of the Symposium on Surface Mining Hydrology, Sedimentology, and Reclamation*. Univ. of Kentucky, Lexington, KY.

Beverton, R.J.H. and S.J. Holt. 1957. On the dynamics of exploited fish populations. *Fish. Invest.* 19:1-533.

Bidgood, B.F. 1974. Reproductive potential of two lake whitefish (*Coregonus clupeaformis*) populations. *J. Fisheries Res. Board of Canada* 31:1631-1639.

Billings, W.D., K.M. Peterson, G.R. Shaver, and A.W. Trent. 1977. Root growth, respiration, and carbon dioxide evolution in an arctic tundra soil. *Arc. Alp. Res.* 9:129-137.

Bingham, G., E.H. Clark II, L.V. Haygood, and M. Leslie (eds.) 1990. *Issues in Wetlands Protection: Background Papers Prepared for the National Wetlands Policy Forum*. The Conservation Foundation, Washington, D.C.

Bishop, R.A. 1981. Iowa's wetlands. *Proc. Iowa Acad. Sci.* 88:11-16.

Blake, N.M. 1980. *Land Into Water - Water Into Land: A History of Water Management in Florida*. Univ. Florida Presses, Tallahassee, FL.

Blumm, M.C. and D.B. Zaleha. 1989. Federal wetlands protection under the Clean Water Act: regulatory ambivalence, intergovernmental tension, and a call for reform. *Univ. Colorado Law Review* 60:695-772.

Bosselman, F.P. 1989. Sweeden's Swamp: the morass of wetland regulation. *Land Use Law* (Mar. 1989):3-6.

Boto, K.G. and W.H. Patrick, Jr. 1979. Role of wetlands in the removal of suspended sediments. *In* P.E. Greeson, J.R. Clark, and J.E. Clark (eds.) *Wetland Functions and Values. The State of Our Understanding.* Proceedings of the National Symposium on Wetlands, Amer. Water Res. Assoc. Minneapolis, MN.

Boyd, W.L. and J.W. Boyd. 1972. Microorganisms in frost scars. *Arc. Alp. Res.* 4:257-260.

Brinson, M.M., B.L. Swift, R.C. Plantico, and J.S. Barclay. 1981. *Riparian Ecosystems: Their Ecology and Status.* U.S. Fish & Wildl. Serv., FWS/OBS-81/17, Washington, D.C.

Brooks, R.P. 1991. Restoration ecology: repairing the damage. *In:* D. J. Decker, M. E. Krasny, G. R. Goff, C. R. Smith, and D. W. Gross (eds.) *Challenges in the Conservation of Biological Resources.* Westview Press, Boulder, CO.

Brower, J.E. and J.H. Zar. 1977. *Field and Laboratory Methods for General Ecology.* Wm. C. Brown, Dubuque, IA.

Brown, L.L. and E.O. Wilson. 1956. Character displacement. *Systematic Zool.* 5:49-64.

Brown, S. 1989. Michigan: an experiment in section 404 assumption. *National Wetlands Newsletter* 11(4):5.

Brown, S.L. 1990. Structure and dynamics of basin forested wetlands in North America. *In:* A.E. Lugo, M.M. Brinson, and S.L. Brown (eds.) *Forested Wetlands. Ecosystems of the World.* Elsevier, Amsterdam.

Brown, S.L., E.W. Flohrschutz, and H.T. Odum. 1984. Structure, productivity, and phosphorus cycling of the scrub cypress ecosystem. *In:* K.C. Ewel and H.T. Odum (eds.) *Cypress Swamps.* Univ. Presses of Florida, Gainesville, FL.

Browne, R.A. 1981. Lakes as islands: Biogeographic distribution, turnover rates, and species composition in the lakes of central New York. *J. Biogeog.* 8:75-83.

Buol, S.W., F.D. Hole, and R.J. McCracken. 1980. *Soil Genesis and Classification.* Iowa State University Press, Ames, IA.

Burke, D.G., E.J. Meyers, R.W. Tiner, Jr., and H. Groman. 1988. *Protecting Nontidal Wetlands.* Amer. Planning Assoc., Chicago, IL.

Burnham, K.P. and D.R. Anderson. 1984. The need for distance data in transect counts. *J. Wildl. Manage.* 48:1248-1254.

Burns, T.P. 1989. Lindeman's contradiction and the trophic structure of ecosystems. *Ecology* 70:1355-1362.

Burns, T. P., A.L. Brenkert, and K. A. Rose. MS. The magnitude of indirect and direct effects in model ecosystems. Unpublished manuscript.

Cahoon, D.R. and J.C. Stevenson. 1986. Production, predation, and decomposition in a low-salinity *Hibiscus* marsh. *Ecology* 67:1341-1350.

Cairns, J., Jr. (ed.) 1980. *The Recovery Process in Damaged Ecosystems.* Ann Arbor Science Publ. (Butterworth), Boston.

Cairns, J., Jr. (ed.) 1988. *Rehabilitating Damaged Ecosystems. Vols. 1 & 2.* CRC Press, Boca Raton, FL.

Callahan, K., G.R. Hath, M. Hutchinson, J.A. Gebrian, J.H. Mills, and H. Withers. 1992. *An Inland Wetland Commissioner's Guide to Site Plan Review.* Connecticut Dept. of Environmental Protection, Hartford, CT.

Caughley, 1977. *Analysis of Vertebrate Populations.* John Wiley & Sons, London.

Chapman, R.N. 1931. *Animal Ecology with Especial Reference to Insects.* McGraw-Hill, New York.

Chapman, V.J. 1960. *Salt Marshes and Salt Deserts of the World.* Interscience, New York.

Chapman, V.J. 1974. Salt marshes and salt deserts of the world. *In:* R.J. Reimold and W.H. Queens (eds.) *Ecology of Halophytes.* Academic Press, New York.

Charles, D.F. (ed.) 1990. *Acidic Deposition and Aquatic Ecosystems: Regional Case Studies.* Springer-Verlag, New York.

Christian, R.R., R.B. Hanson, J.R. Hall, and W.J. Wiebe. 1981. Aerobic microbes and meiofauna. *In:* L.R. Pomeroy and R.G. Wiegert (eds.) *The Ecology of a Salt Marsh.* Springer-Verlag, New York.

Clark, J.R. and J. Benforado (eds.). 1981. *Wetlands of Bottomland Hardwood Forests: Proceedings of a Workshop on Bottomland Hardwood Forests of the Southeastern United States.* Elsevier, New York.

Clements, F.E. 1916. Plant succession: analysis of the development of vegetation. *Publ. Carnegie Inst., Washington* 242:1-512.

Clements, F.E. 1936. The nature and structure of the climax. *Ecology* 24:252-284.

Clewell, A.F. 1988. Bottomland hardwood forest creation along new headwater streams. *In:* J.A. Kusler, M.L. Quammen, and G. Brooks (eds.) *Proceedings of the National Wetland Symposium: Mitigation of Impacts and Losses.* Assoc. State Wetland Managers, Tech. Rep. 3, Berne, NY.

Clewell, A.F. and R. Lea. 1990. Creation and restoration of forested wetland vegetation in the southeastern United States. *In:* J.A. Kusler and M.E. Kentula (eds.) *Wetland Creation and Restoration: The Status of the Science.* Island Press, Washington, D.C.

Clifford, H.T. and W. Stephenson. 1975. *An Introduction to Numerical Classification.* Academic Press, New York.

Clymo, R.S. 1964. The origin of acidity in *Sphagnum* bogs. *Bryologist* 67:427-431.

Cochran, W.G. 1977. *Sampling Techniques.* John Wiley & Sons, New York.

Cochran-Stafira, D.L. and R.A. Anderson. 1984. Diatom flora at kettle-hole bog in relation to hydrarch succession zones. *Hydrobiologia* 109:265-273.

Cody, M.L. 1973. Character convergence. *Ann. Rev. Ecol. Syst.* 4:189-212.

Cohen, A.D., D.J. Casagrande, M.J. Andrejko, and G.R. Best (eds.) 1984. *The Okefenokee Swamp: Its Natural History, Geology, and Geochemistry.* Wetlands Surveys, Los Alamos, NM.

Colinvaux, P. 1982. Towards a theory of history: fitness, niche and clutch of *Homo sapiens. Ecology* 70:393-412.

Colinvaux, P. 1986. *Ecology.* John Wiley & Sons, New York.

Connell, J.H. 1961. The influence of interspecific competition and other factors on the distribution of the barnacle *Chthamalus stellatus. Ecology* 42:710-723.

Connell, J.H. and R.O. Slatyer. 1977. Mechanisms of succession in natural communities and their role in community stability and organisation. *Amer. Nat.* 111:1119-1144.

Conner, W.H. and J.W. Day, Jr. 1982. The ecology of forested wetlands in the southeastern United States. In: B. Gopel, R.E. Turner, R.G. Wetzel, and D.F. Whigham (eds.) *Wetlands: Ecology and Management.* International Scientific Publ., Jaipur, India.

Conner, W.H. and J.W. Day, Jr. 1987. *The Ecology of Barataria Basin, Louisiana: An Estuarine Profile.* U.S. Fish & Wildl. Serv., Biol. Rep. 85(7.13), Washington, D.C.

Conner, W.H., J.R. Toliver, and F.H. Sklar. 1986. Natural regeneration of baldcypress (*Taxodium distichum* (L.) Rich.) in a Louisiana swamp. *Forest Ecology and Management* 14:305-317.

Conservation Foundation. 1988. *Protecting America's Wetlands: An Action Agenda. Final Report of the National Wetlands Policy Forum.* The Conservation Foundation, Washington, D.C.

Cooley, W.W. and P.R. Lohnes. 1971. *Multivariate Data Analysis.* John Wiley & Sons, New York.

Copeland, B.J., R.G. Hodson, S.R. Riggs, and J.E. Easley, Jr. 1983. The Ecology of Albemarle Sound, North Carolina: An Estuarine Profile. U.S. Fish & Wildl. Serv., FWS/OBS-83/01, Washington, D.C.

Cowardin. L.M. 1978. Wetland classification in the United States. *J. For.* 76:666-668.

Cowardin, L.M., V. Carter, F.C. Golet, and E.T. LaRoe. 1979. *Classification of Wetlands and Deepwater Habitats of the United States.* U.S. Fish & Wildl. Serv., FWS/OBS 79-31, Washington, D.C.

Cowardin, L.M., D.S. Gilmer, and L.M. Mechlin. 1981. Characteristics of central North Dakota wetlands determined from sample aerial photographs and ground study. *Wildl. Soc. Bull.* 9:280-288.

Crum, H. 1988. *A Focus on Peatlands and Peat Mosses.* Univ. Michigan Press, Ann Arbor MI.

Cunningham, R.A., W.B. Stoebuck, and D.A. Whitman. 1984. *The Law of Property.* West Publ. Co., St. Paul, MN.

Cutler, M.R. 1978. The mission of wetland science. *In:* P.E. Greeson, J.R. Clark, and J.E. Clark (eds.) *Wetland Functions and Values. The State of Our Understanding.* Proceedings of the National Symposium on Wetlands, Amer. Water Res. Assoc., Minneapolis, MN.

Dabel, C.P. and F.P. Day, Jr. 1977. Structural comparisons of four plant communities in the Great Dismal Swamp. *Bull. Torrey Botan. Club* 104:352-360.

Dahl, T.E. 1990. *Wetland Losses in the United States. 1780's to 1980's.* U.S. Fish & Wildl. Serv., Washington, D.C.

Dahl, T.E., C.E. Johnson, and W.E. Frayer. 1991. *Status and Trends of Wetlands in the Conterminous United States, Mid-1970's to mid 1980's.* U.S. Fish & Wildl. Serv., Washington, D.C.

Damman, A.W.H. and T.W. French. 1987. *The Ecology of Peat Bogs of the Glaciated Northwestern United States: A Community Profile.* U.S. Fish & Wildlife Serv., Biol. Rept. 85, Washington D.C.

Dardeau, M.R., R.F. Modlin, W.W. Schroeder, and J.P. Stout. 1992. Estuaries. *In:* C.T. Hackney, S.M. Adams, and W.H. Martin (eds.) *Biodiversity of the Southeastern United States.* John Wiley & Sons, New York.

Davis, C. 1991. Making no assumptions. *National Wetlands Newsletter* 13(2):6-7.

Day, J.W., Jr., C.A.S. Hall, W.M. Kemp, and A. Yánez-Arancibia. 1989. *Estuarine Ecology.* John Wiley & Sons, New York.

Day, R.T., P.A. Keddy, J. McNeill, and T. Carleton. 1988. Fertility and disturbance gradients: a summary model for riverine marsh vegetation. *Ecology* 69:1044-1054.

Dean, R.G. 1979. Effects of vegetation on shoreline erosional processes. *In:* P.E. Greeson, J.R. Clark, and J.E. Clark (eds.) *Wetland Functions and Values: The State of Our Understanding.* Amer. Water Res. Assoc., Minneapolis, MN.

Decker, D.J. and G.R. Goff (eds.) 1987. *Valuing Wildlife. Economic and Social Perspectives.* Westview Press, Boulder, CO.

Deland, M.R. 1992. No net loss of wetlands: a comprehensive approach. *Natural Resources & Environment* (Amer. Bar Assoc.) 7(1):3-5, 52-53.

Delaune, R.D., R.J. Buresh, and W.H. Patrick, Jr. 1979. Relationship of soils properties to standing crop biomass of *Spartina alterniflora* in a Louisiana marsh. *Estuarine Coastal Marine Sci.* 8:477-487.

Delaune, R.D., R.H. Baumann, and J.G. Gosselink. 1983. Relationships among vertical accretion, coastal submergence, and erosion in a Louisiana Gulf Coast marsh. *J. Sed. Petrol.* 53:147-157.

Dennison, M.S. 1992. This Land Is Your Land? *Planning Advisory Service Memo,* Amer. Planning Assoc. (Oct. 1992):1-4.

Dennison, M.S. 1993. Coastal laws strike a balance. *Environmental Protection* (Apr. 1993):39.

Dial, R.S. and D.R. Deis. 1986. *Mitigation Options for Fish and Wildlife Resources Affected by Port and Other Water-Dependent Developments in Tampa Bay, Florida.* U.S. Fish & Wildl. Serv., Biol. Rept. 86(6), Slidell, LA.

Dickerman, J.A., A.J. Stewart, and R.G. Wetzel. 1986. Estimates of net annual aboveground production: sensitivity to sampling frequency. *Ecology* 67:650-659.

Dierberg, F.E. and P.L. Brezonik. 1983. Nitrogen and phosphorus mass balances in natural and sewage-enriched cypress domes. *J. Appl. Ecology* 20:323-337.

Digby, P.G.N. and R.A. Kempton. 1987. *Multivariate Analysis of Ecological Communities.* Chapman & Hall, London.

Dix, M.R. and S. Denson. 1993. Florida's assumption of federal dredge-and-fill jurisdiction: clearing the permitting stream bed or muddying administrative waters? *The Florida Bar Journal* 67(4):56-59.

Dixon, W.J. and M.B Brown (eds.) 1990. *BMDP Statistical Software*. Univ. California Press, Berkeley, CA.

Dorge, C.L., W.J. Mitsch, and J.R. Wiemhoff. 1984. Cypress wetlands in southern Illinois. In: K.C. Ewel and H.T. Odum (eds.) *Cypress Swamps*. Univ. Presses of Florida, Gainesville, FL.

Dreher, R.G. 1992. EPA recants role in federal oversight. *National Wetlands Newsletter*. 14(5):10-11.

Drew, R.D. and N.S. Schomer. 1984. *An Ecological Characterization of the Caloosahatchee River/Big Cypress Watershed*. U.S. Fish & Wildl. Serv., FWS/OBS-82-58.2, Washington, D.C.

DuBusk, W.F. and K.R. Reddy. 1987. Removal of floodwater nitrogen in a cypress swamp receiving primary wastewater effluent. *Hydrobiologia* 153:79-86.

Duever, M.J., J.E. Carlson, and L.A. Riopelle. 1984. Corkscrew Swamp: a virgin cypress strand. In: K.C. Ewel and H.T. Odum (eds.) *Cypress Swamps*. Univ. Presses of Florida, Gainesville, FL.

Ecological Society of America. 1992. Evaluation of proposed revisions to the 1989 *Federal Manual for Identifying and Delineating Jurisdictional Wetlands*. *Bull. Ecological Society of America* 73: 14-22.

Ehrenfeld, D. 1988. Why put a value on biodiversity? In: E.O. Wilson (ed.) *Biodiversity*. Nat'l Academy Press, Washington, D.C.

Eisler, R. 1985 - 1990. *Contaminant Hazard Reviews*. No. 1-20., U.S. Fish & Wildl. Serv., Biol. Rep. 85(1.1-1.20), Washington, D.C.

Elton, C.S. 1927. *Animal Ecology*. Macmillan, New York.

Elton, C.S. 1946. Competition and structure of ecological communities. *J. Anim. Ecol.* 15:54-68.

Emlen, J.M. 1984. *Population Biology. The Coevolution of Population Dynamics and Behavior*. MacMillan, New York.

Environmental Defense Fund/World Wildlife Fund. 1992. *How Wet Is a Wetland? The Impacts of the Proposed Revisions to the Federal Delineation Manual*. Report by Environmental Defense Fund and World Wildlife Fund (Jan. 16, 1992), reproduced in *Daily Env't Rep. (BNA)*, (Jan. 17, 1992):E1.

Erickson, N.E. and D.M. Leslie, Jr. 1988. *Soil-Vegetation Correlations in Coastal Mississippi Wetlands*. U.S. Fish & Wildl. Serv., Biol. Rep. 89(3), Washington, D.C.

Ewel, K.C. 1990. Swamps. In: R.L. Myers and J.J. Ewel (eds.) *Ecosystems of Florida*. Univ. Central Florida Press, Orlando, FL.

Ewel, K.C. and H.T. Odum (eds.) 1984. *Cypress Swamps*. Univ. Presses of Florida, Gainesville, FL.

Faber, P.A., E. Keller, A. Sands, and B.M. Masser. 1989. *The Ecology of Riparian Habitats of the Southern California Coastal Region: A Community Profile.* U.S. Fish & Wildl. Serv., Biol. Rep. 85(7.27), Washington, D.C.

Feierabend, J.S. 1992. *Endangered Species - Endangered Wetlands: Life on the Edge.* National Wildlife Federation, Washington, D.C.

Federal Interagency Committee for Wetland Delineation. 1989. *Federal Manual for Identifying and Delineating Jurisdictional Wetlands.* U.S. Army Corps of Engineers, U.S. Environmental Protection Agency, U.S. Fish & Wildl. Serv., and U.S.D.A. Soil Conservation Service, Washington, D.C.

Fenchel, T. 1975. Character displacement and coexistence in mud snails (Hydrobiidae). *Oecologia* 20:19-32.

Fenchel, T. and S. Kolding. 1976. Habitat selection and distribution patterns of five species of amphipod genus *Gammarus*. *Oikos* 33:316-322.

Findlay, S., K. Howe, and H.K. Austin. 1990. Comparison of detritus dynamics in two tidal freshwater wetlands. *Ecology* 71:288-295.

Finger, T.R. and E.M. Stewart. 1987. Responses of fishes to flooding regime in lowland hardwood wetlands. *In:* W.J. Matthews and D.C. Heins (eds.) *Community and Evolutionary Ecology of North American Fishes.* Univ. Oklahoma Press, Norman, OK.

Flanagan, P.W. and A.K. Veum. 1974. Relationships between respiration, weight loss, temperature and moisture in organic residues in tundra. *In:* A.J. Holding, O.W. Heal, S.F. MacLean, and P.W. Flanagan (eds). *Soil Organisms and Decomposition in Tundra.* Tundra Biome Steering Committee, Stockholm.

Florida Department of Environmental Regulation. 1991. *Report on the Effectiveness of Permitted Mitigation.* Florida Dept. Envir. Reg., Tallahassee, FL.

Forbes, S.A. 1887. The lake as a microcosm. *Bull. Peoria (Illinois) Sci. Assoc.* [reprinted in *Ill. Nat. Hist. Surv. Bull.* 15:537-550 (1925).]

Foster, D.R. and H.E. Wright, Jr. 1990. Role of ecosystem development and climate change in bog formation in central Sweden. *Ecology* 71:450-463.

Frayer, W.E. and J.M. Hefner. 1991. *Florida's Wetlands. Status and Trends, 1970's to 1980's.* U.S. Fish & Wildl. Serv., Atlanta, GA.

Frayer, W.E., T. Monahan, D. Bowden, and F. Graybill. 1983. *Status and Trends of Wetlands and Deepwater Habitats in the Conterminous United States.* U.S. Fish & Wildl. Serv., Washington, D.C.

Frohring, P.C., D.P. Vorhees, and J.A. Kushlan. 1988. History of wading bird populations in the Everglades: a lesson in the use of historical information. *Colonial Waterbirds* 11:328-335.

Futuyma, D.J. and M. Slatkin (eds.) 1983. *Coevolution.* Sinauer Assoc., Sunderland, MA.

Garbisch, E.W. 1986. Highways and Wetlands: *Compensating Wetland Losses.* U.S. Dept. Transportation, Federal Highway Admin., Rep. No. FHWA-IP-86-22, Washington, D.C.

Garbisch, E.W. 1989. Wetland enhancement, restoration, and construction. *In:* S.K. Majumdar, R.P. Brooks, F.J. Brenner, and R.W. Tiner, Jr. (eds.) *Wetlands Ecology and Conservation: Emphasis in Pennsylvania*. Pennsylvania Academy of Science, Easton, PA.

Garbisch, E.W., T.J. Denbow, G.A. Thunhorst, D.W. Rothman, C.C. Bartoldus, M.L. Kraus, D. Klements, and D.R. MacLean. 1992. *Guidelines for the Development of Wetland Replacement Areas*. Nat'l Coop. Hwy. Res. Prog., Rept. 25-3, Washington, D.C.

Gardner, R. H., O'Neill, R. V., Mankin, J. B. and J. H. Carney 1981. A comparison of sensitivity analysis and error analysis based on a stream ecosystem model. *Ecological Modelling* 12:177-194.

Gates, C.E. 1979. Line transects and related issues. *In:* R.M. Cormack et al.(eds.) *Sampling Biological Populations*. Int'l Coop. Publ. House, Fairfield, VA.

Gates, C.E. 1980. *LINETRAN, a general computer program for analyzing line-transect data*. J. Wildl. Manage. 44:658-661.

Gauch, H.G. 1982. *Multivariate Analysis in Community Ecology*. Cambridge Univ. Press, Cambridge, U.K.

Gause, G.F. 1934. *The Struggle for Existence*. Williams & Wilkins, Baltimore (republished in 1971 by Dover, New York).

Gause, G.F., N.P. Smaragdova, and A.A. Witt. 1936. Further studies of interaction between predators and prey. *Amer. Nat.* 70:1-18.

Geltman, E.A.G. 1989. Regulation of non-adjacent wetlands under section 404 of the Clean Water Act. *New England Law Review* 23:615-643.

Gerlach, S. 1981. *Marine Pollution*. Springer-Verlag, New York.

Gerritsen, J. and H.S. Greening. 1989. Marsh seed banks of the Okefenokee Swamp: effects of hydrologic regime and nutrients. *Ecology* 70:750-763.

Glaser, P.H. 1983. *Eleocharis rostellata* and its relation to spring fens in Minnesota. *Michigan Botanist* 22:19-21.

Glaser, P.H. 1987. *The Ecology of Patterned Boreal Peatlands of Northern Minnesota: A Community Profile*. U.S. Fish & Wildl. Serv., Biol. Rep. 85(7.14), Washington, D.C.

Gleason, H.A. 1917. The structure and development of the plant association. *Bull. Torrey Botan. Club* 44:463-481.

Gleason, H.A. 1926. The individualist concept of the plant association. *Bull. Torrey Bot. Club* 53:7-26.

Gleason, H.A. 1927. Further views on the succession concept. *Ecology* 8:299-326.

Gleason, P.J. (ed.) 1984. *The Environments of South Florida, Present and Past. II.* Miami Geological Society, Miami, FL.

Good, R.E., D.F. Whigham, and R.L. Simpson (eds.) 1978. *Freshwater Wetlands: Ecological Processes and Management Potential*. Academic Press, New York.

Goode, B.N. 1989. In defense of Nationwide Permit 26. *National Wetlands Newsletter* 11(6):4.

Goldman-Carter, J. 1989. Nationwide Permit 26: the wetlands giveaway. *National Wetlands Newsletter* 11(6):4.

Goldman-Carter, J.L. 1992. The unraveling of no net loss. *National Wetlands Newsletter* 14(5):12-14.

Goldsmith, A. and E.H. Clark II. 1990. Nonregulatory programs promoting wetland protection. *In:* G. Bingham, E.H. Clark II, L.V. Haygood, and M. Leslie (eds.) *Issues in Wetlands Protection: Background Papers Prepared for the National Wetlands Policy Forum.* The Conservation Foundation, Washington, D.C.

Gore, A.J.P. (ed.) 1983. *Ecosystems of the World. Vol. 4A, Mires: Swamp, Bog, Fen, and Moor.* Elsevier, Amsterdam.

Gore, J.A. (ed.) 1985. *The Restoration of Rivers and Streams: Theories and Experience.* Butterworth, Boston.

Gosselink, J.G. 1984. *The Ecology of Delta Marshes of Coastal Louisiana: A Community Profile.* U.S. Fish & Wildl. Serv., FWS/OBS-84/09, Washington, D.C.

Gosselink, J.G., W.H. Conner, J.W. Day, and R.E. Turner. 1981. Classification of wetland resources: land, timber, and ecology. *In:* B.D. Jackson and J.L. Chambers (eds.) *Timber Harvesting in Wetlands.* Louisiana State Univ., Baton Rouge, LA.

Gosselink, J.G., L.C. Lee, and T.A. Muir (eds.) 1990. *Ecological Processes and Cumulative Impacts: Illustrated by Bottomland Hardwood Wetland Ecosystems.* Lewis Publ., Boca Raton, FL.

Goulding, M. 1980. *The Fishes and the Forest.* Univ. California Press, Berkeley, CA.

Grant, P.R. 1975. The classical case of character displacement. *Evol. Biol.* 8:237-337.

Grant, R.R., Jr. and R. Patrick. 1970. Tinicum Marsh as a water purifier. *In: Two Studies of Tinicum Marsh.* The Conservation Foundation, Washington, D.C.

Green, R.H. 1979. *Sampling Design and Statistical Methods for Environmental Biologists.* John Wiley & Sons, New York.

Green, R.H. 1980. Multivariate approaches in ecology: the assessment of ecological similarity. *Ann. Rev. Ecol. Syst.* 11:1-14.

Greeson, P.E., J.R. Clark, and J.E. Clark (eds.) 1979. *Wetland Functions and Values: The State of Our Understanding.* Amer. Water Res. Assoc., Minneapolis, MN.

Greig-Smith, P. 1983. *Quantitative Plant Ecology. 3rd Ed.* Blackwell, Oxford.

Guillory, V. 1979. Utilization of an inundated floodplain by Mississippi River fishes. *Fla. Sci.* 42:222-228.

Guthery, F.S. 1981. Playa basins and resident wildlife in the Texas panhandle. *In:* J.S. Barclay and W.V. White (eds.) *Playa Lakes Symposium Proceedings.* U.S. Fish Wildl. Serv., FWS/OBS-81/07, Washington D.C.

Haines, E.B. 1979. Interactions between Georgia salt marshes and coastal waters: a changing paradigm. *In:* R.J. Livingston (ed.) *Ecological Processes in coastal and Marine Systems.* Plenum Press, New York.

Hall, C.A.S. and R.M. Moll. 1975. Methods of assessing aquatic primary productivity. *In:* H. Leith and R.H. Whittaker (eds.) *Primary Productivity of the Biosphere.* Springer-Verlag, New York.

Hammer, D.A. 1992. *Creating Freshwater Wetlands.* Lewis Publ., Boca Raton, FL.

Hardy, A.C. 1924. The herring in relation to its animate environment. *Min. Agri. and Fish., Fishery Invest. (Series 2).* 7:1-57.

Harper, J.L. 1977. *The Population Biology of Plants.* Academic Press, New York.

Harrington, H.F. 1986. Michigan 404 program assumption. *National Wetlands Newsletter* 7(1):10.

Hedgepeth, J.W. 1987. San Francisco Bay: the wetlands besieged. *In: Proceedings of the National Wetland Protection Symposium.* U.S. Fish & Wildl. Serv., FWS/OBS-7897, Washington, D.C.

Hedrick, P.W. 1984. *Population Biology: The Evolution and Ecology of Populations.* Jones & Bartlett, Boston.

Hefner, J.M. 1986. Wetlands of Florida. *In:* E.D. Estevez, J. Miller, J. Morris, and R. Hamman (eds.) *Proceedings of the Conference: Managing Cumulative Effects in Florida Wetlands.* New College Env. Studies Prog., Publ. No. 37. Omnipress, Madison, WI.

Heinselman, M.L. 1970. Landscape evolution and peatland types in the Lake Agassiz Peatlands Natural Area, Minnesota. *Ecol. Monogr.* 40:235-261.

Heltshe, J.F. and N.E. Forrester. 1983. Estimating species richness using jacknife procedure. *Biometrics* 39:1-11.

Hendricks, W.A. 1956. *The Mathematical Theory of Sampling.* Scarecrow, New Brunswick, NJ.

Herdendorf, C.E. 1987. *The Ecology of the Coastal Marshes of Western Lake Erie: A Community Profile.* U.S. Fish & Wildl. Serv., Biol. Rep. 85(7.9), Washington, D.C.

Herdendorf, C.E., C.N. Raphael, and E. Jaworski. 1986. *The Ecology of Lake St. Clair Wetlands: A Community Profile.* U.S. Fish & Wildl. Serv., Biol. Rep. 85(7.7), Washington, D.C.

Higashi, M., B.C. Patten, and T.P. Burns. 1991. Network trophic dynamics: an emerging paradigm in ecosystems ecology. *In:* M. Higashi and T.P. Burns (eds.) *Theoretical Studies of Ecosystems: The Network Perspective.* Cambridge Univ. Press, New York.

Hill, A. B. 1965. The environment and disease: association or causation? *Proc. Royal Soc. Medicine (London).* 58:295-300.

Hill, M.O. 1973. Reciprocal averaging: an eigenvector method of ordination. *J. Ecol.* 61:225-235.

Hill, M.O. and H.G. Gauch. 1980. Detrended correspondence analysis: an improved ordination technique. *Vegetatio* 42:47-58.

Hoffmeister, J.E. 1974. *Land From the Sea, the Geologic Story of South Florida.* Univ. Miami Press, Coral Gables, FL.

Hollands, G.G. and D.W. Magee. 1986. *A Method for Assessing the Functions of Wetlands. In:* J.A. Kusler and P. Riexinger (eds.) *Proceedings of the National Wetlands Assessment Symposium*, Assoc. of State Wetland Managers, Tech. Rep. 1, Berne, NY.

Holling, C.S. 1973. Resilience and stability of ecological systems. *Ann. Rev. Ecol. Syst.* 4:1-24.

Hook, D.D. and R. Lea (eds.) 1989. *Proceedings of the Symposium. Forested Wetlands of the Southern United States.* U.S. Forest Service, Gen. Tech. Rep. SE-50, Washington, D.C.

Horn, H.J. 1966. Measurement of "overlap" in comparative ecological studies. *Amer. Nat.* 100:419-424.

Horner, R.R. and K.J. Raedeke. 1989. *Guide for Wetland Mitigation Projects Monitoring.* Washington State Dept. Transp., Rept. No. WA-RD 195.1, Seattle, WA.

Houck, O.A. 1989. Hard choices: the analysis of alternatives under section 404 of the Clean Water Act and similar environmental laws. *Univ. Colorado Law Rev.*, 60:773-840.

Howarth, R.W. 1988. Nutrient limitation of net primary production in marine ecosystems. *Ann. Rev. Ecol. Syst.* 19: 89-110.

Hubbard, D.E., D.D. Millar, D.D. Malo, and K.F. Higgins (eds.) 1988. *Soil-Vegetation Correlations in Prairie Potholes of Beadle and Deuel Counties, South Dakota.* U.S. Fish & Wildl. Serv., Biol. Rep. 88(22), Washington, D.C.

Huenneke, L.F. and R.R. Sharitz. 1986. Microsite abundance and distribution of woody seedlings in a South Carolina cypress-tupelo swamp. *Amer. Midl. Natur.* 115:328-335.

Huffman, T. 1991. A return to ecological concepts. *National Wetlands Newsletter.* 13(6):10-12.

Hutchinson, G.E. 1949. Circular causal systems in ecology. *Ann. New York Acad. Sci.* 51:221-246.

Hutchinson, G.E. 1959. Homage to Santa Rosalia, or why are there so many kinds of animals? *Amer. Nat.* 93:145-159.

Hutchinson, G.E. 1965. *The Ecological Theater and the Evolutionary Play.* Yale Univ. Press, New Haven, CT.

Hutchinson, G.E. 1967. *A Treatise on Limnology. Vol. 2. Introduction to Lake Biology and Limnoplankton.* Wiley-Interscience, New York.

Hutchinson, G.E. 1978. *An Introduction to Population Ecology.* Yale Univ. Press, New Haven, CT.

Jahn, L.A. and R.V. Anderson. 1986. *The Ecology of Pools 19 and 20, Upper Mississippi River: A Community Profile.* U.S. Fish & Wildl. Serv., Biol. Rep. 85(7.6), Washington, D.C.

Jaworski, E. and C.N. Raphael. 1978. *Fish, Wildlife, and Recreational Values of Michigan's Coastal Wetlands.* Michigan Dept. Nat. Res., Lansing, MI.

Johnson, C.W. 1985. *Bogs of the Northeast*. Univ. Press of New England, Hanover, NH.

Jolly, G.M. 1965. Explicit estimates from capture - recapture data with both death and immigration -stochastic model. *Biometrika* 52:225-247.

Jones, K.L. 1982. Prey patterns and trophic niche overlap in four species of Caribbean frogs. *In*: N.J. Scott (ed.) *Herpetological Communities*. U.S. Fish & Wildl. Serv., Res. Rep. 13, Washington, D.C.

Jones, S. 1992. Future seen in the Crystal River. *National Wetlands Newsletter*. 14(2):12-13.

Jordan, T.E., D.F. Whigham, and D.L. Correll. 1989. The role of litter in nutrient cycling in a brackish tidal marsh. *Ecology* 70:1906-1915.

Jordan, W.R., III, M.E. Gilpin, and J.D. Aber (eds.) 1987. *Restoration Ecology: A Synthetic Approach to Ecological Research*. Cambridge Univ. Press, Cambridge, U.K.

Josselyn, M. 1983. *The Ecology of San Francisco Bay Tidal Marshes: A Community Profile*. U.S. Fish & Wildl. Serv., FWS/OBS-83/23, Washington, D.C.

Juday, C. 1940. The annual energy budget of an inland lake. *Ecology* 21:438-450.

Kadlec, J.A. and W.A. Wentz. 1974. *State-of-the-Art Survey and Evaluation of Marsh Plant Establishment Techniques: Induced and Natural. Vol. 1: Report of Research*. U.S. Army Corps of Engineers, Rept. D-74-9, Vicksburg, MS.

Kantrud, H.A., G.L. Krapu, and G.A. Swanson. 1989. *Prairie Basin Wetlands of the Dakotas: A Community Profile*. U.S. Fish & Wildl. Serv., Biol. Rept. 85(7.28), Washington, D.C.

Kaplan, E.H. 1988. Mangrove communities of Florida and the Carribean. *In: A Field Guide to Southeastern and Carribean Seashores*. Houghton-Mifflin, Boston.

Kemp, W.M. and W.R. Boynton. 1984. Spatial and temporal coupling of nutrient inputs to estuarine primary production: the role of particulate transport and decomposition. *Bull. Marine Sci.* 35: 522-535.

Kentula, M.E., R.P. Brooks, S.E. Gwin, C.C. Holland, A.D. Sherman, J.C. Sifneos, and A.J. Hairston (ed.) 1992. *An Approach to Improving Decision Making in Wetland Restoration and Creation*. Island Press, Washington, D.C.

Kershaw, K.A. and J.H.H. Looney. 1985. *Quantitative and Dynamic Plant Ecology, 3rd ed.* Edw. Arnold, London.

Klopatek, J.M. 1978. Nutrient dynamics of freshwater riverine marshes and the role of emergent macrophytes. *In*: R.E. Good, D.F. Whigham, and R.L. Simpson (eds.) *Freshwater Wetlands. Ecological Processes and Management Potential*. Academic Press, New York.

Knudson, P.E., R.A. Brocchu, W.N. Seelig, and M. Inskeep. 1982. Wave dampening in *Spartina alterniflora* marshes. *Wetlands* 2:87-104.

Kolding, S. and T. Fenchel. 1976. Coexistence and life cycle characteristics of five species of amphipod genus *Gammarus*. *Oikos* 33:323-327.

Kortekaas, W.M.E., E. Van der Maarel, and W.G. Beeftink. 1976. A numerical classification of European *Spartina* communities. *Vegetatio* 33:51-60.

Krebs, C.J. 1989. *Ecological Methodology.* Harper & Row, New York.

Krebs, C.J. 1990. *Ecology: The Experimental Analysis of Distribution and Abundance,* 3rd ed. Harper Collins, New York.

Kruczynski, W.L. 1982. Salt marshes of the northeastern Gulf of Mexico. *In:* R.R. Lewis III (ed.) *Creation and Restoration of Coastal Plant Communities.* CRC Press, Boca Raton, FL.

Kruczynski, W.L. 1990a. Mitigation and the section 404 program: a perspective. *In:* J.A. Kusler and M.E. Kentula (eds.). *Wetland Creation and Restoration: The Status of the Science.* Island Press, Washington, D.C.

Kruczynski, W.L. 1990b. Options to be considered in preparation and evaluation of mitigation plans. *In:* J.A. Kusler and M.E. Kentula (eds.). *Wetland Creation and Restoration: The Status of the Science.* Island Press, Washington, D.C.

Krzanowski, W.J. 1988. *Principles of Multivariate Analysis.* Clarendon Press, Oxford.

Kushlan, J.A. 1974a. Effects of a natural fish kill on the water quality, plankton, and fish population of a pond in the Big Cypress Swamp, Florida. *Trans. Amer. Fish. Soc.* 103:235-243.

Kushlan, J.A. 1974b. Observations on the role of the American alligator (*Alligator mississippiensis*) in the southern Florida wetlands. *Copeia* 1974:993-996.

Kushlan, J.A. 1976. Environmental stability and fish community diversity. *Ecology* 57:821-825.

Kushlan, J.A. 1986. The Everglades: management of cumulative ecosystem degradation. *In:* E.D. Estevez, J. Miller, J. Morris, and R. Hamman (eds.) *Proceedings of the Conference: Managing Cumulative Effects in Florida Wetlands.* New College Env. Studies Prog., Publ. No. 37. Omnipress, Madison, WI.

Kushlan, J.A. 1987. External threats and internal management: the hydrologic regulation of the Everglades. *Environ. Manag.* 11:109-119.

Kushlan, J.A. 1989a. Avian use of fluctuating wetlands. *In:* R.R. Sharitz and J.W. Gibbons (eds.) *Freshwater Wetlands and Wildlife.* DOE Symposium Series No. 61. U.S. Dept. of Energy, Oak Ridge, TN.

Kushlan, J.A., 1989b. Wetlands and wildlife, the Everglades perspective. *In:* R.R. Sharitz and J.W. Gibbons (eds.) *Freshwater Wetlands and Wildlife.* DOE Symposium Series No. 61. U.S. Dept. of Energy, Oak Ridge, TN.

Kushlan, J.A. 1990. Freshwater marshes. *In:* R.L. Myers and J.J. Ewel (eds.) *Ecosystems of Florida.* Univ. Central Florida Press, Orlando.

Kushlan, J.A. 1991. The Everglades. *In:* R.J. Livingston (ed.) *The Rivers of Florida.* Springer-Verlag, New York.

Kushlan, J.A. and T. Jacobsen. 1990. Environmental variability and the reproductive success of Everglades alligators. *J. Herpetol.* 24:176-184.

Kusler, J.A. 1983. *Our National Wetland Heritage. A Protection Guidebook.* Environmental Law Institute, Washington, D.C.

Kusler, J. 1992. The mitigation banking debate. *National Wetlands Newsletter* 14(1):4.

Kusler, J. and H. Groman. 1986. Mitigation: an introduction. *National Wetlands Newsletter* 8(5):2-3.

Kusler, J.A. and M.E. Kentula. 1990. Executive summary. *In:* J.A. Kusler and M.E. Kentula (eds.). *Wetland Creation and Restoration: The Status of the Science.* Island Press, Washington, D.C.

Kusler, J.A., M.L. Quammen, and G. Brooks. 1988. *Proceeding of the National Wetlands Symposium: Mitigation of Impacts and Losses.* Assoc. State Wetland Managers, Tech. Rep. 3, Berne, NY.

Lack, D. 1946. Competition for food by birds of prey. *J. Anim. Ecol.* 15:123-129.

Larsen, J.A. 1982. *Ecology of the Northern Lowland Bogs and Conifer Forests.* Academic Press, New York.

Larson, J.S. 1976. *Models for Assessment of Freshwater Wetlands.* Univ. Massachusetts, Water Resour. Res. Cent., Rep. No. 32., Amherst, MA.

Larson, J.S., M.S. Bedinger, C.F. Bryan, S. Brown, R.T. Huffman, E.L. Miller, D.G. Rhodes, and B.A. Touchet. 1981. Transition from wetlands to uplands in southeastern bottomland hardwood forests. *In:* J.R. Clark and J. Benforado (eds.) *Wetlands of Bottomland Hardwood Forests.* Elsevier, New York.

Leck, M.A. and R.L. Simpson. 1987. Seed bank of a freshwater tidal wetland: turnover and relationship to vegetation change. *Amer. J. Bot.* 74:360-370.

Legendre, L. and P. Legendre. 1983. *Numerical Ecology.* Elsevier, New York.

Leitch, J.A. and L.E. Danielson, 1979. *Social, Economic, and Institutional Incentives to Drain or Preserve Prairie Pothole Wetlands.* Dept. Agric. and Applied Economics, Univ. Minnesota, St. Paul, MN.

Leslie, M. 1990. Mitigation policy. *In:* G. Bingham, E.H. Clark II, L.V. Haygood, and M. Leslie (eds.) *Issues in Wetlands Protection: Background Papers Prepared for the National Wetlands Policy Forum.* The Conservation Foundation, Washington, D.C.

Leslie, M. and E.H. Clark II. 1990. Perspectives on wetlands loss and alteration. *In:* G. Bingham, E.H. Clark II, L.V. Haygood, and M. Leslie (eds.) *Issues in Wetlands Protection: Background Papers Prepared for the National Wetlands Policy Forum.* The Conservation Foundation, Washington, D.C.

Lewis, R.R. (ed.) 1982. *Creation and Restoration of Coastal Plant Communities.* CRC Press, Boca Raton, FL.

Lewis, R.R. 1990. Wetlands restoration/creation/enhancement terminology: suggestions for standardization. *In:* J.A. Kusler and M.E. Kentula (eds.) *Wetland Creation and Restoration: The Status of the Science.* Island Press, Washington, D.C.

Lewis, R.R. 1992. Why Florida needs mitigation banking. *National Wetlands Newsletter* 14(1):7.

Lewis, R.R. and E.D. Estevez. 1988. *The Ecology of Tampa Bay, Florida: A Community Profile.* U.S. Fish & Wildl. Serv., Biol. Rep. 85(7.18), Washington, D.C.

Lindeman, R.L. 1941a. The developmental history of Cedar Creek Bog, Minnesota. *Amer. Midland Nat.* 25:101-112.
Lindeman, R.L. 1941b. Seasonal food-cycle dynamics in a senescent lake. *Amer. Midland Nat.* 26:636-673.
Lindeman, R.L. 1942. The trophic dynamic aspects of ecology. *Ecology* 23:399-418.
Linthurst, R. 1980. An evaluation of aeration, nitrogen, pH and salinity factors affecting *Spartina alterniflora* growth: a summary. *In:* V. Kennedy (ed.) *Estuarine Perspectives.* Academic Press, New York.
Linthurst, R.A. and E. Seneca. 1981. Aeration, nitrogen and salinity as determinants of *Spartina alterniflora* Loisel, growth response. *Estuaries.* 4:53-63.
Livingston, R.J. 1984. *The Ecology of the Apalachicola Bay System: An Estuarine Profile.* U.S. Fish & Wildl. Serv., FWS/OBS-82/05, Washington, D.C.
Livingston, R.J. 1990a. Projected changes in estuarine conditions based on models of long-term atmospheric alteration. *In: The Potential Effects of Global Climate Change in the United States.* U.S. Environmental Protection Agency, Publ. CR-814608-01-0, Washington, D.C.
Livingston, R.J. 1990b. Inshore Marine Habitats. *In:* R.L. Myers and J.J. Ewel (eds.) *Ecosystems of Florida.* Univ. Central Florida Press, Orlando, FL.
Livingston, R.J. 1991. Medium sized rivers: Gulf coastal plain. *In:* C.T. Hackney, S.M. Adams, and W.H. Martin (eds.) *Biodiversity of the Southeastern United States.* John Wiley & Sons, New York.
Livingston, R.J. and O. Loucks. 1979. Productivity, trophic interactions, and food web relationships in wetlands and associated systems. *In:* P.E. Greeson, J.R. Clark, and J.E. Clark (eds.) *Wetland Functions and Values. The State of Our Understanding.* Proceedings of the National Symposium on Wetlands, Amer. Water Res. Assoc., Minneapolis, MN.
Lonard, R.L. and E.J. Clairain. 1986. Identification of methodologies for the assessment of wetland functions and values. *In:* J.A. Kusler and P. Riexinger (eds.) *Proceedings of the National Wetlands Assessment Symposium.* Assoc. of State Wetland Managers, Tech. Rep. 1, Berne, NY.
Lonard, R.T., E.J. Clairain, R.T. Huffman, J.W. Hardy, L.D. Brown, P.E. Ballard, and J.W. Watts. 1981. *Analysis of Methodologies Used for the Assessment of Wetland Values.* U.S. Army Corps of Engineers, Vicksburg, MS.
Long, B.M. 1980. *Soil Survey of Berkeley County, South Carolina.* U.S. Dept. of Agriculture, Soil Conservation Service and U.S. Forest Service. Washington, D.C.
Long, E. R. and L. G. Morgan, 1991. *The Potential for Biological Effects of Sediment-Sorbed Contaminants Tested in National Status and Trends Program.* Nat'l Oceanogr. Atmosph. Admin., Tech. Mem. NOS OMA 52, Washington, D.C.
Lopushinsky, W., and J.A. Max. 1990. Effect of soil temperature on root and shoot growth and on budburst timing in conifer seedling transplants. *New Forests* 4: 107-124.

Lotka, A.J. 1925. *Elements of Physical Biology.* Williams & Wilkins, Baltimore [reprinted: 1956. *Elements of Mathematical Biology.* Dover, New York].

Lubchenko, J. 1978. Plant species diversity in a marine intertidal community: importance of herbivore food preference and algal competitive abilities. *Amer. Nat.* 112:23-39.

Ludden, A.P., D.L. Frink, and D.H. Johnson. 1983. Water storage capacity of natural wetland depressions in the Devils Lake basin of North Dakota. *J. Soil and Water Conserv.* 38:45-48.

Ludwig, J.A. and J.F. Reynolds. 1988. *Statistical Ecology.* John Wiley & Sons, New York.

Lugo, A.E., M.M. Brinson, and S. Brown. 1990. *Ecosystems of the World 15, Forested Wetlands.* Elsevier, Amsterdam.

Lugo, A.E. and S.C. Snedaker. 1974. The ecology of mangroves. *Ann. Rev. Ecol. Syst.* 5:39-64.

Lyr, H. and G. Hoffman. 1967. Growth rates and growth periodicity of tree roots. *In:* J.A.A. Romberger and P. Mikola (eds). *International Review of Forest Research. Vol. 2.* Academic Press, New York.

MacArthur, R.H. 1972. *Geographical Ecology.* Harper & Row, New York.

MacArthur, R.H. and J.H. Connell. 1966. *The Biology of Populations.* John Wiley & Sons, New York.

MacArthur, R.H. and R. Levins. 1964. Competition, habitat selection, and character displacement in a patchy environment. *Proc. Nat. Acad. Sci.* 51:1207-1210.

MacArthur, R.H. and R. Levins. 1967. The limiting similarity, convergence, and divergence of coexisting species. *Amer. Nat.* 101:377-385.

MacArthur, R.H. and E.O. Wilson. 1967. *The Theory of Island Biogeography.* Princeton Univ. Press, Princeton, NJ.

MacFadyen, A. 1957. *Animal Ecology.* Pitman, London.

Mackay, D. 1991. *Multimedia Environmental Models: The Fugacity Approach.* Lewis Publ., Boca Raton, FL.

Mandelker, D. 1985. *NEPA: Law and Litigation.* Callaghan, Deerfield, IL.

Marble, A.D. 1992. *A Guide to Wetland Functional Design.* Lewis Publishers, Boca Raton, FL.

Marinucci, A.C., J.E. Hobbie, and J.V.K. Helfrich. 1983. Effects of litter nitrogen on decomposition and microbial biomass in *Spartina alterniflora. Microbial Ecology* 9:27-40.

Marsh, L.L. and D.R. Acker. 1992. Mitigation banking on a wider plane. *National Wetlands Newsletter* 14(1):8-9.

Martin, A.C., N. Hotchkiss, F.M. Uhler, and W.S. Bourn. 1953. *Classification of Wetlands of the United States.* U.S. Fish & Wildl. Serv., Spec. Sci. Rep. 20, Washington, D.C.

May, R.M. 1975. Patterns of species abundance and diversity. *In:* M.L. Cody and J.M. Diamond (eds.) *Ecology and Evolution of Communities.* Harvard Univ. Press, Cambridge, MA.

McCormick, J. and H.A. Somes, Jr. 1982. *The Coastal Wetlands of Maryland.* Maryland Dept. Natural Resources, Annapolis, MD.

McIntosh, R.P. 1967. An index of diversity and the relation of certain concepts to diversity. *Ecology* 48:392-404.

Metzler, K.J. and R.W. Tiner, Jr. 1992. *Wetlands of Connecticut.* U.S. Fish & Widlife Service, Washington, D.C.

Meyer, J.S., C.G. Ingersoll, L.L. McDonald, and M.S. Boyce. 1986. Estimating uncertainty in population growth rates: jacknife vs. bootstrap techniques. *Ecology* 67:1156-1166.

Miller, R. 1993. *The Long-Term Environmental Impacts and the Costs Associated With Forest Road Crossings of Wetlands in Pennsylvania.* M.S. Thesis. Pennsylvania State University, University Park, PA.

Miller, R.S. 1967. Competition and species diversity. *Brookhaven Symp. Biol.* 22:63-70.

Miller, W.R. and F.E. Egler. 1950. Vegetation of the Wequetequock-Pawcatuck tidal marshes. *Conn. Ecol. Monog.* 20:143-172.

Minshall, G.W., W.S.E. Jensen, and W.S. Platts. 1989. *The Ecology of Stream and Riparian Habitats in the Great Basin Region: A Community Profile.* U.S. Fish & Wildl. Serv., Biol. Rep. 85(7.24), Washington, D.C.

Mitsch, W.J., C.L. Dorge, and J.R. Wiemhoff. 1979. Ecosystem dynamics and a phosphorus budget of an alluvial cypress swamp in southern Illinois. *Ecology* 69:1116-1124.

Mitsch, W.J. and J.G. Gosselink. 1993. *Wetlands, 2nd ed.* Van Nostrand Reinhold, New York.

Möbius, K. 1877. *Die Auster und die Austerwirthschaft.* Wiegundt, Hempel, and Parey, Berlin (in German). Translated by Rice, H.J. 1880. Rept. U.S. Comm. Fisheries: 683-751. [partially reprinted in Kormondy, E.J. (ed.) 1965. *Readings in Ecology.* Prentice-Hall, Englewood Cliffs, NJ].

Moler, P.E. (ed.) 1992. *Rare and Endangered Biota of Florida. Vol. III. Amphibians and Reptiles.* Univ. Press of Florida, Gainesville, FL.

Monk, C.D. 1965. Southern mixed hardwood forests of north central Florida. *Ecol. Monogr.* 35:335-354.

Monk, C.D. 1966. An ecological study of hardwood swamps in north-central Florida. *Ecology* 47:649-654.

Montague, C.L., S.M. Bunker, E.B. Haines, M.L. Pace, and R.L. Wetzel. 1981. Aquatic macroconsumers. *In:* L.R. Pomeroy and R.G. Wiegert (eds.) *The Ecology of a Salt Marsh.* Springer-Verlag, New York.

Montague, C.L. and R.G. Wiegert. 1990. Salt Marshes. *In:* R. L. Myers and J. J. Ewel (eds.) *Ecosystems of Florida.* Univ. Central Florida Press, Orlando, FL.

Moore, P.D. (ed.) 1984. *European Mires.* Academic Press, London.

Moore, P.D. and D.J. Bellamy. 1974. *Peatlands.* Springer-Verlag, Inc., New York.

Moran, M.A., T. Legovic, R. Benner, and R. Hodgson. 1988. Carbon flow from lignocellulose: a simulation analysis of a detritus-based ecosystem. *Ecology* 69:1525-1536.

Morisita, M. 1959. Measuring of interspecific association and similarity between communities. *Mem. Fac. Sci. Kyushu Univ., Ser. E (Biol.)* 3:65-80.
Mueller-Dombois, D. and H. Ellenberg. 1974. *Aims and Methods of Vegetation Ecology.* John Wiley & Sons, New York.
Musselman, L.J., O.L. Nickrent, and G.F. Levy. 1977. A contribution towards a vascular flora of the Great Dismal Swamp. *Rhodora* 79:240-268.
Myers, R.L. 1983. Site susceptibility to invasion by the exotic tree *Melaleuca quinquenervia* in southern Florida. *J. Appl. Ecol.* 20:645-658.
Myers, R.L. 1984. Ecological compression of *Taxodium distichum var. nutans* by *Melaleuca quinquenervia* in southern Florida. *In:* K.C. Ewel and H.T. Odum (eds.) *Cypress Swamps.* Univ. Presses of Florida, Gainesville, FL.
Myers, R.L. and J.J. Ewel. 1990. Problems, prospects, and strategies for conservation. *In:* R.L. Myers and J.J. Ewel (eds.) *Ecosystems of Florida.* Univ. Central Florida Press, Orlando, FL.
Nachlinger, J.L. 1988. *Soil-Vegetation Correlations in Riparian and Emergent Wetlands, Lyon County, Nevada.* U.S. Fish & Wildl. Serv., Biol. Rep. 88(17), Washington, D.C.
NAPAP, 1990. *1989 Annual Report to the President and Congress.* National Acid Precipitation Assessment Program, Washington, D.C.
National Research Council. 1992. *Restoration of Aquatic Ecosystems: Science, Technology, and Public Policy.* National Academy Press, Washington, D.C.
National Wetlands Newsletter. 1986. EPA and FWS sign new §404(q) MOAs with Army. *National Wetlands Newsletter* 8(1):2.
Neilsen, B. and L. Cronin (eds.) 1981. *Estuaries and Nutrients.* Humana Publ., Clifton, NJ.
Nelson, R.W., W.J. Logan, and E.C. Weller. 1983. *Playa Wetlands and Wildlife on the Southern Great Plains: A Characterization of Habitat.* U.S. Fish & Wildl. Serv., FWS/OBS-83/28, Washington, D.C.
Niering, W.A. 1985. *Wetlands.* Alfred A. Knopf, New York.
Niering, W.A. 1991. *Wetlands of North America.* Thomasson-Grant, Charlottesville, VA.
Niering, W.A. and R.S. Warren. 1980. Vegetation patterns and processes in New England salt marshes. *BioScience* 30:301-307.
Nitecki, M.H. (ed.) 1983. *Coevolution.* Univ. Chicago Press, Chicago.
Nixon, S. W. 1980. Between coastal marshes and coastal waters - a review of twenty years of speculation and research on the role of salt marshes in estuarine productivity and water chemistry. *In:* P. Hamilton and K.B. MacDonald (eds.) *Estuarine and Wetland Processes.* Plenum Press, New York.
NOAA. 1990. *Biennial Report to Congress on Coastal Zone Management.* National Oceanic and Atmospheric Administration, Washington, D.C.
Noble, I.R. and R.O. Slatyer. 1979. The effect of disturbance on plant succession. *Proc. Ecol. Soc. Australia* 10:135-145.

Norton, B.G. 1987. *Why Preserve Natural Variety?* Princeton Univ. Press, Princeton, N.J.

Norton, B.G. 1988. Commodity, amenity and morality. The limits of quantification in valuing biodiversity. *In:* E.O. Wilson (ed.) *Biodiversity.* Nat'l Academy Press, Washington, D.C.

Novitzky, R.P. 1979. Hydrologic characteristics of Wisconsin's wetlands and their influence on floods, stream flow, and sediment. *In:* P.E. Greeson, J.R. Clark, and J.E. Clark (eds.) *Wetland Functions and Values: The State of Our Understanding.* Amer. Water Res. Assoc., Minneapolis, MN.

Odum, E.P. 1969. The strategy of ecosystem development. *Science* 164:262-270.

Odum, E.P. 1971. *Fundamentals of Ecology, 3rd ed.* W.B. Saunders, Philadelphia.

Odum, E.P. 1979. The value of wetlands: a heirarchical approach. *In:* P.E. Greeson, J.R. Clark, and J.E. Clark (eds.) *Wetland Functions and Values. The State of Our Understanding.* Proceedings of the National Symposium on Wetlands, Amer. Water Res. Assoc. Minneapolis, MN.

Odum, E.P. and A.A. de la Cruz. 1967. Particulate organic detritus in a Georgia salt marsh-estuarine ecosystem. *In:* G.H. Lauff (ed.) *Estuaries.* Amer. Assoc. Adv. Sci., Washington, D.C.

Odum, H.T. 1983. *Systems Ecology: An Introduction.* John Wiley & Sons, New York.

Odum, H.T., K.C. Ewel, W.J. Mitsch, and J.W. Ordway. 1975. *Recycling Treated Sewage through Cypress Wetlands in Florida.* Univ. Florida Center for Wetlands, Occ. Pub. 1, Gainesville, FL.

Odum, W.E. 1988. Comparative ecology of tidal freshwater and salt marshes. *Ann. Rev. Ecol. Sys.* 19:147-176.

Odum, W.E., J.S. Fisher, and J.C. Pickral. 1979. Factors controlling the flux of particulate organic carbon from estuarine wetlands. *In:* R.J. Livingston (ed.) *Ecological Processes in Coastal and Marine Systems.* Plenum Press, New York.

Odum, W.E. and C.C. McIvor. 1990. Mangroves. *In:* R.L. Myers and J.J. Ewel (eds.) *Ecosystems of Florida.* Univ. Central Florida Press, Orlando, FL.

Odum, W.E., C.C. McIvor, and T.J. Smith. 1982. *The Ecology of Mangroves of South Florida: A Community Profile.* U.S. Fish & Wildl. Serv., FWS/OBS-81/24, Washington, D.C.

Odum, W.E., T.J. Smith, J.K. Hoover, and C. McIvor. 1984. *The Ecology of Tidal Freshwater Marshes of the United States East Coast: A Community Profile.* U.S. Fish & Wildl. Serv., FWS/OBS-83/17, Washington, D.C.

Oosting, H.J. 1956. *The Study of Plant Communities, 2nd ed.* San Francisco, Freeman.

Orloci, L. 1978. *Multivariate Analysis in Vegetation Research.* Dr. W. Junk B.V., The Hague.

Owens-Smith, J., T.P. Burns, B.C. Patten, and K.A. Fox. In Press. Wetlands protection under legal protection. *In:* Patten, B.C. (ed.) *Wetlands and Shallow Continental Water Bodies. Volume II.* SPB Academic Publ., The Hague, Netherlands.

Paine, R.T. 1966. Food web complexity and species diversity. *Amer. Nat.* 100:65-75.
Paine, R.T. 1969. The *Pisaster-Tegula* interaction: prey patches, predator food preference, and intertidal community structure. *Ecology* 50:93-120.
Patten, B.C., T.P. Burns, and M. Higashi. 1989. Network trophic dynamics: the food web of an Okefenokee Swamp aquatic bed marsh. *In:* R.R. Sharitz and J.W. Gibbons (eds.) *Freshwater Wetlands and Wildlife.* DOE Symposium Series No. 61. U.S. Dept. of Energy, Oak Ridge, TN.
Patterson, S.G. 1986. *Mangrove Community Boundary Interpretation and Detection of Areal Changes in Marco Island, Florida: Application of Digital Image Processing and Remote Sensing Techniques.* U.S. Fish & Wildl. Serv., Biol. Rept 86(10), Washington, D.C.
Payne, N.F. and F. Copes (eds.) 1988. *Wildlife and Fisheries Habitat Improvement Handbook.* U.S. Dept. of Agric. Forestry Service, Wildlife & Fisheries, Washington, D.C.
Pearl, R. 1928. *The Rate of Living.* Knopf, New York.
Peet, R.K. 1974. The measurement of species diversity. *Ann. Rev. Ecol. Syst.* 5:285-307.
Pennings, S.C. and R.M. Callaway. 1992. Salt marsh zonation: the relative importance of competition and physical factors. *Ecology* 73:681-690.
Peters, D.S., D.W. Ahrenholtz, and T.R. Rice. 1979. Harvest and value of wetland associated fish and shellfish. *In:* P.E. Greeson, J.R. Clark, and J.E. Clark (eds.) *Wetland Functions and Values. The State of Our Understanding.* Proceedings of the National Symposium on Wetlands, Amer. Water Res. Assoc. Minneapolis, MN.
Peterson, B.J. and R.W. Howarth. 1987. Sulfur, carbon, and nitrogen isotopes used to trace organic matter flow in the salt-marsh estuaries of Sapelo Island, Georgia. *Limnol. Oceanog.* 32:1195-1213.
Pfeiffer, W.J. and R.G. Wiegert. 1981. Grazers on *Spartina* and their predators. *In:* L.R. Pomeroy and R.G. Wiegert (eds.) *The Ecology of a Salt Marsh.* Springer-Verlag, New York.
Phillips, P. 1987. The mitigation muddle. *Urban Land* (Jan. 1987):36-37.
Pianka, E.R. 1989. *Evolutionary Ecology.* Harper Collins, New York.
Pielou, E.C. 1984. *The Interpretation of Ecological Data.* John Wiley & Sons, New York.
Pierce, R.J. 1991. Redefining our regulatory goals. *National Wetlands Newsletter.* 13(6):12-13.
Ping, C.L., J.P. Moore, and M.H. Clark. 1990. Wetland properties of permafrost soils in Alaska. Presented to the VIII International Soil Correlation Meeting: Classification and Management of Wet Soils, Louisiana to Texas, Oct. 6-20, 1990.
Ponnamperuma, F.N. 1972. The chemistry of submerged soils. *Advances in Agronomy* 24: 29-96.
Pomeroy, L.R. and R.G. Wiegert (eds.) 1981. *The Ecology of a Salt Marsh.* Springer-Verlag, New York.

Pomeroy, L.R., W.M. Darley, E.L. Dunn, J.L. Gallagher, E.B. Haines, and D.M. Whitney. 1981. Primary production. In: L.R. Pomeroy and R.G. Wiegert (eds.) *The Ecology of a Salt Marsh.* Springer-Verlag, New York.

Radford, A.E. 1978. Natural area classification systems: a standardized scheme for basic inventory of species, community, and habitat diversity. *In: Proceedings of the National Symposium, Classification, Inventory, and Analysis of Fish and Wildlife Habitat.* U.S. Fish & Wildl. Serv., Washington, D.C.

Rand, G. M. and S. R. Petrocelli. 1985. *Fundamentals of Aquatic Toxicology.* Hemisphere Publ. Corp., New York.

Ranwell, D.S. 1974. The salt marsh to tidal woodland transition. *Hydrobiological Bull. (Amsterdam)* 8:139-151.

Redfield, A.C. 1972. Development of a New England salt marsh. *Ecol. Monogr.* 42:201-237.

Redmond, A. 1992. How successful is mitigation? *National Wetlands Newsletter* 14(1):5-6.

Reed, P.B., Jr. 1988a. *National List of Plant Species That Occur in Wetlands: National Summary.* U.S. Fish & Wildl. Serv., Biol. Rept. 88(24), Washington, D.C.

Reed, P.B., Jr. 1988b. *National List of Plant Species That Occur in Wetlands: California (Region 0).* U.S. Fish & Wildl. Serv., Biol. Rept. 88(26.0), Washington, D.C.

Reed, P.B., Jr. 1988c. *National List of Plant Species That Occur in Wetlands: Northeast (Region 1).* U.S. Fish & Wildl. Serv., Biol. Rept. 88(26.1), Washington, D.C.

Reed, P.B., Jr. 1988d. *National List of Plant Species That Occur in Wetlands: Southeast (Region 2).* U.S. Fish & Wildl. Serv., Biol. Rept. 88(26.2), Washington, D.C.

Reed, P.B., Jr. 1988e. *National List of Plant Species That Occur in Wetlands: South Plains (Region 6).* U.S. Fish & Wildl. Serv., Biol. Rept. 88(26.6), Washington, D.C.

Reed, P.B., Jr. 1988f. *National List of Plant Species That Occur in Wetlands: Northwest (Region 9).* U.S. Fish & Wildl. Serv., Biol. Rept. 88(26.9), Washington, D.C.

Rey, J.H. 1981. Ecological biogeography of arthropods on *Spartina* islands in northwest Florida. *Ecol. Monog.* 51:237-265.

Rheinhardt, R. 1992. A multivariate analysis of vegetation patterns in tidal freshwater swamps of lower Chesapeake Bay, U.S.A. *Bull. Torrey Botan. Club* 119:192-207.

Richardson, C.J. (ed.) 1981. *Pocosin Wetlands.* Hutchinson Ross Publishing Co., Stroudsburg, PA.

Richardson, C.J., D.L. Tilton, J.A. Kadlec, J.P.M. Chamie, and W.A. Wentz. 1978. Nutrient dynamics of northern wetland ecosystems. *In:* R.E. Good, D.F. Whigham, and R.L. Simpson (eds.) *Freshwater Wetlands. Ecological Processes and Management Potential.* Academic Press, New York.

Ricklefs, R.E. 1973. *Ecology.* Chiron Press, Newton, MA.

Robinson, N. 1982. SEQRA's Siblings: Precedents from little NEPA's in The Sister States. *Albany L. Rev.* 46:1155.

Rohlf, F.J. 1985. *BIOM. A Package of Statistical Programs to Accompany the Text Biometry.* Exeter Publ., Setauket, NY.

Rohlf, F.J. 1988. *NTSYS-pc: Numerical Taxonomy and Multivariate Analysis System (rev. ed.).* Exeter, New York.

Rohlf, F.J., J. Kishpaugh, and D. Kirk. 1971. *NT-SYS. Numerical Taxonomy System of Multivariate Statistical Programs.* Tech. Rep., State Univ. of New York at Stony Brook, New York.

Romesburg, H.C. 1984. *Cluster Analysis for Researchers.* Lifetime Learning Publ., Belmont, CA.

Rozas, L.P. and W.E. Odum. 1987. Use of tidal freshwater marshes by fishes and macrofaunal crustaceans along a marsh stream-order gradient. *Estuaries* 10:36-43.

Salvesen, D. 1990. *Wetlands. Mitigating and Regulating Development Impacts.* Urban Land Institute, Washington, D.C.

Sanders, H.L. 1968. Marine benthic diversity: a comparative study. *Amer. Nat.* 102:243-282.

Sather, J.H. and R.D. Smith. 1984. *An Overview of Major Wetland Functions and values.* U.S. Fish & Wildl. Serv., Publ. OBS-84/18, Washington, D.C.

Schalles, J.F. and D.J. Shure. 1989. Hydrology, community structure and productivity patterns of a dystrophic Carolina bay wetland. *Ecol. Monogr.* 59:365-385.

Schlesinger, W.H. 1978. Community structure, dynamics, and nutrient ecology in the Okefenokee cypress swamp-forest. *Ecol. Monog.* 48:43-65.

Schneider, R.L. and R.R. Sharitz. 1986. Seed bank dynamics in a southeastern riverine swamp. *Amer. J. Botany* 73:1022-1030.

Schneller-McDonald, K., L.S. Ischinger, and G.T. Auble. 1989. *Wetland Creation and Restoration: Description and Summary of the Literature.* U.S. Fish & Wildl. Serv., Biol. Rep. No. 89, Washington, D.C.

Schomer, N.S. and R.D. Drew. 1982. *An Ecological Characterization of the Lower Everglades, Florida, and the Florida Keys.* U.S. Fish & Wildl. Serv., FWS/OBS-82/58.1, Washington, D.C.

Scodari, P.F. 1990. *Wetlands Protection: The Role of Economics.* Environmental Law Institute, Washington, D.C.

Seliskar, D.M. and J.L. Gallagher. 1983. *The Ecology of Tidal Marshes of the Pacific Northwest Coast: A Community Profile.* U.S. Fish & Wildl. Serv., Publ. FWS/OBS-82/32, Washington, D.C.

Sharitz, R.R. and J.W. Gibbons (eds.) 1982. *The Ecology of Southeastern Shrub Bogs (Pocosins) and Carolina Bays: A Community Profile.* U.S. Fish & Wildl. Serv., FWS/OBS-82/04, Washington, D.C.

Sharitz, R.R. and W.J. Mitsch. 1993. Southern floodplain forests. *In:* W.H. Martin, S.G. Boyce, and A.C.E. Echternacht (eds.) *Biodiversity of the Southeastern United States: Lowland Terrestrial Communities.* John Wiley & Sons, New York.

Shaw, S.P. and C.G. Fredine. 1956. *Wetlands of the United States. Their Extent, and Their Value for Waterfowl and Other Wildlife*. U.S. Fish & Wildl. Serv., Circular 39, Washington, D.C.

Shelley, P. 1991. Losing our wetlands, science, and credibility. *National Wetlands Newsletter*. 13(6):13-14.

Silverberg, S.M. and M.S. Dennison. 1993. *Wetlands and Coastal Zone Regulation and Compliance*. John Wiley & Sons, New York.

Simberloff, D.S. 1972. Properties of the rarefaction diversity measurement. *Amer. Nat.* 106:414-418.

Simberloff, D.S. 1976. Experimental zoogeography of islands: effects of island size. *Ecology* 57:629-648.

Simberloff, D.S. and E.O. Wilson. 1969. Experimental zoogeography of islands: effects of island size. *Ecology* 50:278-296.

Simberloff, D.S. and E.O. Wilson. 1970. Experimental zoogeography of islands: a two-year record of colonization. *Ecology* 51:934-937.

Simon, S., A. Nugteren, and M. Morris. 1993. Restoring Illinois' wetlands. *Illinois Nat. Hist. Repts.* 320:5-6.

Simpson, R.L., R.E. Good, M.A. Leck, and D.F. Whigham. 1983. The ecology of freshwater tidal wetlands. *BioScience* 33: 255-259.

Simpson, R.L., D.F. Whigham, and R. Walker. 1978. Seasonal patterns of nutrient movement in a freshwater tidal marsh. *In:* R.E. Good, D.F. Whigham, and R.L. Simpson (eds.) *Freshwater Wetlands. Ecological Processes and Management Potential*. Academic Press, New York.

Sipple, W.S. 1988. *Wetland Identification and Delineation Manual. Volumes I and II*. U.S. Environmental Protection Agency, Office of Wetlands Protection, Washington, D.C.

Sklar, F.H. 1985. Seasonality and community structure of the backswamp invertebrates in a Louisiana cypress-tupelo wetland. *Wetlands* 5:69-83.

Smith, A.G., J.H. Stroud, and J.B. Gollop. 1964. Prairie pothole marshes. *In:* J.P. Linduska (ed.) *Waterfowl Tomorrow*. U.S. Fish & Wildl. Serv., Washington, D.C.

Smith, E.P. and G. van Belle. 1984. Nonparametric estimation of species richness. *Biometrics* 40:119-129.

Smith, L.M. and J.A. Kadlec. 1983. Seed banks and their role during drawdown of a North American marsh. *J. Applied Ecol.* 20:673-684.

Smock, L.A., E. Gilinsky, and D.L. Stoneburner. 1985. Macro-invertebrate production in a southeastern United States backwater stream. *Ecology* 66:1491-1503.

Sneath, P.H.A. and R.R. Sokal. 1973. *Numerical Taxonomy*. W.H. Freeman, San Francisco.

Snedaker, S. and S. Brown. 1981. Water quality and mangrove ecosystem dynamics. U.S. Environmental Protection Agency, EPA-600/4-81-002, Gulf Breeze, FL.

Soil Survey Staff. 1990. *Keys to Soil Taxonomy, 4th ed.* Soil Management Support Services, Tech. Monog. No. 6. Virginia Polytechnic Institute and State University, Blacksburg, VA.

Soil Survey Staff. 1975. *Soil Taxonomy: A Basic System for Soil Classification.* U.S. Dept. Agric., Soil Conservation Service, Agricultural Handbook No. 436, Washington, D.C.

Sokal, R.R. and F.J. Rohlf. 1981. *Biometry. The Principles and Practice of Statistics in Biological Research, 2nd ed.* W.H. Freeman, New York.

Sokolove, R.D. and P.D. Huang. 1992. Privitization of wetland mitigation banking. *Natural Resources & Environment* (Amer. Bar Assoc.) 7(1):36-38, 68-69.

Sousa, W.P. 1979. Experimental investigation of disturbance and ecological succession in a rocky intertidal algal community. *Ecol. Monogr.* 49:227-254.

Southwood, T.R.E. 1977. Habitat, the template for ecological strategies? *J. Anim. Ecol.* 46:337-365.

Southwood, T.R.E. 1978. *Ecological Methods, 2nd ed.* Chapman & Hall, London.

Stavens, R. and A. Jaffe. 1990. Unintended impacts of public investments on private decisions: the depletion of forested wetlands. *Amer. Econ. Rev.* 80:337-352.

Steeman-Nielsen, E. 1952. The use of radioactive carbon (C^{14}) for measuring organic production in the sea. *J. Cons. Perm. Int. Explor. Mer.* 18:117-140.

Stegman, J.L. 1976. Overview of current wetland classification and inventories in the United States and Canada: *In:* J.H. Sather (ed.) *National Wetland Classification and Inventory Workshop Proceedings-1975.* U.S. Fish & Wildl. Serv., Washington, D.C.

Stewart, R.E. and H.A. Kantrud. 1972. *Vegetation of Prairie Potholes, North Dakota, in Relation to Quality of Water and Other Environmental Factors.* U.S. Geol. Surv., Prof. Pap. 585-D, Washington, D.C.

Stout, J.P. 1984. *The Ecology of Irregularly Flooded Salt Marshes of the Northeastern Gulf of Mexico: A Community Profile.* U.S. Fish & Wildl. Serv., FWS/OBS-81/24, Washington, D.C.

Stowe, L.G. and M.J. Wade. 1979. The detection of small-scale patterns in vegetation. *J. Ecol.* 67:1047-1064.

Suter, G. W. II 1993. *Ecological Risk Assessment.* Lewis Publ., Boca Raton, FL.

Suter, G. W., II, and J. M. Loar. 1992. Weighting the ecological risk of hazardous waste sites. *Environ. Sci. Technol.* 26:432-438.

Sykes, P.A., Jr., 1983. Recent population trend of the snail kite in Florida and its relation to water levels. *J. Field Ornith.* 54:237-246.

Tanner, J.T. 1978. *Guide to the Study of Animal Populations.* Univ. Tennessee, Knoxville, TN.

Tanner, W.F. 1960. Florida coastal classification. *Trans. Gulf Coast Assoc. Geol. Soc.* 10:259-266.

Tansley, A.G. 1920. The classification of vegetation and the concept of development. *Ecology* 8:118-149.

Tansley, A.G. 1935. The use and abuse of vegetational concepts and terms. *Ecology* 16:284-307.

Taylor, T.J., N.E. Erickson, R. Tumlison, J.A. Ratzlaff, and K.D. Cunningham. 1984. *Groundwater Wetlands of the Cimarron Terrace, Northcentral Oklahoma.* Dept. of Zoology, Oklahoma State University, Stillwater, OK.

Teal, J.M. 1957. Community metabolism in a temperate cold spring. *Ecol. Monogr.* 27:283-302.

Teal, J.M. 1962. Energy flow in the salt marsh ecosystem of Georgia. *Ecology* 43:614-624.

Teal, J.M. 1986. *The Ecology of Regularly Flooded Salt Marshes of New England: A Community Profile.* U.S. Fish & Wildl. Serv., Biol. Rept. 85(7.4), Washington, D.C.

Thayer, G.W. (ed.) 1992. *Restoring the Nation's Marine Environment.* Maryland Sea Grant College, College Park, MD.

Thibodeau, F.R. and B.D. Ostro. 1981. An economic analysis of wetlands protection. *J. Environ. Manage.* 12:19-30.

Thunhorst, G.W. 1992. *Wetland Planting Guide for the Northeastern United States.* Environmental Concern, St. Michaels, MD.

Tilman, D. 1977. Resource competition between planktonic algae: an experimental and theoretical approach. *Ecology* 58:338-348.

Timbrell, J. A. 1982. *Principle of Biochemical Toxicology.* Taylor & Francis, Ltd., London.

Tiner, R.W., Jr. 1984. *Wetlands of the United States: Current Status and Recent Trends.* U.S. Fish & Widlife Service, Washington, D.C.

Tiner, R.W., Jr. 1985a. *Wetlands of New Jersey.* U.S. Fish & Widlife Service, Washington, D.C.

Tiner, R.W., Jr. 1985b. *Wetlands of Delaware.* U.S. Fish & Wildl. Serv./Delaware Dept. Nat. Res. & Environ. Control, Dover, DE.

Tiner, R.W., Jr. 1988. *Field Guide to Nontidal Wetland Identification.* Maryland Dept. of Natural Resources, Water Resources Administration, Annapolis, MD and U.S. Fish & Wildl. Serv., Newton Corner, MA.

Tiner, R.W., Jr. 1989. Wetland boundary delineation. *In:* S.K. Majumdar, R.P. Brooks, F.J. Brenner, and R.W. Tiner, Jr. (eds). *Wetlands Ecology and Conservation: Emphasis in Pennsylvania.* Pennsylvania Acad. Sci., Lafayette College, Easton, PA.

Tiner, R.W., Jr. 1990. Use of high-altitude aerial photography for inventorying forested wetlands in the United States. *Forest Ecol. Manage.* 33/34:593-604.

Tiner, R.W. 1991a. The concept of a hydrophyte for wetland identification. *BioScience* 41:236-247.

Tiner, R.W. 1991b. How wet is a wetland? *Great Lakes Wetlands Newsletter* 2: 1-4,7.

Tiner, R.W. 1991c. *Maine Wetlands and Their Boundaries.* Maine Dept. of Economic and Community Development, Augusta, ME.

Tiner, R.W. 1993a. Using plants as indicators of wetlands. *Proc. Acad. Nat. Sci., Philadelphia* 144:240-253.
Tiner, R.W. 1993b. The primary indicators method - a practical approach to wetland recognition and delineation in the United States. *Wetlands* 13:50-64.
Tiner, R.W. 1993c. *Field Guide to Coastal Wetland Plants of the Southeastern United States.* Univ. Massachusetts Press, Amherst, MA.
Tiner, R.W., Jr. and J.T. Finn. 1986. *Status and Recent Trends of Wetlands of Five Mid-Atlantic States: Delaware, Maryland, Pennsylvania, Virginia, and West Virginia.* U.S. Fish & Wildl. Serv./U.S. Environmental Protection Agency, Philadelphia, PA.
Tiner, R.W., Jr. and P.L.M. Veneman. 1987. *Hydric Soils of New England.* Cooperative Extension Service Bull. C-183R, Univ. of Massachusetts, Amherst, MA.
Tomovic, R. 1963. *Sensitivity Analysis of Dynamic Systems.* McGraw-Hill, New York.
Tonn, W.M. and J.J. Magnuson. 1982. Patterns in the species composition and richness of fish assemblages in northern Wisconsin lakes. *Ecology* 63:1149-1166.
Townsend, C.R., A.G. Hildrew, and J.E. Francis. 1983. Community structure in some southern English streams: the influence of physicochemical factors. *Freshwater Biol.* 13:521-544.
Twilley, R.R., A.E. Lugo, and C. Patterson-Zucca. 1986. Litter production and turnover in basin mangrove forests in southwest Florida. *Ecology* 67:670-683.
U.S. Army/U.S Environmental Protection Agency. 1989. Memorandum of agreement between the Department of the Army/Environmental Protection Agency concerning the determination of the geographic jurisdiction of the Section 404 program and the application of the exemptions under 404(f) of the Clean Water Act. January 19, 1989.
U.S. Army Corps of Engineers. 1987. *Corps of Engineers Wetlands Delineation Manual.* U.S. Army Corps of Engineers, Environmental Laboratory, Tech. Rept. Y-87-1, Vicksburg, MS.
U.S. Environmental Protection Agency. 1989. *Risk Assessment Guidance for Superfund. Vol. II, Environmental Evaluation Manual.* U.S. Environmental Protection Agency, Interim Final. EPA/540/1-89/001, Washington, D.C..
U.S. Environmental Protection Agency. 1990. *Wetlands: Region 4 Implementation and Management of the Section 404 Wetlands Program.* Rept. of Audit E1h7F8-04-0331-0100208, U.S. Environmental Protection Agency, Atlanta, GA.
U.S. Environmental Protection Agency. 1991a. Ecological Assessment of Superfund Sites: An Overview. *Eco Update,* 1(2), EPA Publ. 9345.0-05I, Washington, D.C.
U.S. Environmental Protection Agency. 1991b. *Summary of Advanced Identification Projects Under Section 230.80 of the 404(b)(1) Guidelines.* U.S. Environmental Protection Agency, Washington, D.C.
U.S. Environmental Protection Agency. 1992a. *Final Report to Congress on § 319 of the Clean Water Act (1989).* U.S. Environmental Ptorection Agency, EPA-506/9-90, (WH-553), Washington, D.C.

U.S. Environmental Protection Agency. 1992b. The role of Natural Resource Trustees in the Superfund process. *Eco Update* 1(3), EPA Publ. 9345.0-05I, Washington, D.C.

U.S. Environmental Protection Agency. 1992c. *Framework for Ecological Risk Assessment.* U.S. Environmental Protection Agency, EPA/630/R-92/001, Washington, D.C.

U.S. Environmental Protection Agency. 1992d. Developing a work scope for ecological assessments. *Eco Update*, 1(4), EPA Publ. 9345.0-05I, Washington, D.C.

U.S. Environmental Protection Agency. 1992e. *Fact Sheet. Rookery Bay Wetlands Advanced Identification Project, Collier County, Florida.* U.S. Environmental Protection Agency, Atlanta, GA.

U.S. Fish and Wildlife Service. 1980. *Habitat Evaluation Procedures (HEP).* U.S. Fish & Wildl. Serv., Ecol. Services Manual 101, Washington, D.C.

U.S. Fish and Wildlife Service. 1989. *National Wetlands Priority Conservation Plan.* U.S. Fish & Wildl. Serv., Washington, D.C.

U.S. Fish and Wildlife Service. 1990. *Wetlands: Meeting the President's Challenge (Wetlands Action Plan).* U.S. Fish & Wildl. Serv., Washington, D.C.

U.S. General Accounting Office. 1988. *Wetlands - The Corps of Engineers' Administration of the Section 404 Program.* U.S. General Accounting Office, Washington, D.C.

U.S. Soil Conservation Service. 1991. *National List of Hydric Soils.* National Technical Committee for Hydric Soils, U.S. Department of Agriculture, Washington, D.C.

U.S. Soil Conservation Service. 1992. *Field Handbook. Chapter 13: Wetland Restoration, Enhancement, and Creation.* U.S. Dept. of Agric., Soil Conservation Service, Washington, D.C.

Utida, S. 1957. Cyclic fluctuations of population density intrinsic to a host-parasite system. *Ecology* 38:442-449.

Valiela, I. 1984. *Marine Ecological Processes.* Springer-Verlag, New York.

Van Cleve, K., and D. Sprague. 1971. Respiration rates in the forest floor of birch and aspen stands in interior Alaska. *Arc. Alp. Res.* 3:17-26.

Vandermeer, J.H. 1972. Niche theory. *Ann. Rev. Ecol. Syst.* 3:107-132.

van der Valk, A.G. 1981. Succession in wetlands: a Gleasonian approach. *Ecology* 62:688-696.

van der Valk, A.G. 1985. Vegetation dynamics of prairie glacial marshes. *In:* J. White (ed.) *The Population Structure of Vegetation.* Dr. W. Junk, Publ., Dordrecht, The Netherlands.

van der Valk, A.G. 1989. *Northern Prairie Wetlands.* Iowa State Univ. Press, Ames, IA.

van der Valk, A.G. and C.B. Davis. 1978. The role of seed banks in the vegetation dynamics of prairie glacial marshes. *Ecology* 59:322-335.

Veneman, P.L.M. and R.W. Tiner. 1990. *Soil-Vegetation Correlations in the Connecticut River Floodplain of Western Massachusetts.* U.S. Fish & Wildl. Serv., Biol. Rept. No. 90(6), Washington, D.C.

Verhulst, P.F. 1838. Notice sur la loi que la population suit dans son accroissement. *Correspondences Math. Phys.* 10:113-121.
Vince, S.W., S.R. Humphry, and R.W. Simons. 1989. *The Ecology of Hydric Hammocks: A Community Profile.* U.S. Fish & Wildl. Serv., Biol. Rept. No. 85, Washington, D.C.
Volterra, V. 1926. Fluctuations in the abundance of a species considered mathematically. *Nature* 118:558-560.
Want, W.L. 1993. *Law of Wetlands Regulation.* Clark, Boardman, Callaghan, New York [updated yearly].
Wauer, R.H. 1977. Significance of Rio Grande riparian system upon the avifauna. *In:* R.R. Johnson and D.A. Jones (eds.) *Importance, Preservation, and Management of Riparian Habitat. A Symposium.* U.S. Forest Service, Gen. Tech. Rept. RM-43, Washington, D.C.
Welch, B.L., R.B. Whitlatch, and W.F. Bohlen. 1982. Relationship between physical characteristics and organic carbon sources as a basis for comparing estuaries in southern New England. *In:* V.S. Kennedy (ed.) *Estuarine Comparisons.* Academic Press, New York.
Weinstein, M.P. 1979. Shallow marsh habitats as primary nurseries for fishes and shellfish, Cape Fear River, North Carolina. *Fish. Bull.* 77:339-357.
Weinstein, M.P., S.L. Weiss, and M.F. Walters. 1980. Multiple determinants of community structure in shallow marsh habitats, Cape Fear River Estuary, North Carolina. *Marine Biol.* 58:227-243.
Weller, M.W. 1981. *Freshwater Marshes.* Univ. Minnesota Press, Minneapolis, MN.
Wentworth, T.R., G.P. Johnson and R.L. Kologiski. 1988. Designation of wetlands by weighted averages of vegetation data: a preliminary evaluation. *Water Resources Bull.* 24:389-396
Wharton, C.H., W.M Kitchens, E.C. Pendleton, and T.W. Sipe. 1982. *The Ecology of Bottomland Hardwood Swamps of the Southeast: A Community Profile.* U.S. Fish & Wildl. Serv., FWS/OBS-81/37, Washington, D.C.
Wharton, C.H., V.W. Lambou, J.Newsom, P.V. Winger, L.L Gaddy, and R. Mancke. 1981. The fauna of bottomland hardwoods in southeastern United States. *In:* J.R. Clark and J. Benforado (eds.) *Wetlands of Bottomland Hardwood Forests.* Elsevier, Amsterdam.
Whitney, D.M., A.G. Chalmers, E.B. Haines, R.B. Hanson, L.R. Pomeroy, and B. Sherr. 1981. The cycles of nitrogen and phosphorus. *In:* L.R. Pomeroy and R.G. Wiegert (eds.) *The Ecology of a Salt Marsh.* Springer-Verlag, New York.
Whittaker, R.H. 1962. Classification of natural communities. *Botan. Rev.* 28:1-239.
Whittaker, R.H. 1967. Gradient analysis of vegetation. *Biol. Rev.* 42:207-264.
Whittaker, R.H. 1970. The biochemical ecology of higher plants. *In:* E. Sondheimer and J.B. Simeone (eds.) *Chemical Ecology.* Academic Press, New York.
Whittaker, R.H. 1975. *Communities and Ecosystems. 2nd ed.* MacMillan, New York.

Whittaker, R.J. 1991. Small-scale pattern: an evaluation of techniques with an application to salt marsh vegetation. *Vegetatio* 94:81-94.

Wiebe, W.J., R.R. Christian, J.A. Hansen, G. King, B. Sherr, and G. Skyring. 1981. Anaerobic respiration and fermentation. *In:* L.R. Pomeroy and R.G. Wiegert (eds.) *The Ecology of a Salt Marsh.* Springer-Verlag, New York.

Wiegert, R.G. 1962. The selection of an optimum quadrat size for sampling the standing crop of grasses and forbs. *Ecology* 43:125-129.

Wiegert, R.G and B.J. Freeman. 1990. *Tidal Salt Marshes of the Southeast Atlantic Coast: A Community Profile.* U.S. Fish & Wildl. Serv., Biol. Rep. 85(7.29), Washington, D.C.

Wiegert, R.G. and L.R. Pomeroy. 1981. The salt marsh ecosystem: a synthesis. *In:* L.R. Pomeroy and R.G. Wiegert (eds.) *The Ecology of a Salt Marsh.* Springer-Verlag, New York.

Wiegert, R.G., L.R. Pomeroy, and W.J. Wiebe. 1981. Ecology of salt marshes: an introduction. *In:* L.R. Pomeroy and R.G. Wiegert (eds.) *The Ecology of a Salt Marsh.* Springer-Verlag, New York.

Wilcher, L.S. and R.W. Page. 1990. A victory on many fronts. *National Wetlands Newsletter* 12(2):2, 5-7.

Wilcox, D.A. 1986. The effects of deicing salts on vegetation in Pinhook Bog, Indiana. *Canadian J. Botany* 64:865-874.

Wilen, B.O. and R.W. Tiner. 1993. Wetlands of the United States. *In:* D. Whigham, D. Dykyjova, and S. Hejny (eds.) *Wetlands of the World: Inventory, Ecology and Management. Vol. I.* Kluwer Acad. Publ., Dordrecht, Netherlands.

Willard, D., M. Leslie, and R.B. Reed. 1990. Defining and delineating wetlands. *In:* G. Bingham, E.H. Clark II, L.V. Haygood, and M. Leslie (eds.) *Issues in Wetlands Protection: Background Papers Prepared for the National Wetlands Policy Forum.* The Conservation Foundation, Washington, D.C.

Williams, W.T. 1971. Principles of clustering. *Ann. Rev. Ecol. Syst.* 2:303-326.

Windell, J.T., B.E. Willard, D.J. Cooper, S.Q. Foster, C.F. Knud-Hanson, L.P. Rink, and G.N. Kiladis. 1986. *An Ecological Characterization of Rocky Mountain Montane and Subalpine Wetlands.* U.S. Fish & Wildl. Serv., Biol. Rep. 86(11), Washington, D.C.

Winkler, M.G. 1988. Effect of climate on development of two Sphagnum bogs in south-central Wisconsin. *Ecology* 69:1032-1043.

Winter, L. 1990. Sununu pulled rank. *National Wetlands Newsletter* 12(2):3, 7-8.

Winter, T.C. and M.R. Carr. 1980. *Hydrologic Setting of Wetlands in the Cottonwood Lake Area, Stutsman County, North Dakota. U.S. Geological Survey, Water Resources Investigations* 80-99, Reston, VA..

Winter, T.C., and M.R. Llamas (eds.) 1993. Hydrogeology of wetlands. *J. Hydrology* 141:1-269.

Wolf, R., L.C. Lee, and R.R. Sharitz (eds.) 1986. Wetland creation and restoration in the United States from 1970 to 1985: an annotated bibliography. *Wetlands* 6(1):1-88.

Wood, L.D. 1989. The forum's recommendation to delegate section 404 to the states: a bad deal for wetlands. *National Wetlands Newsletter* 11(4):2-3.

Wood, L.D. 1990. Section 404 delegation: a rebuttal to Governor Kean. *National Wetlands Newsletter* 12(1):2-3.

Wright, J.C. 1959. Limnology of Canyon Ferry Reservoir. II. Phytoplankton standing crop and primary production. *Limnol. Oceanogr.* 4:235-245.

Zachritz, W.H. and J.W. Fuller. 1993. Performance of an artificial wetlands filter treating facultative lagoon effluent at Carville, Louisiana. *Water Environ. Res.* 65:46-52.

Zagata, M.D. 1985. Mitigation by "banking" credits - a Louisiana pilot project. *National Wetlands Newsletter* 7(3):9-11.

Zallen, M. 1992. The mitigation agreement - a major devlopment in wetland regulation. *Natural Resources & Environment* (Amer. Bar Assoc.) 7(1):19-21, 60-62.

Zampella, R.A., G. Moore, and R.E. Good. 1992. Gradient analysis of pitch pine (*Pinus rigida* Mill.) lowland communities in the New Jersey Pinelands. *Bull. Torrey Botan. Club* 119:253-261.

Zedler, J.B. 1982. *The Ecology of Southern California Coastal Salt Marshes: A Community Profile.* U.S. Fish & Wildl. Serv., FWS/OBS-81/54, Washington, D.C.

Zedler, J.B. 1988. Restoring diversity in salt marshes: can we do it? *In:* E.O. Wilson (ed.) *Biodiversity.* Nat'l Academy Press, Washington, D.C.

Zedler, J.B. and C.S. Norby. 1986. *The Ecology of Tijuana Estuary, California: An Estuarine Profile.* U.S. Fish & Wildl. Serv., Biol. Rep. 85(7.5), Washington, D.C.

Zedler, P.H. 1987. *The Ecology of Southern California Vernal Pools: A Community Profile.* U.S. Fish & Wildl. Serv., Biol. Rep. 85(7.11), Washington, D.C.

Zoltai, S.C., F.C. Pollett, J.K. Jeglum, and G.D. Adams. 1975. Developing a wetland classification for Canada. *Proc. North Amer. For. Soils Conf.* 4:497-511.

Appendix

Federal and State Wetland Agencies and Offices

U.S. ARMY CORPS OF ENGINEERS DIVISION OFFICES*

*Please note that some states are within the jurisdiction of more than one divisional office because Corps' divisions are organized by watershed area and not by state boundary.

New England Division
Main, Vermont, Connecticut, New Hampshire, Massachusetts, Rhode Island

424 Trapelo Road
Waltham, MA 02254-9149
(617) 647-8778

North Atlantic Division
New York, Pennsylvania, Virginia, Maryland, Delaware, West Virginia, Vermont

90 Church Street
New York, NY 10007-2979
(212) 264-7500

South Atlantic Division
Virginia, North Carolina, South Carolina, Georgia, Florida, Alabama, Mississippi, Tennessee, Puerto Rico, U.S. Virgin Islands

Room 313
77 Forsyth Street, S.W.
Atlanta, GA 30335-8801
(404) 331-6715

Ohio River Division
Pennsylvania, Virginia, West Virginia, Kentucky, Tennessee, Ohio, Indiana, Illinois, Alabama, North Carolina, Georgia

P.O. Box 1159
Cincinnati, OH 45201-1159
(513) 684-3010

North Central Division
North Dakota, Minnesota, South Dakota, Iowa, Missouri, Illinois, Wisconsin, Michigan, Indiana, Ohio

536 South Clark Street
Chicago, IL 60605-6319
(312) 353-6319

Lower Mississippi Valley Division
Louisiana, Mississippi, Tennessee, Arkansas, Missouri, Illinois, Kentucky

P.O. Box 80
Vicksburg, MS 39180-0080
(601) 631-5052

Missouri River Division
Montana, North Dakota, South Dakota, Wyoming, Nebraska, Iowa, Colorado, Kansas, Missouri

P.O. Box 103
Downtown Station
Omaha, NE 68101-0103
(402) 221-7208

Southwestern Division
Colorado, Kansas, Missouri, New Mexico, Oklahoma, Arkansas, Texas, Louisiana

1114 Commerce Street
Dallas, TX 75242-0216
(214) 767-2510

North Pacific Division
Washington, Oregon, Idaho, Montana, Wyoming, Nevada, Alaska

P.O. Box 2870
Portland, OR 97208-2870
(503) 326-3768

South Pacific Division
Oregon, California, Arizona, New Mexico, Colorado, Utah, Wyoming, Idaho, Nevada

630 Sansome Street Room 720
San Francisco, CA 94111-2206
(415) 705-2405

Pacific Ocean Division
Hawaii

Building 230
Fort Shafter, HI 96858-5440
(808) 438-9258

U.S. ENVIRONMENTAL PROTECTION AGENCY NATIONAL AND REGIONAL OFFICES

Environmental Protection Agency
Office of Wetlands Protection
401 M Street, S.W.
Washington, DC 20460
(202) 475-7799

EPA Region I
Connecticut, Massachusetts, Maine, New Hampshire, Rhode Island, Vermont

JFK Federal Building
Boston, MA 02203

EPA Region II
New Jersey, New York, Puerto Rico, Virgin Islands

26 Federal Plaza
New York, NY 10278

EPA Region III
Delaware, Maryland, Pennsylvania, Virginia, West Virginia, District of Columbia

841 Chestnut Street
Philadelphia, PA 19107

EPA Region IV
Alabama, Florida, Georgia, Kentucky, Mississippi, North Carolina, South Carolina, Tennessee

345 Courtland Street, N.E.
Atlanta, GA 30365

EPA Region V
Illinois, Indiana, Michigan, Minnesota, Ohio, Wisconsin

230 South Dearborn Street
Chicago, IL 60604

EPA Region VI
Arkansas, Louisiana, New Mexico, Oklahoma, Texas

1201 Elm Street
Dallas, TX 75270

EPA Region VII
Iowa, Kansas, Missouri, Nebraska

726 Minnesota Avenue
Kansas City, KS 66101

EPA Region VIII
Colorado, Montana, North Dakota, South Dakota, Utah, Wyoming

One Denver Place
999 18th Street, Suite 1300
Denver, CO 80202

EPA Region IX
Arizona, California, Hawaii, Nevada, American Samoa, Guam, Trust Territories of the Pacific

214 Fremont Street
San Francisco, CA 94105

EPA Region X
Alaska, Idaho, Oregon, Washington

1200 6th Avenue
Seattle, WA 98101

OTHER FEDERAL OFFICES

U.S. Fish and Wildlife Service
Department of Interior
Division of Habitat
1849 C Street, N.W.
Washington, D.C. 20240
(703) 358-2201

Federal Emergency Management Agency (FEMA)
Federal Insurance Administration
Federal Center Plaza
500 C Street, S.W.
Washington, D.C. 20472
(202) 646-2774

National Park Service
Department of Interior
Office of Land Resources
1849 C Street, N.W.
Washington, D.C. 20240
(202) 208-5881

National Oceanic & Atmospheric Administration
Department of Commerce
Ocean and Coastal Resource Management Office
Universal Building South
1825 Connecticut Avenue, N.W.
Washington, D.C. 20235
(202) 673-5138

STATE WETLAND AGENCIES AND OFFICES

Alabama

Department of Economic and Community Affairs
P.O. Box 2939
Montgomery, AL 36105
(205) 284-8774

Field Operations Division
Alabama Department of Environmental Management
1751 Con. W.L. Dickinson Drive
Montgomery, AL 36130
(205) 271-7700

Environmental Scientist
Alabama Department of Environmental Management
2204 Perimeter Road
Mobile, AL 36615
(205) 479-2336

Alaska

Department of Environmental Conservation
2330 Hospital Drive
Juneau, AK 99811
(907) 465-2600

Arkansas

Department of Pollution Control and Ecology
8001 National Drive
Little Rock, AR 72219
(501) 562-7444

Arkansas Soil and Water Commission
1 Capitol Mall, Suite 2D
Little Rock, AR 72201
(501) 371-1611

California

Federal Programs Manager
California Coastal Commission
45 Fremont St., Suite 2000
San Francisco, CA 94105
(415) 904-5200

Wetlands Task Force
California Coastal Commission
640 Capitola Road
Santa Cruz, CA 95062
(408) 479-3511

California Coastal Conservancy
1330 Broad, Suite 1100
Oakland, CA 94612
(510) 658-5254

Colorado

Department of Health
Water Quality Control Division
4210 East 11th Avenue
Denver, CO 80220
(303) 331-4756

Connecticut

Division of Inland Water Resource Management
Department of Environmental Protection
Room 207, State Office Building
165 Capital Avenue
Hartford, CT 06106
(203) 566-7280

Delaware

Department of Natural Resources and Environmental Control
89 King's Highway
P.O. Box 1401
Dover, DE 19903
(302) 739-4691

Florida

Division of Water Management
Department of Environmental Regulation
Twin Towers Office Building
2600 Blair Stone Road
Tallahassee, FL 32399-2400
(904) 488-0130

Department of Natural Resources
3900 Commonwealth Blvd.
Tallahassee, FL 32399
(904) 488-1554

State Resources Management Bureau
Department of Community Affairs
Howard Building
2571 Executive Center Circle, East
Tallahassee, FL 32399
(904) 488-9210

Georgia

Marsh and Beach Section
Coastal Resources Division
Department of Natural Resources
One Conservation Way
Brunswick, GA 31523
(912) 264-7218

Hawaii

Office of State Planning
Office of the Governor
P.O. Box 2540
Honolulu, HI 96811
(808) 587-2833

Illinois

Water Resources Division
Illinois Department of Transportation
2300 S. Dirksen Parkway
Springfield, IL 62707-8415
(217) 782-3488

Department of Environmental Conservation
524 S. 2nd Street
Springfield, IL 62706
(217) 782-3715

Indiana

Department of Natural Resources
Division of Water
2475 Directors Row
Indianapolis, IN 46241
(317) 232-4160

Department of Environmental Management
105 S. Meridian
P.O. Box 6015
Indianapolis, IN 46206-6015
(317) 232-8603

Iowa

Wildlife Bureau
Department of Natural Resources
Wallace State Office Building
East 9th and Grand
Des Moines, Iowa 50319
(515) 281-6156

Governmental Liaison Bureau
Department of Natural Resources
Wallace State Office Building
East 9th and Grand
Des Moines, Iowa 50319
(515) 281-8973

Kansas

Bureau of Water
Department of Health and Environment
Forbes Field, Bldg. 17
Topeka, KS 66620
(913) 296-5500

Kentucky

Division of Water
18 Reilly Road
Frankfort, KY 40601
(502) 564-3410

Louisiana

Department of Natural Resources, Wildlife and Fisheries
Ecological Studies Section
P.O. Box 15570
Baton Rouge, LA 70895
(504) 342-9274

Maine

Division of Natural Resources
Department of Environmental Protection
State House, Station 17
Augusta, ME 04333
(207) 289-2111

State Planning Office
184 State Street
Augusta, ME 04330
(207) 289-3261

Maryland

Nontidal Wetlands Division
Water Resources Administration
Department of Natural Resources
Tawes State Office Building, E-2
Annapolis, MD 21401
(301) 974-3841

Tidal Wetlands Division
Department of Natural Resources
Tawes State Office Building, D-4
Annapolis, MD 21401
(301) 974-3871

Chesapeake Bay Critical Area Commission
275 West Street, Suite 320
Annapolis, MD 21401
(301) 974-2426

Massachusetts

Division of Wetland and Waterways Regulations
Department of Environmental Protection
1 Winter Street
Boston, MA 02108
(617) 767-5518

Michigan

Land and Water Protection Section
Land and Water Management Division
Department of Natural Resources
P.O. Box 30028
Lansing, MI 48909
(517) 335-2694

Great Lakes Shorelands Management Section
Land and Water Management Division
Department of Natural Resources
P.O. Box 20038
Lansing, MI 48909
(517) 373-1950

Minnesota

Protected Waters and Wetlands Permit Program
Department of Natural Resources
500 Lafayette Road - Box 32
St. Paul, MN 55155-4032
(612) 296-4800

Board of Water and Soil Resources
155 South Wabash St., Suite 104
St. Paul, MN 55107
(612) 296-0879

Mississippi

Coastal Management Section
Bureau of Marine Resources
Department of Wildlife Fisheries & Parks
2620 Beach Boulevard
Biloxi, MS 39531

Montana

Department of Fish, Wildlife and Parks
1420 East Sixth Avenue
Helena, MT 59620
(406) 444-2544

Water Quality Bureau
A-206 Cogswell Building
Helena, MT 59620
(406) 444-2406

New Hampshire

Department of Environmental Services
Wetlands Bureau
P.O. Box 2008
Concord, NH 03301
(603) 271-2147

New Jersey

Bureau of Freshwater Wetlands
Division of Coastal Resources
Department of Environmental Protection
CN-401
501 East State Street
Trenton, NJ 08625
(609) 984-0853

New Mexico

Surface Water Quality Bureau
Environmental Improvement Division
Department of Health and Environment
1190 St. Francis Drive
Santa Fe, NM 87503
(505) 827-2804

New York

Freshwater Wetlands Program Manager
Division of Fish & Wildlife
Department of Environmental Conservation
50 Wolf Road
Albany, NY 12233
(518) 457-9713

Bureau of Marine Habitat
Marine Regulatory Division
Department of Environmental Conservation
Building 40
SUNY at Stony Brook
Stony Brook, NY 11749
(516) 751-7900

Coastal Erosion Section
Department of Environmental Conservation
50 Wolf Road
Albany, NY 12233
(518) 457-3157

North Carolina

Division of Coastal Management
Department of Environment, Health, & Natural Resources
P.O. Box 27687
Raleigh, NC 27611
(919) 733-2293

North Dakota

North Dakota Water Commission
900 East Boulevard
Bismarck, ND 58505
(701) 224-2750

Ohio

Wildlife Division
Department of Natural Resources
Fountain Square
Columbus, OH 43224
(614) 265-6305

Oregon

Wetlands Program Manager
Division of State Lands
775 Summer Street, N.E.
Salem, OR 97310
(503) 378-3805

Pennsylvania

Division of Rivers and Wetlands Conservation
Department of Environmental Resources
Environmental Review Section
Technical Assistance and Education Section
P.O Box 8761
Harrisburg, PA 17105-8761
(717) 541-7803

Division of Waterway and Storm Waters Management
Bureau of Dams and Waterway Management
Department of Environmental Resources
One Ararat Boulevard, Room 149
Harrisburg, PA 17110
(717) 541-7904

Rhode Island

Division of Freshwater Wetlands
Department of Environmental Management
291 Promenade Street
Providence, RI 02908-6820
(410) 277-6820

South Carolina

Executive Director
South Carolina Coastal Council
AT&T Capital Center
1201 Main Street, Suite 1520
Columbia, SC 29201
(803) 737-0881

Permit Administrator
South Carolina Coastal Council
4130 Faber Place
Charleston, SC 29405
(803) 744-5838

Tennessee

Division of Water Pollution Control
Department of Health and Environment
344 Cordell Hull Bldg.
Nashville, TN 37219
(615) 741-2275

Texas

Water Quality Division
Texas Water Commission
P.O. Box 13087, Capitol Station
Austin, TX 78711
(512) 463-8412

Utah

Wildlife Resources Division
Department of Natural Resources
1596 W. North Temple
Salt Lake City, UT 84116
(801) 533-9333

Vermont

Wetlands Coordinator
Water Quality Division
Department of Environmental Conservation
10-North Building
103 South Main Street
Waterbury, VT 05676

Virginia

Habitat Management Division
Marine Resources Commission
P.O. Box 756
Newport News, VA 23607
(804) 247-2200

Washington

Wetlands Management and Planning Section
Department of Ecology
State of Washington
P.O. Box 47600
Olympia, WA 98504-7600
(206) 459-6790

West Virginia

Wildlife Division
Department of Natural Resources
Charleston, WV 26305
(304) 348-2771

Wisconsin

Water Regulation Section
Bureau of Water Regulation and Zoning
Department of Natural Resources
P.O. Box 7921
Madison, WI 53707
(608) 266-7360

Wyoming

Water Quality Division
Department of Environmental Quality
Herschler Building 4W
122 West 25th Street
Cheyenne, WY 82002
(307) 777-7081

List of Abbreviations and Acronyms

1987 Manual	1987 Corps of Engineers Wetlands Delineation Manual
1989 Manual	1989 Federal Manual for Identifying and Delineating Jurisdictional Wetlands
ADID	Advance Identification of Wetlands
ANOVA	Analysis of Variance
APA	Administrative Procedures Act
AWQC	Ambient Water Quality Criteria
BAT	Best Available Technology
BCF	Bioconcentration Factor
BLM	Bureau of Land Management
BMP	Best Management Practices
BOD	Biological Oxygen Demand
CEQ	Council on Environmental Quality
CERCLA	Comprehensive Environmental Response, Compensation and Liability Act
CMP	Coastal Zone Management Program
COA	Correspondence Analysis
COC	Contaminant of Concern
Corps	U.S. Army Corps of Engineers
CWA	Clean Water Act
CZMA	Coastal Zone Management Act
DCA	Detrended Correspondence Analysis
DOI	U.S. Department of the Interior
DOJ	U.S. Department of Justice
EA	Environmental Assessment
EIS	Environmental Impact Statement

EO	Executive Order
EPA	U.S. Environmental Protection Agency
ERA	Ecological Risk Assessment
ESA	Endangered Species Act
FAC	Facultative
FACU	Facultative Upland
FACW	Facultative Wetland
FEMA	Federal Emergency Management Agency
FmHA	Farmer's Home Administration
FONSI	Finding of No Significant Impact
FSA	Food Security Act of 1985
FWCA	Fish and Wildlife Coordination Act
FWS	U.S. Fish and Wildlife Service
GAO	U.S. General Accounting Office
HEP	Habitat Evaluation Procedure
HU	Habitat Unit
MMPA	Marine Mammal Protection Act
MOA	Memorandum of Agreement
MOU	Memorandum of Understanding
NAS	National Academy of Sciences
NEPA	National Environmental Policy Act
NHPA	National Historic Preservation Act
NMFS	National Marine Fisheries Service
NOAA	National Oceanographic and Atmospheric Administration
NOAEL	No Observed Adverse Effect Level
NPDES	National Pollutant Discharge Elimination System
NPS	National Park Service
NTCHS	National Technical Committee for Hydric Soils
NWI	National Wetlands Inventory
OBL	Obligate Wetland
OTA	Office of Technology Assessment
PCA	Principle Components Analysis
QA	Quality Assurance
RA	Regional Administrator
RCRA	Resource Conservation and Recovery Act

RGL	Regulatory Guidance Letter
RHA	Rivers and Harbors Act
ROD	Record of Decision
SAMP	Special Area Management Plan
SARA	Superfund Amendments and Reauthorization Act of 1986
SCS	Soil Conservation Service
SOF	Statement of Findings
UPGMA	Unweighted Pair-Group Using Arithmetic Means
UPL	Obligate Upland
USDA	U.S. Department of Agriculture
WET	Wetland Evaluation Technique

Index to Scientific and Common Names

Abies, 87
  amabilis, 202
  concolor, 202
  lasiocarpa, 202
Acer, 81
  negundo, 97
  rubrum, 81, 87
  saccharinum, 97
Acris, 57
Acrostichum, 144
Actitis macularia, 107
Agelaius phoeniceus, 107, 141
Agkistrodon, 57
  piscivorus, 84, 114
Aix sponsa, 84
Ajaia ajaja, 145
Alces, 57
Alder, 76, 92, 93, 189
  speckled, 87, 97
Algae, 51, 81, 113, 131, 139
  blue-green, 60, 135, 138-139
  green, 139
Alkali weed, spreading, 138
Allenrolfea occidentalis, 104
Alligator, 94, 107, 111, 114, 126, 141
  mississippiensis, 57, 84, 141

Alnus, 76
  rugosa, 87, 97
Amaranthus cannabinus, 138
Ambystoma, 57
  cingulatum, 56
  tigrinum, 56, 118
Amia calva, 83
Ammocrypta, 125
Ammodramus maritimus, 141
Amphipod, 83, 117, 140, 145
Amphiuma, 57
Anas, 57, 118
  acuta, 107, 108, 118
  affinis, 118
  americana, 118
  collaris, 118
  crecca, 118
  discors, 107, 118
  fulvigula, 114
  platyrhynchos, 107, 118
  strepera, 118
Andromeda, 76
Andropogon, 81
  virginicus, 104
Anemone, 145
Anopheles, 56, 125

Anser, 57
  albifrons, 108
Anthrocnemum subterminale, 37
Aramus guarauna, 84
Arctophila fulva, 103
Ardea, 57
  herodias, 114
Arethusa bulbosa, 121
Aristida stricta, 104
Arrowhead, 103, 104, 116, 138
Arrowleaf, 113
Arrowwood, 87
  southern, 165, 166
Arum, arrow, 103
Ash, 76, 81, 87
  black, 88, 97
  carolina, 97
  green, 97, 165, 166
  white, 97, 163
Asimina triloba, 163
Asio flammeus, 123
Aster, 138
Atherinidae, 140
Auriparus flaviceps, 99
Avens, white, 163
Avicennia, 130
  germinans, 143
Avocet, 57
Aythya, 57, 118
  collaris, 107
  valisineria, 118
Azolla, 81

Baccharis, 134
Bacopa, 104
Balanus, 37, 145

Barnacle, 37, 145
Bass, 127
  largemouth, 56
  smallmouth, 56
  striped, 56
Bay, 76
  loblolly, 87, 92
  swamp, 81, 87
  sweet, 81, 87, 92
Beakrush, 59, 103, 104
Bear, 94
  black, 84, 89, 145
  polar, 261
Beaver, 57, 89, 99, 107, 118, 126
Beech, 87
  American, 163
Betula, 76
  nigra, 97
  pumila, 121
Bidens, 104
Birch, 76, 93
  bog, 121
  river, 97
  yellow, 88
Bittern, American, 107, 118
  least, 107
Bivalve, 145
Blackbird, red-winged, 107, 118, 141
  yellow-headed, 107, 118
Blackfly, 107
Bladderwort, 104, 112, 116, 121
Blechnum serrulatum, 81
Blueberry, 92, 121
  highbush, 165, 166
Bluejoint, 103

Boar, wild, 84
Bobcat, 94
Botaurus lentiginosus, 107
Bowfin, 83
Boxelder, 97
Branta, 57
   canadensis, 108, 118
Brevoortia, 140
Bromeliad, 81
Bubulcus ibis, 141
Bucephala, 57
Bufo, 57, 318
Bullhead, 83
Bulrush, 62, 103-104, 116
   saltmarsh, 138
   soft-stemmed, 138
Bunchberry, 163
Burreed, 104, 116
Buteo, 57
Butorides virescens, 84
Butterfly, monarch, 40
   viceroy, 40
Buttonbush, 81, 87, 92-93, 97, 113
Buttonwood, 143-144

Cabomba caroliniana, 104
Caddisfly, 125
Caenis, 125
Cajeput, 80 (*See* Paperbark)
Calamagrostis canadensis, 103
Calluna vulgaris, 121
Calopogon tuberosus, 121
Cambarus, 89
Campephilus principalis, 84
Canvasback, 118

Carex, 103, 116
   limosa, 121
   oligosperma, 121
   pauciflora, 121
Carpinus caroliniana, 97
Carya aquatica, 97
Castor, 57
   canadensis, 89
Cataptrophorus semipalmatus, 118
Catbird, grey, 99
Catfish, 56
Catharus fuscens, 99
Cattail, 24, 102-104, 113, 116, 138
   narrow-leaved, 138
Cedar, 76, 87, 94
   Atlantic white, 92
   northern white, 87, 97
   white, 87-88, 189
Celtis, 97
   laevigata, 81
Centropomus undecimalis, 145
Cephalanthus occidentalis, 81
Cetacea, 261
Chamaecyparis, 76
   thyoides, 87
Chamaedaphne, 76
   calyculata, 121
Chara, 57, 105, 124
Charadrius vociferus, 107
Chelydra, 57
Chen, 57
Cherry, black, 97
Chilonius niger, 118
Chironomid, 89, 117
Chironomus, 125
Chlidonius niger, 107

Chrysemys, 57
Chthamalus, 37
Cistothornus palustris, 141
Cladium jamaicense, 103
Cladocera, 117, 127
Cladopodiella fruitans, 121
Clam, 125
Claytonia virginica, 163
Clemmys, 57
Cliftonia monophylla, 88
Conocarpus erecta, 143
Contopus virens, 99
Conuropsis carolinensis, 84
Coot, 57, 118
  American, 108
Copepod, 117, 127, 145
Cordgrass, 130
  big, 138
  California, 138
  salt-meadow, 44
  saltwater, 44
  smooth, 129, 137-138, 142, 144
Cormorant, 145
Cornus, 92
  canadensis, 163
  stolonifera, 87
Cottongrass, 103
Cottonmouth, 84
Cottonwood, 97, 98
Coyote, 118
Crab, 145
  fiddler, 145, 306, 318
Cranberry, 121, 191
Crane, 57
  sandhill, 108, 111, 123
  whooping, 108

Crassostrea, 145
Crayfish, 83, 89, 107, 114, 125, 314, 317
Creeper, Virginia, 163
Cressa truxillensis, 138
Croaker, 56
Crocodile, American, 57, 140, 145
Crocodylus acutus, 57, 141, 145
Crotalus adamanteus, 94
Crowberry, 121
Crustacean, 125
  amphipod, 36
Ctenium aromaticum, 104
Culaea inconstans, 117
Cuscuta salina, 138
Cutgrass, 103-104
Cypress, 24, 38, 76, 78, 80-85, 87, 92, 94, 97
  bald, 77, 79
  dwarf, 76, 81
  pond, 79, 80
Cyprinid, 114
Cyprinodontidae, 140
Cyrilla racemiflora, 88, 92

Daphnea pulex, 32
Decapod, 140, 145
Deer, 84, 89, 94, 118
  white-tailed, 84, 114
Dendroica palmarum, 123
  pinus, 99
Deschampsia caespitosa, 139
Diatom, 47, 132, 135, 139
Diospyros virginiana, 163
Distichlis, 134
  spicata, 104, 137-138, 273

Dodder, salt, 138
Dogwood, 92
  red-osier, 87, 93
  silky, 93
Dorosoma cepedianum, 127
Dove, mourning, 99, 108
Dragonfly, 114
Drepanocladus, 121
Drosera intermedia, 121
  rotundifolia, 121
Drum, 56
Drymarchon corais, 145
Dryocopus lineatus, 84
Duck, 57, 118, 141
  mottled, 114
  redhead, 118
  ring-necked, 107, 118
  ruddy, 107, 118
  wood, 84, 99, 118
Duckweed, 81
Dumetella carolinensis, 99

Eagle, 57
  bald, 145, 312
Egret, 57, 140, 145
  cattle, 141
  great, 114
  snowy, 114
Egretta, 57
  alba, 114
  caerulea, 114
  thula, 114
  tricolor, 114
Eichornia crassipes, 81
Elanoides, 57
Elder, 93

Eleocharis, 103
Elm, 93, 97
  american, 87-88, 97
Empetrum, 121
Ephemerella, 125
Eretmochelys imbricata, 145
Erianthus giganteus, 113
Eriocaulon, 59
  compressum, 81
Eriophorum angustifolium, 103, 121
  spissum, 103
  vaginatum, 121
Etheostoma, 125
Eudocimus albus, 114, 145
Eurycea, 57

Fagus grandifolia, 87, 163
Falco peregrinus, 145
Falcon, American peregrine, 145
Fanwort, 104
Felis concolor, 84
  concolor coryi, 145
Fern, 81
  cinnamon, 165
  mangrove, 144
  marsh, 103
  resurrection, 81
  water, 81
Fetterbush, 81, 87, 92
Fir, 87
  balsam, 88, 189
  Pacific silver, 202
  subalpine, 202
  white, 202
Fire flag, 104

Fish, salmonid, 125
  sciaenid, 140
Flamingo, 57
Flatworm, 125, 145
Flea, water, 32
Fly, 83, 123, 306
  deer, 114, 125
  horse, 125
  psyllid, 123
  tipulid, 123
Forestiera acuminata, 93
Fox, 118
Foxtail, 113
Fraxinus, 76
  americana, 97, 163
  caroliniana, 81, 87, 97
  nigra, 87, 97
  pennsylvanica, 97
  profunda, 80
Frog, 57, 318
  chorus, 118
  grass, 40
  leopard, 118
Fulica, 57
  americana, 108

Gadwall, 118
Gallinula, 57
  chloropus, 108
Gallinule, 57
  common, 108
Gambusia affinis, 83, 114
Gar, 83
Gaultheria procumbens, 163
Gaylussacia, 92
Geothlypis trichas, 99

Geum canadense, 163
Glasswort, 138
  common, 138
  Parsh's, 37
  perennial, 138
Gnat, 114
Gnat-catcher, blue-gray, 84
Godwit, marbled, 118
Goldenrod, spring flower, 93
Goose, 57, 108, 141
  Canada, 118
Gordonia lasianthus, 87
Grass, 81, 87, 102-104, 116, 124, 134, 138, 140
  alkali, 137
  black, 44, 137
  cotton, 121
  panic, 81
  pendant, 103
  reed canary, 103
  salt, 137-138
  salt meadow, 273
  spike, 273
  torpedo, 110
  white, 165-166
Grebe, 118
  pied-billed, 108
Greenbriar, 92
Grus, 57
  americana, 108
  canadensis, 108, 111
Gull, 57, 141
Gum, 76
  black, 80, 87, 97
  sweet, 87, 97, 165

Habenaria blephariglottis, 121
   clavellata, 121
Hackberry, 81, 97
Hairgrass, tufted, 138
Haliaetus, 57
   leucocephalus, 145
Hare, 118, 123
Haw, black, 163
Hawk, 57
Heath, 121
Heather, 121
Hemlock, 162, 189
   eastern, 163, 202
   western, 202
Hemp, salt marsh, 138
Heron, 40, 57, 84, 89, 108, 111, 140, 145
   great blue, 114
   green-backed, 84
   little blue, 114
   tricolored, 114
Herring, 49
Hibiscus palustris, 138
Hickory, water, 97
High-tide bush, 138
Hog, 84
Holly, 81, 87, 92, 113
   American, 163, 165
Honeysuckle, Japanese, 163
Hornbeam, American, 97
Huckleberry, 92
Hyacinth, water, 81
Hydrobia, 36
Hyla, 57
   andersoni, 94
Hyssop, water, 104

Ibis, 84, 99, 108, 111
   white, 114, 145
Ictalurus, 83
Icterus galbula, 99
Ilex, 87
   cassine, 81
   coriacea, 92
   glabra, 81, 92
   opaca, 163
   verticillata, 87
Insect, 140
Iodine bush, 104
Isopod, 83, 117, 140
Itea virginica, 81
Iva, 134
   frutescens, 138
Ixobrychus exilis, 107

Juncus, 103, 134
   acutus, 138
   balticus, 137
   gerardi, 44, 137
   roemerianus, 138

Kalmia, 76
   cuneata, 93
   politifolia, 121
Killdeer, 107, 118
Kite, 57
   snail, 114

Lachnanthes caroliniana, 81, 104
Lagodon, 140
Laguncularia racemosa, 143
Larch, 189

Larix, 76
  laricina, 87, 97
Laurel, 92, 121
Lavender, California sea, 138
Leatherleaf, 121, 189
Ledum, 76
  groenlandicum, 121
  palustre, 121
Leech, 125
Leersia oryzoides, 103
Lemming, 123
Lepidochelys kempi, 145
Lepisosteus, 83
Lepomis, 114, 127
Lettuce, water, 81
Lily, water, 110, 112, 116, 124
Limnodrilus, 125
Limonium californicum, 138
  nashii, 138
Limosa fedoa, 118
Limpet, 40
Limpkin, 84
Lindera benzoin, 87
Liquidambar styraciflua, 87
Liriodendron tulipifera, 163
Littorina, 40
Liverwort, 121
Lizard tail, 81
Lobster, spiny, 145
Lonicera japonica, 163
Loosestrife, rough-leaf, 93
Lopodytes cucullatus, 118
Ludwigia repens, 105
Lutjanus apodus, 145
Lutra canadensis, 145
Lymnaea, 125

Lynx rufus, 94
Lyonia lucida, 81, 87, 92
  mariana, 163
Lysimachia asperulaefolia, 93

Magnolia virginiana, 81
Maianthemum canadense, 163
Maidencane, 103-104, 110, 113
Malaclemys terrapin, 140
Mallard, 107, 118
Mammal, marine, 261
Manatee, West Indian, 141, 145, 263
Mangrove, 24, 26, 45, 68, 130-131, 138, 142-145, 152, 191
  black, 143
  red, 143
  white, 143
Maple, 81
  red, 81, 87-88, 92, 97, 165-166, 189
  silver, 97
Mayflower, Canada, 163, 165-166
Mayfly, 114, 125
Megalops atlanticus, 145
Melaleuca, 38, 45
  quinquenervia, 80
Melanerpes carolinus, 99
Meleagris gallopavo, 99
Melospiza georgiana, 107, 123
  melodia, 107
Menhaden, 56, 140
Mephitis mephitis, 145
Merganser, hooded, 118
Mergus, 57
Mesquite, screwbean, 98

Micropterus, 127
Microtus, 141
Midge, 89, 107, 125
Mink, 84, 114, 118, 145
Minnow, 114
  fathead, 117
Mitchella repens, 163
Modiolus, 125
Mollusk, gastropod, 140
Moose, 57, 123
Mosquito, 56, 107, 114, 125
Mosquitofish, 83, 114
Moss, 119, 121
  peat, 92, 120
  spanish, 81
Mouse, 118
Mugil cephalus, 145
Mugilidae, 140
Muhlenbergia fillipes, 113
Muhly, 113
Mullet, 57, 140, 145
Muskellunge, 56
Muskrat, 107, 118, 141
  roundtail, 114
Mussel, 125
Mustela vison, 84, 114, 145
Mycteria americana, 84, 145
Myocastor, 57
  coypus, 141
Myrica cerifera, 81
Myriophyllum, 124
Mytilus, 125

Najas, 112
Needlerush, black, 138, 142, 144
Neofiber alleni, 114, 141

Nephrolepsis exaltata, 81
Nerodia, 57, 83, 114
  fasciata, 140
  fasciata taeniata, 145
Nightshade, bittersweet, 163
Nitrobacter, 53
Nitrosomonas, 53
Nuphar luteum, 112
Nutria, 57, 141
Nycticorax, 57
Nymphaea, 57, 59, 104, 116
Nyssa, 76
  aquatica, 80
  sylvatica var. *biflora*, 80

Oak, bur, 97
  laurel, 87, 97
  northern pin, 87
  overcup, 87, 97
  white, 87, 163
Odocoilus virginiana, 84
Oligochaete, 57, 83, 127
Ondatra, 107
  zibethicus, 118, 141
Orchid, 82, 121
Orioles, 99
Oryzmyzes, 107
Oryzomys, 141
Osmunda cinnamomea, 81
Osprey, 57
Ostracod, 127
Ostrea, 145
Otter, 99, 107
  river, 145
  sea, 261
Ovenbird, 99

Owl, barred, 89, 99
  great gray, 123
  short eared, 123
Oxyura jamaicensis, 107, 118
Oyster, 145

Palaemonetes, 114
Palila, 260
Panaeus duorarum, 145
Pandion, 57
Panicum, 81, 124
  hemitomon, 103
  repens, 110
  virgatum, 138
Panther, 84
  Florida, 114, 145, 263
Panulirus argus, 145
Paperbark, 38, 80
Parakeet, Carolina, 84
Paramecium, 36
Parsnip, water, 138
Parthenocissus quinquefolia, 163
Partridgeberry, 163
Pawpaw, 163
Peewee, wood, 99
Pelican, 145
Pelicanus, 145
Peltandra sagittaefolia, 93
  virginicum, 103
Pepperbush, sweet, 165-166
Perch, yellow, 56
Percina, 125
Persea, 76
  borbonia, 92
  palustris, 81, 87
Persimmon, 163

Phalacrocorax, 145
Phalaris arundinacea, 103
Phalarope, Wilson's, 118
Phalaropus tricolor, 118
Phragmites, 316
  australis, 102
Physa, 125
Phytoplankton, 49, 126, 132-133, 147
Picea, 76
  engelmannii, 202
  glauca, 97, 202
  mariana, 121
  rubens, 163, 202
  sitchensis, 202
Pickerelweed, 103-104, 110, 113, 138
Pickleweed, 37
Picoides pubescens, 99
Pike, northern, 56
Pimephales promelas, 117
Pine, 77, 81, 210
  eastern white, 202
  jack, 202
  loblolly, 87, 163, 165
  lodgepole, 202
  longleaf, 31, 88, 163, 201-202
  pond, 88, 92, 94
  ponderosa, 202
  scotch, 121
  slash, 80, 87
  white, 163, 189
Pinfish, 140
Pinnipedia, 261
Pintail, 107-108, 118

Index to Scientific and Common Names 419

Pinus, 77
  banksiana, 202
  contorta, 202
  elliottii, 80
  palustris, 202
  ponderosa, 202
  rigida, 31, 88, 163, 201-202
  serotina, 88
  strobus, 163, 202
  sylvestris, 121
  taeda, 87, 163
Pipewort, 59, 81
Pisaster, 39
Pisidium, 125
Pistia, 81
Planera aquatica, 97
Planertree, 97
Platanus occidentalis, 97
Podilymbus podiceps, 108
Poeciliidae, 140
Pogonia ophioglossides, 121
Polioptila caerulea, 84
Polygonium, 103
Polypodium polypodies, 81
Pomacea paludosa, 114
Pondweed, 116
Pontedaria cordata, 103, 138
Pophyrula, 57
Poplar, 87, 92
  balsam, 97
  tulip, 163
Populus balsamifera, 87, 97
  deltoides, 97
  fremontii, 98
Porzana carolina, 107
Potamogeton, 57, 124

Prawn, 107, 114
Privet, swamp, 93
Procambarus, 89, 318
  alleni, 84, 114, 145
Prosopis pubescens, 98
Protonotaria citrea, 84
Prunus serotina, 97
Pseudacris, 57
  woodhousei, 118
Pseudemys, 57, 114
Puccinellia, 137
Pumpkin ask, 80

Quercus alba, 163
  bicolor, 87
  laurifolia, 87, 97
  lyrata, 87, 97
  macrocarpa, 97
  palustris, 87

Rabbit, marsh, 94
  swamp, 99
Raccoon, 84, 89, 99, 107, 145, 314
Rail, 118
  clapper, 141
  Virginia, 107
Rallus limicola, 107
  longirostris, 141
Rana, 57, 318
  pipiens, 118
Rat, 141
  rice, 107
Rattlesnake, 94
Redroot, 81, 104
Redbay, 92

Reed, 104
  common, 102, 103
Rhizophora, 130
  mangle, 143
Rhynchospora, 59
  alba, 103, 121
Rice, wild, 102
Rosa multiflora, 163
  palustris, 81
Rose, multiflora, 163
  swamp, 81
Rose-mallow, 138
Rostrhamus, 57
  sociabilis, 114
Rumex, 104
Rush, 62, 103
  baltic, 137
  black, 113
  spiny, 138

Sacaton, 104
Sagittaria, 103, 138
Salamander, tiger, 118
Salicornia, 104, 138
  subterminalis, 138
  virginica, 37, 138
Salix, 87
  caroliniana, 81, 113
  nigra, 34, 81
Salmon, 56
Salt marsh hay, 137
Saltwort, 104
Sandpiper, spotted, 107
Sarracenia purpurea, 121
Saururus cernuus, 81

Sawgrass, 45, 103-104, 110, 112-113
Scaup, lesser, 118
Schoenus nigricans, 113
Scirpus, 103
  acutus, 116
  cespitosus, 103
  fluviatilis, 116
  robustus, 138
  validus, 116, 138
Sciurus carolinensis, 94
  niger, 84
Sea blite, 104
  California, 138
Sea urchin, 145
Sedge, 102-104, 116, 121
Seiurus aurocapillus, 99
  noveboracensis, 99, 123
Shad, gizzard, 127
Shieldwort, arrowleaf, 93
Shrew, 118
Shrimp, pink, 145
Silverside, 140
Sirenia, 261
Sium suave, 138
Skunk, striped, 145
Smartweed, 103, 110
Smilax, 40
  laurifolia, 92
Snail, 107, 117, 123, 125, 145
  apple, 114
  mud, 36
  periwinkle, 40
Snail darter, 259

Snake, 107, 111
  Atlantic salt marsh, 145
  indigo, 145
  salt marsh, 140
  water, 40, 57, 83, 114
Snapper, mangrove, 145
Snook, 145
Solanum dulcamara, 163
Solidago verna, 93
Sora, 107, 123
Sparganium, 104
  eurycarpum, 116
Sparrow, savannah, 118
  seaside, 141
  song, 107
  swamp, 107, 123
Spartina, 26, 130, 134, 138, 140
  alterniflora, 44, 129-130, 137-138, 141
  cynosuroides 138
  foliosa, 138
  patens, 44, 137-138, 273
Spatterdock, 112
Sphaerium, 125
Sphagnum, 92, 103, 119-123
  capillifilium, 120
  cuspidatum, 120
  magellanicum, 120
  major, 120
  papillosum, 120
Spicebush, 87
Spider, 140
Spikerush, 103-104, 113
Spirodela, 124
  polyrhiza, 81
Sponge, 145

  freshwater, 35
Spoonbill, roseate, 145
Sporobolus airoides, 104
Spring beauty, 163
Spruce, 76, 205
  black, 121, 189
  Engelmann, 202
  red, 163, 202
  sitka, 202
  white, 88, 97, 202
Squirrel, 84, 89
  fox, 84
  gray, 94
Staggerbush, 163
Starfish, 39, 145
Stickleback, brook, 117
Stizostedion lucioperca, 127
Stonewort, 116
Stork, 108
  wood, 84, 114, 145
Strix nebulosa, 123
  varia, 89, 99
Suaeda, 104
  californica, 138
Sunfish, 114, 127
Sus scrofa, 84
Swallow, 141
Switchgrass, 138
Sycamore, American, 97
Sylvilagus floridanus, 94

Tabanus, 125
Tamarack, 76, 87, 88, 97, 121, 189
Tamarisk, 98, 124
Tamarix pentandra, 98
Taricha, 57

Tarpon, 145
Taxodium, 76
  ascendens, 79
  distichum, 77
Tea, Labrador, 121
Teal, blue-winged, 107, 118
  green-winged, 118
Telmatodytes palustris, 107
Tern, 57
  black, 107, 118
Terrapin, diamondback, 140
Thalia genticulata, 104
Thelypteris kunthii, 81
  palustris, 103
Thrush, wood, 99
Thuja occidentalis, 87, 97
Tillandsia, 81
  usneoides, 81
Titi, 88, 92
  black, 88
Toad, 40, 57
Toadstool, 22
Topminnow, 140
Tree frog, pine barrens, 94
Trichechus manatus, 34, 56, 125, 141, 145
Tsuga canadensis, 163, 202
  heterophylla, 202
Tubifex, 127, 312
Tule, 103
Tunicate, 145
Tupelo, 81
  water, 80, 87, 97
Turkey, 99
Turtle, 35, 107, 111, 114
  alligator snapping, 40

Atlantic ridley sea, 145
freshwater, 37, 40
hawksbill, 145
painted, 57
pond, 57
snapping, 57
Typha, 57, 138
  angustifolia, 103, 138
  latifolia, 102

Uca, 145, 318
Ulmus americana, 87
Ursus americanus, 84, 145
Utricularia, 104, 113
  cornuta, 121

Vaccinium, 76
Veery, 99
Verdin, 99
Viburnum dentatum, 87
  prunifolium, 163

Walleye, 56, 127
Walnut, 38
Warbler, Canada, 99
  palm, 123
  pine, 99
  prothonotary, 84
Water lily, 59, 104
Water moccasin, 57
Waterthrush, northern, 99, 123
Wax myrtle, 81
Weasel, 36, 118, 123
Wicky, white, 93
Willet, 118

Willow, 81, 87, 91-93, 97, 113, 121, 189
  primrose, 104
  virginia, 81
Wilsonia canadensis, 99
Winterberry, 87
Wintergreen, 163
Wiregrass, 138
Wolffia, 124
Woodpecker, downy, 99
  ivory-billed, 84
  pileated, 84, 89
  red-bellied, 99
  red-cockaded, 263
Woodwardia virginica, 81
Worm, 125
  annelid, 145
  polychaete, 140
Wren, long-billed marsh, 107
  marsh, 118, 141

Xanthocephalus xanthocephalus, 107

Yellowthroat, common, 99

Zenaida macroura, 108
Zizania aquatica, 102

Index

ADVANCE IDENTIFICATION
    OF WETLANDS (ADID)
  Generally, 262, 303
  ADID Process, 262
  Rookery Bay ADID, 263

BALD CYPRESS TREES, 79
BIOCOENOSIS, 19
BIOGEOCHEMICAL CYCLE
  Carbon Cycle, 53
  Hydrologic Cycle, 54
  Nitrogen Cycle, 52
  Phosphorus cycle, 53
  Wetlands and, 52
BIOMES, 20
BIOSPHERE, 21
BIOTIC DIVERSITY, 19
BOGS
  Birds, 123
  Dominant plants, 121
  Fauna, 122
  Generally, *See* Fig. 3.12, 119
  Geophysical Characteristics, 120
  Hydrology, 122
  Mammals, 123
  Peat Moss, 120
  Quaking Bogs, 120
  Soils, 121
  Sphagnum, 120
  Unusual Characteristics, 123
  Vegetation, 120
  Wetland Criteria Characteristics, 120
  Wetland Values, 123

CIRCULAR 39 CLASSIFICATION, 6
CLEAN WATER ACT, 256
  National Pollutant Discharge
    Elimination System (NPDES)
    Permit, 256
  Non-Point Source Pollution, 256
  See also Section 404 Program
CLIMAX
  Disclimax, Formation of, 44
  Ecological Succession and, 43
  Everglades, Disclimax of 45
  Fire Disclimax, 45
  Superorganism Concept and, 44
COASTAL ZONE MANAGEMENT
    ACT, 71, 217, 257
COMMUNITY, 21
  Continuum Concept, 21
  Measurement of Population
    Dynamics, 40
    Field Data Collection, 41

      Life History Parameters, 41
      Life tables, *See* Table 2.1, 41
      Life Tables, Ecosystem
         Management and, 41
      Mortality Curves, 41
      Survivorship Curves, 41
  Population structure of, 21, 31
  Species Richness, 31
COMPETITION
  Allelopathy, 38
  Animal adaptations
      Armor, 40
      Camouflage, 40
      Chemical Defenses, 40
      Escape Mechanisms, 40
      Mimicry, 40
      Schooling Behavior, 40
  Between paperbark tree and natural cypress, 38
  Coevolution, 40
  Competitive exclusion, 36
  Coupled Oscillation Hypothesis, 39
  Exploitation Concept, 33
  Interference Concept, 33
  Interspecific competition, 33, 36
  Intraspecific competition, 33
  K-selected species, 34
  Logistic Growth Equation, 33
  Lotka-Volterra Competition Models, *See* Fig. 2.4, 33, 38
  Morisita's Index of Similarity, 37
  Niche Overlap and, 37
  Niche Partitioning, 36
  Plant Adaptations
      Chemical Defenses, 40
      Structural Defenses, 40
  Predator-Prey Relationships, 38
  R-selected Species, 36
  Resource Overlap in Freshwater
      Turtles, 37
  Wetland Communities and, 37
COMPREHENSIVE ENVIRONMENTAL RESPONSE, COMPENSATION, AND LIABILITY ACT, 304
CONSERVATION FOUNDATION, 9
CONSTRUCTED WETLAND
  Sewage Treatment and, 62
CORPS OF ENGINEERS WETLANDS DELINEATION MANUAL, *See* WETLAND IDENTIFICATION AND DELINEATION
COUNCIL ON COMPETITIVENESS, 9
COUPLED OSCILLATION HYPOTHESIS, 39
COWARDIN CLASSIFICATION, *See* Figs. 1.1, 5.6, 6, 190,
CYBERNETICS, 20
CYPRESS SWAMPS
  Bald Cypress Trees, 79
  Birds, 84
  Cypress Domes, 77, 84
  Cypress Riverine Swamps, 79
  Cypress Strands, 78
  Fringing Cypress Swamps, 79
  Generally, 77
  Geophysical Characteristics, 77
  Groundwater Recharge and, 62
  Mammals, 84
  Okefenokee Swamp, *See* Fig. 3.1, 78
  Tupelo Trees, 80
  Vegetation
      Black Gum Tree, 80
      Paperbark, 80
      Pumpkin Ash, 80

Slash Pine, 80
Understory, 81
Wetland Criteria Characteristics
 Fauna, 83
 Hydrology, 82
 Soils, 82
 Unusual Characteristics, 85
 Vegetation, 79
Wetland Values, 84

DEEPWATER HABITATS, 123
 Generally, 159
 Ecological Characteristics, 124
 Fauna, 125
 Geophysical Characteristics, 124
 Hydrology, 125
 Rivers and Streams, 123
 Soils, 125
 Typical plants, 124
DENITRIFICATION, 53
DREDGE AND FILL PERMITS
 Administrative Appeals Mechanism,
  Lack of, 243
 Application Form, 228
 Application Information, 229
 Application Process, 228
 Challenges to Issuance of Permits,
  248
 Conflict Resolution, 234
 Corps Responses to Permit
  Applications, 234
 Cumulative Impacts, 231
 Dredged material, defined, 214
 Duration of, 238
 EPA Veto Authority, 241
 Exemptions, 240
 Fill material, defined, 215
 Form and Conduct of Hearings, 236
 General Permits, 238
 Issuance of Permit, 237
 Nationwide Permits, 238
 Nationwide Permit 26, 240
 New Jersey State Coastal Wetlands
  Regulation, 270
 New York State Tidal Wetland
  Regulation, 269
 Notice and Comment, 232
 Notice of Hearing, 236
 Other Agencies, Role of, 238
 Permit Denials, Remedies for, 243
 Project Changes After Permit
  Approval, 231
 Public Hearings, 235
 Public Interest Review, 237
 State and Local Wetland Regulations,
  267
 Who Must Apply? 228
DUCKS UNLIMITED, 73

ECOLOGICAL RISK ASSESSMENT
 Generally, 304
 Assessment Endpoints, 306
 Bioconcentration Factor, 312
 CERCLA sites, 305
 Contaminants of Concern (COC), 311
 Data Limitations, 312
 Defined, 305
 East Fork Poplar Creek Case Study,
  316
 Ecological Resources, Risks to, 304
 Ecological Risk Assessors, 305
 Effects Assessment, 313
 Exposure Assessment, 311
 Framework for, 308
 Goals of, 306
 Lead Agency, Role of, 305
 Natural Resource Survey, 305

Natural Resource Trustees, Role of, 305
Phases of, *See* Fig. 9.1, 309
Problem Formulation, 308
Quality Assurance Plans, 307
Quotient Methods, 311, 315
Risk Characterization, 314
Risk to Biota, Assessment of, 306
Site Remediation, Risks from, 305
Social and Political Concerns, Role of, 307
Uncertainty in Characterization of Risk, 306
Wetlands and, 316
Wetland ERAs, Special Problems with, 316-17

ECOLOGICAL SUCCESSION
Autogenic and Allogenic Succession, 45
Climax and, 43
Facilitation Concept, 46
Food Web, 49
Generally, 21, 43
Horizontal Succession and Wetlands, *See* Fig. 2.5, 47
Parameters of, 43
Plant Communities and, 43
Primary Succession, Rate of, 45
Wetlands and, 48
Wetland Food Pyramids, 49
Wetland Succession, 44

ECOLOGIST
Defined, 19
Evolutionary ecologists, 19

ECOLOGY
Defined, 18
Competition Concept, 33

ECOSYSTEM
Generally, 19
Abiotic Components of, 20
Autochthonous Matter, 51
Biogeochemical Cycles, 52
Biomass Production by Plants, 51
Biotic Components of, 20
Ecological Efficiency of, 50
Energy Production and, 50
Fundamental Niche Concept, 23
Hydraulic Model, 51
Limits of, 20
Niche Theory, 23
Primary Productivity, 51
Production, Rate of, 50
Realized Niche Concept, 23
Scope, Problem of, 20
Wetland Productivity, 51

ELTON, CHARLES, 48

EMERGENCY WETLANDS RESOURCES ACT, 9, 72

ENDANGERED SPECIES ACT
Critical habitat, 259
Generally, 16, 57, 71, 258
Habitat Conservation Plans, 302
Palila v. Hawaii Department of Land & Natural Resources, 260
Takings of Endangered Species, 259
Takings of Plants, 260
Tennessee Valley Authority v. Hill, 259

ENERGY AND WATER DEVELOPMENT APPROPRIATIONS ACT, 158

ESTUARINE WETLANDS
Absorption of nutrients and toxins, 131
Categories of, 129
Defined, 128

428 Wetlands

Flood Damage Control and, 130
Food Webs, 129
Functions and values, 130
Human Activities, Influence of, 129
Losses of, 150
Mangrove Swamps, 142
Nitrogen Cycle in, 133
Productivity of, 129, 131
Salinity, 129
Salt Marshes, 134
Sea Level Rise, Losses from, 151
Shoreline Erosion and, 130
    See also MANGROVE
    SWAMPS, SALT MARSHES
EUTROPHICATION, 62
EVERGLADES, 111
Birds, 114
Dominant Plants, 113
Everglades National Park, 115
Fauna, 114
Generally, *See* Fig. 3.10, 111-12
Geophysical Characteristics, 111
Hydrology, 113
Lake Okeechobee, 113
Soils, 113
Everglades, Unusual Characteristics, 115
Vegetation, 112
Wetland Criteria Characteristics, 112
Wetland Values, 114

FARMED WETLANDS, 159
FEDERAL MANUAL FOR
    IDENTIFYING AND
    DELINEATING
    JURISDICTIONAL
    WETLANDS,
    See WETLAND IDENTIFICATION
    AND DELINEATION
FEDERAL WATER POLLUTION
    CONTROL ACT, *See*
    CLEAN WATER ACT, 213
FEN, *See* Fig. 3.8, 104, 106
FISH AND WILDLIFE
    COORDINATION ACT,
Generally, 217
Wetland Mitigation Under, 281
FLOODPLAIN FORESTS
Beavers, 99
Birds, 99
Characteristic Fauna, 99
Geophysical Characteristics, 95
Hydrology, 98
Soils, 98
Trees, 97
Unusual Characteristics, 100
Vegetation, 96
Wetland Criteria Characteristics, 96
Wetland Values, 99
FLOODPLAIN MARSH
Fauna, 110
Generally, See Fig. 3.9, 108-09
Geophysical Characteristics, 109
Hydrology, 110
Soils, 110
Unusual Characteristics, 111
Vegetation, 110
Wet Prairies, 110
Wetland Criteria Characteristics, 110
Wetland Values, 111
FOOD CHAINS, 20
FOOD SECURITY ACT OF 1985, 224
FOOD WEBS, 49
FORBES, S.A., 19

FRESHWATER WETLANDS
Defined, 74
See BOGS, CYPRESS SWAMPS,
EVERGLADES,
FLOODPLAIN FORESTS,
FLOODPLAIN MARSHES,
PALUSTRINE
HARDWOOD SWAMPS,
PALUSTRINE MARSHES,
PRAIRIE POTHOLES,
SHRUB SWAMPS

GREAT DISMAL SWAMP, 85

HABITAT
Classification of, 23
Microhabitat, 23
Organism Concept and, 22
HERONS, 84
HYDRIC SOIL
Defined, 168
Generally, 52
Hydric Entisols, 207
Identification Using 1987 Corps
Manual, 173
Methods for Identifying, 172
Misidentification Problems, 172
Soil Conservation Service Hydric Soil
Mapping Units, 173
Soil Wetland Indicators, 168
Technical Criteria for, See Table 5.4,
168
Wetland Soils, Phosphorous Content,
53
HYDROPHYTIC VEGETATION
Common Plant Adaptations, See
Table 5.3, 167
Dominant Plant Recognition, See
Table 5.2, 162

FAC- and FACU Dominant Plants,
Examples of, See Table 5.1,
163
FAC Neutral Option, 167
Facultative, 161
Facultative Upland, 161
Facultative Wetland, 161
Generally, 52, 161
Obligate Upland, 161
Obligate Wetland, 161
Other Hydrophytic Vegetation
Indicators, 167
Regional Wetland Plant Lists, See Fig.
5.2, 164
Vegetation Inconclusive, 167

IZAAK WALTON LEAGUE, 73

JURISDICTIONAL WETLAND
Defined, 157
Generally, 5
See also SECTION 404 PROGRAM,
WETLAND
JURISDICTION

LITERATURE CITED, 352-83

MACROCONSUMERS, 20
MANGROVE SWAMPS
Birds, 145
Dominant Plants, 144
Fauna, 144
Fishes, 145
Generally, 142
Geophysical Characteristics, 143
Hydrology, 144
Losses of, 152
Mammals, 145
Mollusks, 145

Soils, 144
Vegetation, 143
Wetland Criteria Characteristics, 143
Wetland Values, 147
MARINE MAMMAL PROTECTION ACT
  Generally, 71, 261
  Marine Mammal Defined, 261
  Taking of Marine Mammals, 261
MARINE SANCTUARIES ACT, 261
MARSHES,
  Generally, 100
  Everglades, 111
  Floodplain Marshes, 108
  Palustrine Marshes, 100
  Prairie Potholes, 115
  See also EVERGLADES, FLOODPLAIN MARSHES, PALUSTRINE MARSHES, PRAIRIE POTHOLES
MICROCONSUMERS, 20
MICROCOSM, 19
MIGRATORY BIRD TREATY ACT, 307
MITIGATION BANKING, *See* WETLAND MITIGATION
MOLLISOLS, 208
MOBIUS, K., 19

NATIONAL AUDUBON SOCIETY, 73
NATIONAL ENVIRONMENTAL POLICY ACT
  Cumulative Impacts, 255
  Environmental Assessments, 254
  Environmental Impact Statement, 254
  Generally, 16, 71, 214, 254, 304
  Scoping Process, 254
  Wetland Mitigation Under, 281

NATIONAL ESTUARINE RESEARCH RESERVES PROGRAM, 72
NATIONAL HISTORIC PRESERVATION ACT, 16
NATIONAL LIST OF PLANT SPECIES THAT OCCUR IN WETLANDS, 161
NATIONAL MARINE FISHERIES SERVICE, 11
NATIONAL TECHNICAL COMMITTEE FOR HYDRIC SOILS, 168
NATIONAL TECHNICAL CRITERIA FOR HYDRIC SOILS, *See* Table 5.4, 171
NATIONAL WETLANDS INVENTORY MAPS, 188
NATIONAL WETLANDS POLICY FORUM, 9
NATIONAL WETLANDS PRIORITY CONSERVATION PLAN, 9, 72
NATIONAL WILDLIFE FEDERATION, 73
NATIONAL WILDLIFE REFUGE SYSTEM, 72
NATURE CONSERVANCY, THE, 73
"NAVIGABLE WATERS"
  Interpretation of, 10
  Meaning of, 215
NITRIFICATION, 52
NO NET LOSS OF WETLANDS POLICY, 9
NUTRIENT CYCLES, 20

ODUM, E.P., 19, 43

ODUM, H.T., 20
OKEFENOKEE SWAMP, 59, 85
ORDINATION
  Generally, 27
  Modern Techniques, 29
  PCA Ordination, 30
  Ordination Techniques, 27
ORGANISM
  As Ecosystem Unit, 22

PALUSTRINE HARDWOOD
    SWAMPS
  Bay Swamp, 87
  Dominant Plants, 87
  Fauna, 89
  Fires, 88
  Generally, *See* Fig. 3.2, Plate I, 85
  Geophysical Characteristics, 86
  Hydrology, 88
  Soils, 88
  Trees, 87
  Understory, 87
  Unusual Characteristics, 90
  Vegetation, 86
  Wetland Criteria Characteristics, 86
  Wetland Values, 89
PALUSTRINE MARSHES
  Generally, *See* Fig. 3.6, 100-101
  Beaver, 104
  Birds, 107
  Dominant Plants, 102
  Everglades, 100
  Fauna, 107
  Fens, 104
  Geophysical Characteristics, 102
  Hydrology, 106
  Savannahs, 104
  Soils, 105
  Unusual Characteristics, 108

Vegetation, 102
Wetland Criteria Characteristics, 102
Wetland Values, 108
PHOTOSYNTHESIS, 20
POCOSINS
  Generally, 90
  Vegetation, *See* Fig. 3.4, 92
PONDS AND LAKES
  Fauna, 127
  Generally, 126
  Geophysical Characteristics, 126
  Hydrology, 127
  Soils, 126
  Vegetation, 126
POPULATION ECOLOGY
  Generally, 21
  Mathematical Models and, 22
PRAIRIE POTHOLES
  Birds, 18
  Dominant Plants, 116
  Fauna, 117
  Flood control and, 64
  Generally, *See* Fig. 3.11, 115
  Geophysical Characteristics, 116
  Groundwater Recharge and, 62
  Hydrology, 117
  Mammals, 118
  Soils, 116
  Unusual Characteristics, 119
  Vegetation, 116
  Waterfowl, 118
  Wetland Criteria Characteristics, 116
  Wetland Values, 118
PREDATOR-PREY
    RELATIONSHIPS
  Estuarine wetlands and, 39
  Models for, 39
  Predator, Definition of, 38

PROBLEM WETLANDS
  Alaskan Permafrost Wetlands, 204
  "Cyclical Wetlands", 203
  FAC- and FACU Plant Species, 202
  Generally, 199
  Groundwater Wetlands, 202
  Hydrologically Difficult Wetlands
    Generally, 210
    Hydrologically Altered
      Wetland Systems, 210
    Seasonally Saturated Pine,
      210
    Temporarily Flooded Floodplain
      Wetlands, 210
    Seasonally Saturated Wet Meadows,
      210
  Identification under 1987 Manual,
    200
  Playa Wetlands, 202
  Prairie Pothole Wetlands, *See* Fig. 6.2,
    202-03
  Problematic Field Conditions, 211
  Problematic Hydric Soils, 206
  Problematic Wetland Plant
    Communities, 200
  Seasonal Wetlands, 202
  Types of, 200
  Vernal Pools, 203
PROBLEMATIC FIELD
    CONDITIONS
  Difficult Topography, 211
  Disturbed Areas, 212
  Generally, 211
  Man-induced Wetlands, 212
  Rocky Areas, 211
PROBLEMATIC HYDRIC SOILS
  Entisols, 206
  Hydric Vertisols, 206
  Mollisols, 206
  Spodosols, 206
PRODUCERS, 20

QUADRAT METHOD
  Compared with Releve Method, 26
  Inflection Point, 25
  Line Transects, Sampling by, 26
  Problems with, 26
  Quadrat shape, 26

REGULATORY FRAMEWORK
  Overview, 10, 213
  See SECTION 404 PROGRAM
REGULATORY TAKINGS
    CHALLENGES
  Generally, 244
  Harm-Prevention Analysis, 246
  Lucas v. South Carolina Coastal
    Council, 246
  Takings Test, 245
  Wetlands Cases and, 247
RELEVE METHOD
  Association Concept, 25
RESOURCE CONSERVATION
    AND RECOVERY ACT,
    304
RESTORATION AND CREATION
    OF WETLANDS
  Generally, 319-20
  Aerial Photographs, 328
  Budget Estimates, 332
  Case Study of, 341-47
  Construction Plans, 331
  Construction Sequence, 332
  Cost Considerations, 348
  Costs, *See* Table 10.1, 350
  Creation Defined, 320
  Development of Conceptual Plans,
    325

Earth-Moving Activities, 332
Functional Assessment Methods, 323-24
Funding for Innovative Projects, 325
Grading of Wetland Basin, *See* Fig. 10.2, 333
Habitat Evaluation Procedures, 339
Hydrology, 328
Implementation, 334
In-kind Mitigation, 323
Landscape Context, 331
Location, 326
Maintenance, 334
Monitoring, 336
Monitoring Data, 337
Nursery Stock, 334
On-site Mitigation, 323
Plant Establishment, 333
Plant Species Selection, 335
Process of, 322
Professional Survey, 332
Project Management, 349
Publications, 320
Rationale for, 320
Restoration Defined, 320
Site Morphometry and Soils, 329
Site Objectives, 324
Site Selection, *See* Fig. 10.1, 325, 327
Soil and Substrate Preparation, 332
Status of Science and Technology, 320
Suggested Monitoring Protocol, 338
Tidal Salt Marshes, *See* Fig. 2.8, 59
Time Delays, 331
Topography and Geology, 328
Transplants, 335
Vegetation Measurements, 338
Water Quality Samples, 340
Wetland Permit Sequencing, 349

RIVER AND HARBORS ACT, 71, 214, 255
ROOKERY BAY NATIONAL ESTUARINE SANCTUARY, 72

SALT MARSH
   Generally, *See* Fig. 4.1, 137
   Algae, 135
   Animals, 140
   Birds, 141
   Characteristic Fauna, 140
   Defined, 134
   Dominant Plants, 134, 138
   Fish and Wildlife Service Classification, 135
   Geophysical Characteristics, 135
   Hydrology, 139
   Mammals, 141
   Soils, 139
   Unusual Characteristics, 141
   Vegetation, 136
   Wetland Classification of, 135
   Wetland Criteria Characteristics, 136
   Wetland Values, 141
SECTION 404 PROGRAM
   Advance Identification of Wetlands, 262
   Background, 71, 213
   Creation of, 214
   Discharges Into Waters of the United States, 223
   Draining of Wetlands, 215
   Dredge and Fill Permits, 227
   Dredged Material, Definition of, 214
   Exemptions, 218
   Fill Material Defined, 215
   Impact of Coastal Zone Management Act, 257

Impact of Endangered Species Act, 258
Impact of Marine Mammal Protection Act, 261
Impact of National Environmental Policy Act, 254
Impact of Other Federal Environmental Laws, 253
Impact of River and Harbors Act, 255
Jurisdiction Over "Adjacent" Wetlands, 219
Jurisdiction Over "Isolated" Wetlands, 220
Jurisdictional Determinations, 217
"Navigable Waters", Meaning of, 215
Permit Process, 216
Regulatory Takings Challenges, 244
Special Area Management Plans, 264
State Assumption of the 404 Program, 226
"Waters of the United States", 215
Wetland Jurisdiction, 216
Wetland, Regulatory Definition, 5

SHRUB SWAMPS
  Characteristic Fauna, 94
  Dominant Plants, 90, 93
  Evapotranspiration, 94
  Geophysical Characteristics, 90
  Hydrology, 93
  Pocosins, 90
  Soils, 93
  Unusual Characteristics, 94
  Vegetation, 91
  Wetland Criteria Characteristics, 91
  Wetland Values, 94
  Willow shrub swamp, Fig. 3.3, 91

SIERRA CLUB, 73
SLOUGHS, See CYPRESS SWAMPS, 78

SPECIAL AREA MANAGEMENT PLANS (SAMPs)
  Generally, 264, 302
  Environmental Impact Statements, 266
  National Environmental Policy Act, Impact on, 266
  Purpose and Development of SAMPs, 264
  SAMP for New Jersey's Hackensack Meadowlands District, 266
  Wetland Mitigation and, 266

SPECIES
  Biological Species, 22
  Species, Concept of, 22
  Species Diversity
    Mark-and-Recapture Techniques, 32
    Population Size and, 32
    Shannon-Weaver Measure, 32
  Bootstrap Procedure, 32
  Jackknife Estimate of, 31
  Species Richness
    Mathematical Measurements of, 31
    Rarefaction Procedure, 31

SPODOSOLS, 207
SUPERFUND AMENDMENTS AND REAUTHORIZATION ACT OF 1986, 304
SWAMP LANDS ACT, 70
SWAMPS
  Defined, 76
  Bottomland Hardwood Swamps, 95
  Cedar Swamps, 87
  Cypress Swamps, 77
  Floodplain Forests, 95

Palustrine Hardwood Swamps, 77, 85
Shrub Swamps, 76
Swamp Forests, 76
Swamp Trees, 76

TAKINGS, *See* REGULATORY
   TAKINGS CHALLENGES
TANSLEY, A.G., 19

U.S. CONSTITUTION
   Fifth Amendment, 14
   Fourteenth Amendment, 14

VERNAL POOL,
   Generally, *See* Fig. 6.3, 205
   Phases of Development, 204
VERTISOLS, 209

WATER TUPELO, 80
"WATERS OF THE UNITED
   STATES"
   Meaning of, 10, 154, 215
WETLAND CASES
   Avoyelles Sportsmen's League, Inc. v.
      Marsh, 223, 252
   Bersani v. Deland, 242
   Bersani v. EPA, 290
   Bersani v. U.S. EPA, 242
   Ciampitti v. United States, 248, 274
   Deltona Corp. v. United States, 247
   Florida Rock Industries, Inc. v.
      United States, 247
   Friends of the Crystal River v. U.S.
      Environmental Protection
      Agency, 227
   Gazza v. New York State Department
      of Environmental
      Conservation, 271

Golden Gate Audubon Society v. U.S.
   Army Corps of Engineers,
   251
Hoffman Homes, Inc. v.
   Administrator, United States
   Environmental Protection
   Agency, 222
Hough v. Marsh, 251
James City County, Va. v. U.S.
   Environmental Protection
   Agency, 242
Leslie Salt Co. v. United States, 221,
   241, 252
Loveladies Harbor v. United States,
   247
Mall Properties, Inc. v. Marsh, 234
Missouri Coalition for the
   Environment v. Corps of
   Engineers, 231
National Wildlife Federation v.
   Hanson, 250
Newport Galleria Group v. Deland,
   242, 249
North Carolina Wildlife Federation v.
   Woodbury, 250
Resources Defense Council, Inc. v.
   Callaway, 215
Russo Development Corp v. Thomas,
   231
Save Our Community v. EPA, 224
Sierra Club v. Sigler, 251
Stoeco Dev., Ltd. v. Dep't of the
   Army Corps of Engineers,
   252
Tabb Lakes v. United States, 222
Tull v. United States, 250
United States v. Lee Wood
   Contracting, Inc., 220

436   Wetlands

United States v. Riverside Bayview
    Homes, 219
United States v. Zanger, 241
Zabel v. Tabb, 214
WETLAND CLASSIFICATION
  Based on Wetland Types, 6
  Basin Morphology and Water, 75
  Dominant Plant Type, 24, 75
  Computer Programs for, 27
  Fish and Wildlife Service
      Nomenclature, 75
  Floodplain Wetlands, 75
  Island Biogeography Theory, 26
  Palustrine Wetlands, 75
  Quadrat Method, *See* Fig. 2.1, 25
  Relevé Method, 24
  Species-Area Curves, Use of 26
  Wetland Characteristics, 16
  Wetland, EPA definition of, 156, 160
WETLAND DATA SOURCES
  Generally, 187
  Aerial Photographs, 194
  Hydric Soil Lists, 193
  National Wetlands Inventory Maps,
      188
  Recorded Hydrologic Data, 197
  Soil Surveys 193
  Topographic Maps, 188
  Definition of, 3
WETLAND ECOLOGY
  Classification by Cluster Analyses, 27
  Community Classification, 24
      Modern Procedures for, 27
  Competition Concept, 24, 32
  Detritus, 51
  Ecological Succession and, 24
  Ecosystem Degradation and, 24
  Evapotranspiration, 54
  Generally, 21, 23

Hydroperiod, 54
Ordination and, 24, 27
Plant Adaptations, 37
Population Changes and, 24
Water Budget, 54
WETLAND FUNCTIONS AND
    VALUES
  Amphibian Species, 56
  Animal Reproduction and
      Development, 55
  Anthropocentric Values, 54
  Decrease Runoff Velocity, 65
  Endangered Species Habitat, 57
  Flood and Erosion Control, 63
  Generally, 54, 153
  Groundwater Recharge and Water
      Quality, 61
  Hunting of Waterfowl, 61
  Intrinsic Value, Question of, 54
  Nonadjacent Wetlands, 56
  Plant Productivity, *See* Fig. 2.8, 59
  Recreation, 61
  Reduced Turbidity, 61
  Removal of Nutrients, 61
  Reptile Species, 57
  Sediment Retention, 61
  Sewage treatment, Use for, 62
  Spawning and Feeding Grounds for
      Fish, 56
  U.S. Department of the Interior
      Definition, 55
  Water Recharge, 62
  Wetland Bird Species, 57
  Wetland Mammals, 57
  Wetland Plants, 57
WETLAND HYDROLOGY
  1987 Manual Definition, 173
  Anaerobic Conditions, 173

"Growing Season", Conditions During the 173
Hydrophytic Vegetation, Establishment of, 173
FAC Neutral Option, 175
Growing Season Dates, According to 1987 Manual, 174
Hydrology Indicators, 175
Period of Inundation, 173
Secondary Indicators, 175
WETLAND IDENTIFICATION AND DELIENATION
1987 Corps Manual, Generally, 13, 16, 225
1987 Corps Manual, Purpose of, 153-54
1987 Corps Manual, Requirements of, 176
1987 Corps Manual, Three Parameter Approach of, 160
1989 Manual, Generally, 12, 153, 224
1989 Manual, Ban on Use of, 158
1989 Manual, Controversy Over, 12, 15
1989 Manual, Development of, 156
1989 Manual, Inadequacies of, 225
Aerial Photographs, Use of 177
Atypical Situations, 186
Completed Data Form, Example of, See Fig. 5.3, 182
Comprehensive Determination Method, 179, 185
Corps Delineation Procedures, 178
Data Sources, 179, 187
Field Procedures for, 175
Guidance for Field Work, 177
National Wetlands Inventory Maps, Use of, 177
Process Defined, 175

Positioning of Transects, See Fig. 5.4, 184
Problem Areas, 187
Problem Wetlands, 199
Routine Determination Methods, 178-79
Three Parameter Test, 199
Timing of Site Inspections, 177
WETLAND INDICATORS
Generally, 159
Hydric Soil, 168
Hydrophytic vegetation, 159, 161
Wetland Hydrology, 173
Wetland Plants, 161
WETLAND JURISDICTION
1989 Manual, 224
Draining of Wetlands, 224
Interstate Commerce and, 218
Migratory Bird Link to Interstate Commerce, 221
Jurisdictional Determinations, 16
WETLAND LOSSES
Causes of, 69
Clearcutting of Coastal Wetlands, See Fig. 4.10, 152
Construction of Dikes, Dams, Levees and Seawalls, 69
Discharge of Pollutants, 69
Drainage for Crop or Timber Production, 69
Drainage for Mosquito Control, 69
Dredging for Stream Channelization, 69
Florida, 68
Generally, 10, 14, 67, 150
Indirect Threats, 69
Peat Mining, 69
Sea Level Rise and, 151

WETLAND MITIGATION

1990 Memorandum of Agreement, 283
Advantages of Mitigation Banks, 301
Avoidance, 294
Compensatory Mitigation, 295
    Alternatives for Compensatory Mitigation, 292
    Compensatory Mitigation Proposal, 287
Controversial Issues, 282
Corps' Regulations, 283
Corps/EPA 404(b)(1) Mitigation Guidelines, 286
Corps/EPA Joint MOA, 286
Creation, 292
Current Mitigation Policies, 282
Council on Environmental Quality Definition, 293
Council on Environmental Quality, Role of, 281
Effectiveness of, 280
Enhancement, 292
EPA Guidelines, 284
EPA Veto Authority, 290
Exchange, 292
Fish and Wildlife Coordination Act and, 281
Fish and Wildlife Service, Role of, 284
Forms of Mitigation, 283, 291
Functional Values, 288
    Functional Values, Consideration of, 296
    Functional Values, Fish and Wildlife Service Evaluation of, 285
Generally, 278

Habitat Evaluation Procedures, 285
In-kind Compensatory Mitigation, Preference for, 296
Minimization, 294
Mitigation Alternatives, 291
Mitigation Banking, 298
    Concept of, 299
    Disadvantages of Mitigation Banks, 301
    Mitigation Banking Habitat Units, 300
    Mitigation Banking Recommendations, 302
    Objectives of Mitigation Banking, 299
Monitoring, 297
National Environmental Policy Act and, 281
National Marine Fisheries Service, Role of, 286
Offsite Mitigation, 288
On-site Mitigation, 292
On-site Compensatory Mitigation, Preference for, 296
Origin of Wetland Mitigation Policy, 281
Other Federal Agencies, Role of, 284
"Practicable Alternatives", 287, 288, 291
Preservation, 292
Regulatory Framework, 280
Restoration, 292
Sequencing Requirement, 287, 293
Sequencing Requirement, Exceptions to, 296
Sweedens Swamp Case, 290
Tenneco LaTerre Wetland Mitigation Bank, 299
"Unacceptable Adverse Effect", 289

Wetland Evaluation Technique, 285
WETLAND PLANTS, 57
  Decomposition of, 59
  Emergent plants, 57
  Facultative Wetland Species, 57
  Obligate Wetland Species, 57
  Seed Bank, 59
  Submersed Plants, 57
WETLAND PROTECTION
  American Values and Wetlands, 14
  Conventional Classification of
      Wetlands, 8
  Federal Acquisition Programs, 72
  Fish and Wildlife Service, Wetland
      Definition, 5, 154-55
  Jurisdictional Authority, 11
  Local Ordinances, 72
  Misperceptions About Wetlands, 3
  The Nature Conservancy, Role of 73
  Permit Process, 11
  Property Rights and Wetlands, 14
  Public Attitudes About, 13
  Public Education and, 73
  State Laws, 71
  Wetland Conservation, 67
  Wetland Valuation, 66
  Wetland Values, 3, 8
  See also SECTION 404 PROGRAM
WETLAND REGULATORY
      OFFICES
  Army Corps of Engineers, 384-87
  Environmental Protection Agency,
      388-90
  Other Federal Offices, 391
  State Wetland Offices, 392-405
WETLAND SCIENCE
  Interplay with Regulation, 13
  Wetland, Scientific Definitions of, 4